Reimagining the Historian in Victorian England

Elise Garritzen

Reimagining the Historian in Victorian England

Books, the Literary Marketplace, and the Scholarly Persona

palgrave
macmillan

Elise Garritzen
Department of Philosophy,
History and Art
University of Helsinki
Helsinki, Finland

ISBN 978-3-031-28460-1 ISBN 978-3-031-28461-8 (eBook)
https://doi.org/10.1007/978-3-031-28461-8

Cover illustration: mikroman6/Getty Images

This Palgrave Macmillan imprint is published by the registered company Springer Nature Switzerland AG
The registered company address is: Gewerbestrasse 11, 6330 Cham, Switzerland

Acknowledgments

Writing acknowledgments for a book which unearths how a scholarly persona was fashioned in prefaces, acknowledgments, and other paratexts is a somewhat intimidating task. With all imaginable prefatorial originality and wittiness already done, the essence nonetheless remains: the great pleasure of acknowledging the many debts I have accumulated along the way. I owe my greatest gratitude to Markku Peltonen, whose interest in this project has gone well beyond the intricacies of Samuel Rawson Gardiner; he has responded to my boundless queries with great acumen, his comments and guidance have been invaluable, and collegiality unfailing.

I have been fortunate to receive encouragement throughout this project. I am grateful to Herman Paul who has been generous with his time and expertise. I would like to thank Anthony Grafton for his warm hospitality during my Fulbright fellowship at Princeton University. Leslie Howsam's suggestions and criticism have been indispensable. I am also indebted to Angus Mitchell, Petteri Norring, and Laura Tarkka who gave their time to read and comment on draft chapters of this book. The constructive suggestions of the two anonymous readers helped me to make the final revisions to the manuscript.

I have had the privilege to present papers based on this project on numerous occasions. I owe thanks to the participants of seminar series in Helsinki, Leiden, and Princeton for their thought-provoking suggestions. My particular thanks also go to the audiences in the Creativity and Commerce workshop at Edinburgh University (2014), the SHARP conference in Paris (2016), the Scientific Persona and its Incarnations conference organized by Kirsti Niskanen and the wonderful SPICE project at the University of Stockholm (2017), and the Thanks for Typing conference at Oxford (2018). My research has benefitted from the invitations to contribute to several collective ventures which have pushed my thinking further. I want to thank Helen Kingstone and Trev Broughton, whose "Born 1819" project encouraged me to think about persona in terms of generations, Herman Paul who gave me an opportunity to rethink gendered personae, and the members of the "Irish Women's Writing (1880–1920) Network" who inspired me to explore women's history writing as a collaborative venture.

This book owes its existence to generous support from ASLA-Fulbright Finland, the Alfred Kordelin Foundation, Princeton University, the University of Helsinki, the Kone Foundation, and the Jenny and Antti Wihuri Foundation. It was the funding from the Academy of Finland that allowed me to bring this project to completion. I am also indebted to the help and advice from librarians and archivists in the following institutions: Bodleian Library, British Library, Cambridge University Library, David M. Rubenstein Library, Firestone Library, Hull History Centre, John Rylands Library, Jesus College Archive, King's College Archive Centre, Oxford University Press Archive, Syracuse University Library Special Collections, University of Reading Special Collections, Warwickshire County Archive, Wren Library, and Åbo Akademi University Library. I also want to thank Lucy Kidwell and the editors at Palgrave Macmillan.

The history department at the University of Helsinki has proved to be an inspiring academic home for me. I am grateful beyond words to Anna Koivusalo and Soile Ylivuori for their intellectual companionship. A big thank you goes to Mirkka Lappalainen for collegiality—and for enduring my historiographical nerdiness. Warm thanks are also due to Heikki Mikkeli and Kari Saastamoinen for their support and insightful comments. I have enjoyed the conversations with numerous colleagues: thank you Cesare Cuttica, Matti Hannikainen, Kaarlo Havu, Josephine Hoegaerts, Eva Johanna Holmberg, Reetta Hänninen, Mikko Immanen, Mari Isoaho, Niklas Jensen-Eriksen, Markku Kekäläinen, Laura Kolbe, Janne Lahti, Anu Lahtinen, Antti Lepistö, Adriana Luna-Fabritius, Risto Marjomaa, Jaakko Sivonen, Antti Taipale, Henrika Tandefelt, Maiju Wuokko, and Tupu Ylä-Anttila.

Finally, my gratitude goes to my family; my parents and my sister, whose love and support have been unfaltering. I dedicate this book to Hans who is the paratext that makes my life complete. He not only read every chapter of this book and asked the right and most difficult questions, but often accompanied me on archive journeys near and far. There are no words to express the appreciation I have for the enthusiasm which he shares with me for nineteenth-century historians and their quirks.

Contents

Abbreviations

BCWZ	Bath Chronicle and Weekly Gazette
BL	British Library
CUL	Cambridge University Library
EHR	English Historical Review
HHC	Hull History Centre
JRL	John Rylands Library
KING'S CAM	King's College Archive Centre
NC	Freeman's History of the Norman Conquest of England
NLI	National Library of Ireland
ODNB	Oxford Dictionary of National Biography
OEH	Freeman's Old English History for Children
OUP	Oxford University Press Archives
PMG	Pall Mall Gazette
RUB	Rubinstein Library
SR	Saturday Review
WCA	Warwickshire County Archive

List of Figures

CHAPTER 1

Introduction: From Rhetorical Diarrhea to a Branch of Science

It was the Golden Jubilee of Queen Victoria and the event inspired the Victorians to celebrate their national achievements in science, literature, and art. But where was history? Why were historians' accomplishments not recognized with similar pomp, asked Augustus Jessopp, a curate and enthusiastic amateur historian and antiquary.[1] He was so perturbed about this negligence that he stole the opening pages of a book review he was working on to praise the great innovation of history as science:

> The recognition of history as a science, and the winning for her a throne on which she may take her seat without fear of supercilious slight or contemptuous comparison, is a triumph won for a cause, say rather for a great idea; and the greater the toilers, and the more magnanimous they are, the greater will be their joy at the result which their labours have brought about. The historians have worked so loyally for history that through them we have learnt to understand and believe in a science of history. But it has been a long fight, and it is not yet quite ended.[2]

Many agreed with Jessopp; the *Daily Telegraph* proclaimed fourteen years later how the scientification of history had finally reached its goal and the "historic spirit" had replaced "the spirit of irresponsible dilettantism."[3] This transformation of history from a Romantic, literary pursuit into a branch of science riveted Victorians from Oxbridge colleges to London dinner tables and provincial newspapers. The commentators

E. Garritzen, *Reimagining the Historian in Victorian England*, https://doi.org/10.1007/978-3-031-28461-8_1

might have questioned the desirability of the change but admitted that they did not miss the moral preaching of Carlyle or the "rhetorical diarrhoea" of Macaulay.[4] From this acceptance of history's novel status followed a set of complicated questions. What did scientific history mean, what did it entail, and what constituted "a scientific historian" whose existence Jessopp rallied? Who was eligible to make decisions about historians' "physiognomy" and on what grounds? These were not insignificant concerns, as at stake was the entitlement to speak on behalf of the past. These are also the questions I ask while I trace how a *scholarly persona* of a scientific historian emerged out of these debates. The Victorians would not have recognized the concept as such, but the novelty of a word does not mean the novelty of the thing. The reconceptualization of a "historian" was premised on an idea of a collective ideal type that consisted of a cluster of virtues, skills, character traits, emotions, and manners that were deemed critical for the production of reliable historical knowledge. This collective model came close to what is today called a scholarly persona. The myriad attempts to construct such a persona open a fascinating window onto the Victorian world, where science and scholarship enjoyed great popularity and where the virtues, character traits, and manners of scientists were table talk, as the public had no intention of divorcing the quality of scientific discoveries from the quality of their discoverers, despite scientists' proclamations of disembodied knowledge.

The broad argument which this book pursues is that historians employed the scholarly persona to establish history as a branch of science and a unique discipline. The idea of history as science caught historians' imagination in the late 1850s when the thrust for professionalization and academic specialization was growing in England. Historians, though, began to relate to themselves as professionals only toward the end of the century, and rather than constructing a persona of a professional, they forged a persona that endorsed history as a discipline. Their disciplinary project had a peculiar English flavor as it also developed outside universities and learned societies. This, I hold, had a dramatic impact on the crafting of the persona. According to the conventional narrative of Western historiography, the emergence of history as a discipline was an outcome of three interlinked forces—scientification, professionalization, and institutionalization—and led to the establishment of new departments and chairs, pedagogical innovations such as the research seminar or dissertation work, and the founding of professional journals and societies. These institutional venues created ample opportunities for historians to cultivate their persona.[5] In England, by contrast, institutional development was slow. The ideals of liberal education at Cambridge and Oxford embraced character formation rather than the training of professional skills, and the public investment in history was more modest than elsewhere in Europe, where history was needed to consolidate the rising nation states. Even the Royal Historical Society was populated by amateurs until the end of the century, and the British Academy and the Institute of Historical Research were early twentieth-century innovations.[6] To compensate the slender institutional opportunities for staging the persona, I argue that historians converted books into a sphere where they fashioned and performed the persona of a scientific historian. As the ensuing chapters show, historians used cover design, size, titles, footnotes, typography, and other *paratexts* to construct, negotiate, and display their persona. If the modernizing literary conventions of sciences erased authors from the narratives, as Lorraine Daston and Peter Galison have proposed, personalities and personae flourished in the paratexts.[7] By ascribing agency to book parts, my aim is to make an important shift in the way we perceive the formation of history into a modern discipline.

Books were a powerful channel for promoting the persona as history and related subjects covered 12–15 percent of all published titles in Britain between 1870 and 1919. Victorians, distraught from the constant political, economic, social, and religious turmoil of their era, sought solace from histories that foregrounded steady progress and *The Times* reported in 1882 how "all the world" was reading history.[8] The adoption of books for fashioning a persona means that the persona was forged at the intersection of historians, publishers, and readers. Publishers were instrumental in deciding what kind of history got published and what kinds of paratexts were used in history books, whereas approval from readers was critical for authorizing the persona. Public support was not evident. Books can become popular only if readers find them relevant, but the multi-volume histories which symbolized scholarliness for the historians perplexed the general reader, who found them plain dull reading. The persistent complaint was that historians had replaced instructiveness and elegance with a dry chronicling of facts and, unless they found a happy medium between history as science and history as art, they would lose the ordinary reader—and the *raison d'être* of historical research. Francis Warre-Cornish, schoolmaster with a scholarly bent, formulated this message elegantly as he urged historians to understand that

> The Muse of history must not be excluded from literature, and turned over to science. If Clio ... condescends to be dull, she will become a Danaid, not a Muse, and the public will not come to her empty or over-full pitcher. And if the public does not read, the publisher will not print; and so Clio is starved out.[9]

Historians, enticed by their publishers, realized that they could please the different communities of readers only by diversifying their repertoire. Accordingly, the persona manifested itself in many types of histories, from schoolbooks to exhaustive special studies and articles in the *English Historical Review* (*EHR*).

Here, then, is where I pick up the storyline to investigate how historians mobilized book parts to establish history as a discipline and the scientific historian as its representative and authoritative producer of truthful historical knowledge. I approach paratexts as textual, material, and visual representations of the persona, as spaces where historians crafted the persona, and as convenient pretexts for publishers, reviewers, journalists, and many others to participate in the construction of the physiognomy of a proper historian. Since neither paratexts nor scholarly personae were produced and consumed in a vacuum, it is necessary to shift our gaze between the history books and the world that surrounded them to tease out broader social, cultural, and scholarly factors that were embedded in paratexts and personae alike. By bringing into dialog historiography, persona studies, history of science, history of humanities, history of knowledge, book history, and cultural history, I open new interpretive avenues; through the prisms of paratexts and scholarly personae, I interrogate the role and meanings which epistemic and moral virtues held in the nineteenth-century scholarly culture, the social organization and hierarchies of scholarly communities, the management of scholarly reputations, and the commercialization of historical knowledge.

Impersonal Science and Scholarly Persona

"Even scientists are men of flesh and blood" with "thoughts and passions," Jessopp wrote in the Jubilee eulogy, with one stroke undermining the appeals to disinterested scholarship.[10] As Steven Shapin argues, science is not disembodied; virtues and personal qualities have been pertinent to the making, maintenance, and authority of knowledge since the early modern period.[11] The link between the qualities of the knower and the quality of knowledge permeated Victorian scholarly culture and the vocabulary of virtues and personal characteristics was present wherever history was talked about. Virtues served epistemic, moral, and social purposes as they defined the standards for truthful knowledge, placed ethical demands on historians' minds and bodies, and assisted in regulating scholarly communities. The pervasiveness of the virtue lexicon bordered on banality as it circulated in satire, journalism, and advertisements, where almost every history book was sold with promises of its writer's industriousness, learning, fairness of judgment, impartiality, accuracy, thoroughness, conscientiousness, meticulousness, and clarity of style. That said, historians themselves drew heavily on virtues and character traits in their persona project. The *scholarly persona* as an analytical concept provides tools for investigating the applications of the virtue discourse in different temporal and disciplinary contexts as it pivots on scientific practitioners and practices and foregrounds the social and cultural construction of knowledge and knowers. As a concept, it owes a debt to the cultural turn in the history of science, but the recent trends of individualization, self-promotion, and self-fashioning in academia have propelled it into unforeseen interdisciplinary popularity in history, cultural studies, and sociology.

Lorraine Daston and Otto Sibum set the interpretive frame for persona studies in a seminal essay where they defined a scholarly persona as a collective *ideal* type of what it takes to be a scientist. It is an intermediate between individual identities and institutional templates and creates a collective with shared and recognizable features. The relationship between an individual and a persona is more integrated than just a role one assumes: the persona "shapes the individual in body and mind," enabling one to be a scientist in a discernible way.[12] Personae are, as Herman Paul maintains in his tremendously influential work, schematic models of changing repertoires of virtues, skills, character, traits, and conduct. They combine new elements and existing components which are borrowed and modified to create a persona that matches with the goods that are considered worthwhile to pursue at a specific historical moment.[13] The late-Victorian historians offer a textbook example of a collective attempt to refashion a scholarly persona at a moment when the perimeters of history writing were changing. As they reassessed the goals and status of historical research, they mixed and matched old, new, and revised components to erect a model that set normative and ethical requirements for a scientific historian and helped to manage the cohesion of the scholarly community. Historians' generous use of existing components emphasizes the continuity, not a dramatic break, between the Romantic and scientific modes of history.

Scores of case studies and theoretical appraisals of the scholarly persona in different spatial, temporal, and disciplinary contexts have recently appeared. They follow roughly two main trajectories of either zooming in on the actual personae—their constituents, mutability, and relationship to competing personae—or on the individuals who appropriate and perform the personae.[14] The two approaches have been framed as opposites, but I concur with Paul and Heini Hakosalo, who have questioned this. According to Hakosalo, the perspectives are "complementary rather than incompatible," and, as Paul observes, "personae and performances are like the foci of an ellipse: they presuppose each other."[15] Since this book pivots on the role the persona played in the emergence of history as an academic discipline, its emphasis is on the collective construction and promotion of the persona and in its constituents, but it also considers to a lesser degree how individuals emulated and adopted the ideal persona.

While I draw inspiration from these scholarly debates and from the rich literature on personae, I broaden the perspective in four significant ways as I explore how the persona was forged in *paratexts*, how it was appropriated by *English* historians, how their *overlapping responsibilities* as scholars, educators, and entrepreneurs determined the persona and, finally, how a cadre of *non-expert commentators* contributed to the forming of the persona. As I clarify these four expansions in the following pages of this introduction, it becomes evident that my book contributes to a move toward a more complex understanding of the interplay between personae and their manifestations and their reception both inside and outside the specialist domain.

Personae in Books

Lecture halls, seminar rooms, laboratories, fields, archives, libraries, offices, professors' private studies, and anniversary dinners have been identified as venues where personae are produced, reproduced, and enacted in rituals, words, voices, bodily practices, comportment, and appearances. Similarly, scientific notes and journals, methodological handbooks, Festschrift, commemorative statues, obituaries, autobiographies, memoirs, and ego-documents have emerged as sites for forging and performing the persona.[16] Quite strikingly, no systematic attention has been paid to perhaps the most common manifestation of scientific expertise: the book. By using approaches from the history of the book and the history of authorship, I demonstrate how books as culturally and socially specific textual, material, and aesthetic objects can make an appropriate stage for molding personae. As I uncover the traces of historians' personae in books, I displace the traditional historiographical emphasis on interpretations and philosophies of history and refocus attention on the materiality of history books. As D. F. McKenzie has proposed, when talking about the "sociology of texts," a book's materiality is embedded in its cultural, social, and economic contexts, and readers engage with the materiality just as they do with the text, searching for clues and ascribing meanings to the different aspects of a book.[17] Paratexts in history books were steeped in Victorian values, and in this sense they were not a blank canvas for historians to portray the persona. Rather, the persona became discernible within the contextual entanglements and the wider webs of meaning which the books encapsulated.

Books were a natural habitat for constructing a persona in the late-Victorian era because print culture acquired a mass audience and held a supreme position before radio, cinema, and other twentieth-century communication innovations challenged its dominance. Because of this "textualization" of the world, debates about books tended to be debates about society itself.[18] This applies to the historians who channeled their anxieties about the changing social hierarchies into the criticism of the craze for easily digestible books or the reading practices of the "vulgar public." The bookishness of the nineteenth century was a sum of larger trends that altered the literary marketplace and shaped authorship, publishing, and reading. On the one hand, the introduction of mechanized presses, machine-made paper, and railway distribution and the removal of the so-called taxes on knowledge enabled the production and supply of cheaper books. On the other, the increasing literacy, affluence, and leisure created audiences that previously had not had access to printed material. Publishers responded to the changes by customizing books to different socioeconomic groups and communities of readers.[19] Although the uniformity of an Enlightenment readership was a nostalgic illusion, the reality was that the late-Victorian historians faced a notably fragmented readership with mixed expectations of history books.

Book parts granted historians an opportunity to describe and promote their persona to claim authority for the scientific historian. Distributing awareness of the persona was critical because a persona can become a collective ideal type only if it is recognized and accepted by both expert and non-expert communities. It is because of this that Kirsti Niskanen, Mineke Bosch, and Kaat Wils have proposed that personae strive in a dynamic relationship between scientific expertise and its performances.[20] A growing body of literature has investigated the performative side of science and demonstrated how the carefully choreographed staging of scientific expertise in front of an audience produces and reproduces what constitutes a reliable scientist, scientific practice, or knowledge.[21] Meanwhile, the performativity of scientific publications remains relatively uncharted territory. Books, however, do not appear in print vested with reliability and scholarly significance, but these must be grafted both on them and their authors through rhetorical and performative moves.[22] The enactment of the persona in history books was geared to bestowing authority on both the individual historian and the discipline as a collective.

In terms of bookish performance, Stefan Hilgartner has found several significant differences between printed and face-to-face acts. First, books lack the unintended messages which body language betrays, and second, readers may evaluate a published performance in unexpected temporal and geographical contexts.[23] These points demand some modification. As Chapter 10 argues, books were considered reflections of historians' bodies and appearances and historians and readers alike construed them as expressions of the author's habitus. Books, however, were not without unexpected elements, as histories were accidentally issued with incorrect bindings, missing pages, blurry images, and other errors that could affect the performance. Furthermore, if readers revisited the performance in print, so did historians. The nineteenth-century print culture encouraged the production of revised editions, and historians seized the opportunity to readjust their texts and the performances of their persona.[24] Historiographers, though, have overlooked how the first edition was often only an initial step in a meandering process of constant textual alterations. As historians edited their texts, they also corrected or removed earlier paratexts and added new ones. The presentation of the persona in paratexts was thus mutable. Since multiple editions circulated simultaneously, readers could encounter different variants of that persona.

The presentation of the persona in paratexts entailed collaboration between historians and their publishers. As historians of science have shown, the print culture shaped Victorian sciences profoundly as the changes in publishing, printing, and reading endorsed the multiplication of formats of scientific publications and compelled scientists to reconsider their relationship with different reading communities to retain their scientific authority.[25] History was by no means an exception. Leslie Howsam has emphasized in her pathbreaking work how publishers made the decisions on what got published and in which format, in this way influencing the course of history writing in the late-Victorian period.[26] Publishers regulated the use of paratexts as well, and it has been common to consider paratexts such as size, paper quality, and typography as a publisher's sphere.[27] This is understandable because at least the Victorians credited publishers for the books' aesthetic and material qualities and the standard publishing contracts assigned the responsibility of these paratexts to the publisher. Nevertheless, the publisher's role should not be exaggerated. As numerous examples in this book show, the material paratexts were also often compromises between historians' scholarly concerns and publishers' commercial calculations. While publishers defined the size of the book and the budget for its production, they were otherwise willing to attend to many of the historians' wishes about the style, quality, and content of the paratexts. For example, the authorial attributions on the title page were almost exclusively historians' interpretations of their normative persona, as Chapter 2 demonstrates. Hence, rather than drawing on the conventions of book history and measuring the difference in authorial intention and editorial intervention, I regard the relationship between historians and publishers as a symbiosis. Historians needed publishers to usher their books into the marketplace and publishers needed historians to produce content and lend their names to their catalogs of educational histories. The paratexts with the highest marketing potential and production costs were the primary concern of the publishers and this left a significant paratextual space for historians to use freely for crafting and displaying the persona.

Persona of an English Historian

Historiographers have explored Victorian historians' ideological affinities and narrative strategies, the definitions of history as "science," and the gendering of history.[28] In terms of the scholarly persona, the historiographical emphasis has been on German, Austrian, and Dutch historians. These case studies provide an intriguing parallel to my book, as they help to pinpoint similarities and differences in national historiographical cultures. For German historians, the persona was a cartographical tool for drawing an imaginary disciplinary map according to the competing epistemic and moral commitments which the different groups of historians embraced.[29] If the Germans used the persona to organize the discipline, their English colleagues would apply it to erect a boundary between the scientific historians and their non-expert contestants. A major gap between the scientific and picturesque historians began to take shape in 1860, and writers such as James Anthony Froude were accused of sacrificing facts and truth for dramatic effect.[30] Indeed, there was not a single reality of what a historian meant or how history was written and, for instance, the porousness of the boundary between history and antiquarianism alarmed historians, as Chapter 4 shows. The diversity of history writing manifested itself in the stock of alternative personae and, consequently, in historians' disciplinary boundary work as they warded off the undesirable models to accentuate their own virtuosity. The coterie of scientific historians itself was too small to generate competing persona constellations or to disperse into disciplinary subgroups as happened in Germany. The differences in the scientific persona lay in nuances, and it was a unified and official enough instrument for historians to deploy to edge out those who emulated the wrong kind of persona. This near-institutional character of the persona was critical because competing knowledge systems can be marginalized and the boundaries policed only when the "official" mechanisms of methods and procedures are established.[31] This consistency, however, did not mean that the persona was fixed; as Chapter 2 and Chapter 3 illustrate, historians adjusted their persona according to the shifting institutional conditions and narrative and scholarly ideals.

Although historians utilized the persona to strengthen their disciplinary status, it would be misleading to talk about a battle between professionalism and amateurism. On the one hand, "amateur" is too imprecise a label because Victorians, fascinated by taxonomies, applied more refined categories such as antiquaries, genealogists, biographers, and gentleman scholars, who all came with their own model personae. On the other, the meaning of the"professional" scientist remained unfixed despite the gradual raise of disciplinary professionalization.[32] Historians avoided calling themselves professionals and the persona was not aimed at establishing professional working conditions but at validating scientific history as the authoritative form of historical inquiry, and history as a recognizable discipline with a distinct methodology, epistemology, ethics, and research practices. To quote Robert Townsend, who has written about the professionalization of history in America, "discipline" is constituted "as an organized body of knowledge" and a profession "as an organized form of work."[33] The English historians recognized this difference and oriented their persona project accordingly. Consequently, I refrain from talking about "professional" historians and instead use either "historians" or "scientific historians" when I refer to the members of the community that advocated history as science. This comes with its own challenges, as there was a difference between the empirically oriented "science of history" which Jessopp wrote about and the "scientific history" which carried connotations of positivism and was rejected by the disciplinary elite.[34] However, the semantic distinctions were not watertight. For example, John Horace Round, an epitome of the inductive method, used "scientific history" when referring to the proponents of empirical research.[35] Hence, and without feasible alternatives, I prefer the "scientific," as it does not presuppose a modern scientific profession. Its clumsiness is a reminder of the indefinite character of history during the later Victorian period.

The emergence of history as a discipline rested on its scientification, as historians realized in the 1850s how backward their practices were compared with their continental brethren. They caught the idea of history as science from Henry Thomas Buckle's *History of Civilization in England* (1857–1861), but since they rejected Buckle's positivist philosophy, they found an alternative model for their scientific aspirations from Germany and Leopold von Ranke: empiricism, the inductive method, and an exhaustive study of primary sources were in tune with their individualistic mindset.[36] The persona encapsulated this growing absorption in methods, original authorities, and epistemic and moral virtues, and glued together the informal scholarly community which existed at least as much in the minds of its geographically dispersed members as it did among the colleges of Oxford and Cambridge. Historians collaborated through correspondence and flagged their membership by acknowledging one another's work. Despite the physical distance, historical knowledge was constructed through social interaction and the community was built on intellectual kinship. The persona, as its paratextual renditions show, was used as a regulative tool that ensured the cohesion of the community by stating what was expected of the individual members in terms of virtues, methods, skills, and comportment. The inductive method, for instance, was evoked to claim a privileged ownership of the past, but the method-talk fostered a sense of community as well because the method was removed—at least seemingly—from the political, religious, economic, and social divisions which split nineteenth-century societies.

Because of the social aspect of the persona, I would like to argue that conduct and manners were even more critical components of it than have so far been recognized. Considering how much moral weight Victorians placed on countenance,[37] it is unsurprising that observance of the implied code of scholarly conduct was ingrained in the persona. The comportment of each historian either reaffirmed or undermined the credibility and dignity of the persona, and in this sense the historians' community resembled the early modern Republic of Letters where conduct was perceived as an articulation of right and wrong scientific procedures. Polite manners mirrored epistemology and ethics warranting the reliability of knowledge.[38] Books were no exception to this. Since paratexts were an expression of the collective persona, historians composing them were like Erving Goffman's corporate teammates who were bound by loyalty that ensured the coherence of their separate performances and their compliance with the persona. In such corporate performances, "the good reputation of one practitioner depends on the conduct of the others."[39] Hence, monitoring conduct was a common strategy among historians, but Chapter 3 shows how they made allowances for scholarly etiquette if their public image benefitted from a strategic concealment of improper conduct. The rhetoric and actual practices could differ when historians protected the dignity of their persona. These discrepancies are revealing of the ideals on which historians built their discipline.

A Composite Persona: A Historian, an Educator, an Entrepreneur, and a Middle-Class Man

Two images of a nineteenth-century historian dominate the historiographical scene: the Rankean scholar and a politically committed commentator—or a public intellectual according to Stefan Collini's terminology.[40] But to be a historian was more than that. Between these seemingly opposite poles remain uncovered areas that call for closer attention, for they help to reveal tensions that tell much about historians' values and preferences. To unearth historians' multiple obligations, I draw on Camille Creyghton, who has pointed out how nineteenth-century French historians forged personae that were "composite wholes," sums of their responsibilities as scholars and pedagogues, each with unique requirements, virtues, and skills which they mobilized and reinterpreted when they switched from one pose to another.[41] English historians were scholars, educators, and entrepreneurs, and they pursued goods that ranged from acquiring and disseminating historical knowledge to engaging in political debates and making a profit. The scholarly dimension formed the bedrock of the persona, but since some of the goods which historians pursued contradicted one another, they interpreted the virtues flexibly, prioritized them according to the task at hand, and ascribed multiple meanings to them. For instance, Chapter 9 explores how the elasticity enabled them to resolve the frictions that arose when they composed promotional paratexts that thrived on hyperbole which impaired the virtues of honesty and altruism that they cherished as scholars.

The composite persona also accommodated those commitments that were embedded in Victorian identity markers such as gender, class, ethnicity, religious belief, sexuality, or disability. Mineke Bosch, Kirsti Niskanen, and others have argued that such embodiments can limit access to the available personae. This has generated interest especially in the gendered manifestations and inequalities inscribed in personae.[42] The persona of a late-Victorian historian surely was that of a middle-class Englishman, but one should avoid oversimplification. First, it would be careless to argue that historians supported a unified vision of manliness, for men as dissimilar as Oscar Browning, Montagu Burrows, and Edward Freeman embraced their particular views about masculinity. Second, the manliness of the persona did not mean that women were barred from history. It is still common to recite the feminist pioneers who professed that the professionalization and institutionalization of sciences led to a deliberate exclusion of women from scientific pursuits. While this early scholarship made a valuable contribution by recovering the voices of women scientists, I agree wholeheartedly with Heather Ellis, who has observed how highlighting women's marginalization quite ironically presented women "enjoying less agency than they actually had."[43] The late-Victorian women historians present an excellent opportunity to reconsider women's place in Victorian scholarship. Chapter 2 discusses the representations of gender on title pages and Chapter 7 examines women as politically radicalized historians, but otherwise I have deliberately avoided confining women to one separate chapter with a title around the "gendered" or "feminine" persona. Instead, I have integrated them into the overall narrative to make visible how those women who attended to the inductive method were admitted into the scholarly community and how their historical inquiries were treated as valuable contributions to the common pool of historical knowledge. This does not intend to naively downplay the structural obstacles which women encountered, but to emphasize the fluidity of the gender boundaries during the latter half of the nineteenth century.

Class was another important, yet lesser studied factor in nineteenth-century scholarly personae. The class aspect had two implications for historians. In the first place, the persona shared qualities with the moral *character* of the middle-class man, as both were invested with traits such as industriousness, honesty, integrity, and politeness. In the context of the persona, the practical definition of these moral habits derived from historical pursuits and the training of such qualities was also preparation for being a better citizen. This gave additional weight to historical studies and to the persona. Second, historians mobilized the persona to claim a distinct social position as part of the broader movement of the liberal professions to institute themselves along the old professions of law, medicine, and religion. Historians were preoccupied by the ambivalence of their social standing and Edward Freeman encapsulated this in a self-reflecting essay which he wrote for the American journal *Forum*. His words betray both the challenges which historians faced when trying to describe their status and their eagerness to establish a suitable social category for a "historian":

> I belong to no profession; I can hardly be said to belong to any class. But I have points of contact with several classes. At once a professor in Oxford and a justice of peace in Somerset, I do not feel that I am exactly a country gentleman; still less do I feel that I am exactly an Oxford don; I suppose I am not a 'literary man,' because I have never lived by writing; I suppose I am not a political character, because I have never sat in Parliament. But I feel that I have enough common with all these classes and with other classes as well to understand all of them, without exactly belonging to any of them.[44]

When the *Pall Mall Gazette* published a résumé of the essay, it tellingly called this section "Of no class and profession."[45] It was obvious that the traditional three-class model was unable to accommodate the new liberal professions and that the horizontal fracturing of the middle class demanded more nuanced categories. Unsurprisingly, interpreting the symbols that denoted the status of the new professions was popular among the class-conscious Victorians.[46] The persona project fueled such meditation, and therefore the representations of the persona in paratexts derived both from historians' scholarly needs and from the concurrent social developments. After all, as Chapter 10 argues, the sobriety of the bindings in historical studies imitated the gravity of the plain three-piece suit, unofficial costume of the hard-working Victorian middle-class man. Historians' implicit status anxiety formed an important context for the persona project—and for the materiality of their books.

Non-Expert Commentators on the Persona

Who is entitled to construct, or participate in a construction of, a scholarly persona? So far, the persona has been discussed as the property of scientists or institutions, such as funding bodies and science administrators, who influence the configuration of personae and regulate access to them.[47] However, I would like to suggest that there can be a wider cast of commentators who participate in the debates about personae and offer modifications and alternate models which require a response from the scholarly elite. As historians came in many guises, the campaign to make history a branch of science involved a chorus of voices. Although the leading historians such as Lord Acton, Mandell Creighton, Edward Freeman, Samuel Rawson Gardiner, John Richard Green, John Robert Seeley, and William Stubbs figure prominently in this book, I also give voice to those who found themselves on the outskirts or outside of the scholarly community. Instead of treating Jessopp and his kin as the useful "others" or mere targets for the scientific historians' exclusionary moves, I add depth to my argument about the plurality of opinions which the disciplinary undertaking inspired by recognizing their engagement in the debates about the persona.

Another important body of commentators consisted of reviewers and essayists. Book reviews in iconic periodicals are standard source material for Victorian literary and cultural studies, but newspaper reviews, especially in the regional papers, are an untapped source, even though local editors have been identified as influential figures who introduced books to audiences in different parts of the country. As Andrew Hobbs convincingly argues, "provincial newspapers were not on the fringes of nineteenth-century print culture, but at the very heart of it."[48] The regional newspapers were instrumental in spreading news about scientific history and historians beyond the large cities and university towns, and most people encountered historians secondhand through the reviews. The reviewers' opinions were anything but insignificant for the persona project.

A striking feature in the newspaper reviews is the urge to make sense of the changes which history was undergoing and to map the anatomy of a scientific historian. The reviews were deluged with virtues, character traits, skills, and habits as the writers drew portraits of historians. These brief sketches resonated with the rise of modern celebrity culture which animated interest in scientists' life stories. In addition to the biographical portrayals, scores of reviews were prefixed with brief overviews of recent developments in history. The writers identified similarities and differences between the Romantic and scientific modes of history and drew genealogies to trace the origins of scientific history. The hereditary lineage descended from Hallam, Macaulay, and Milman to Freeman, Stubbs, and Seeley and confirmed that history had evolved, not through a radical rupture, but through continuity and steady progress, just like history itself apparently had progressed from the earliest times to the present day. The writers dropped these names without explanatory labels, as they clearly saw their audience to be cultivated enough to both draw inferences without additional handles and appreciate these historiographical surveys. This was not just some cut-and-paste journalism or an inferior version of the London press. Recognition of the input of provincial reviewers significantly enriches the picture of Victorian history culture.

Historians could be irritated by this attention. They courted public approval for the persona but expected the "unlearned" to accept their version of it without a murmur. They feared that the proliferation of history books in the popular press would create an illusion of history as an "easy" topic that did not demand any special skills and knowledge from its practitioners and was therefore open for anyone to pursue and comment on. They were also dismayed about the unflattering picture which the newspapers and magazines painted of them as mere pedants. The wide circulation of such images put them into disrepute and caused evil to history "as a branch of knowledge," as Freeman expressed the concern.[49] The rectification of these "misconceptions" became a key target of historians' persona project. Although they realized that reputations were conferred by the public, they nonetheless hoped to be able to shape their public disciplinary image. Their attempts to control the representations of the persona in scholarly and non-scholarly outlets is a theme that runs through my book, and historians' reputational distress suggests that it is a topic that should deserve even more attention in persona studies with a historical scope.

Persona in Paratexts

Ever since Gérard Genette introduced the term "paratext" in his pioneering work *Seuils* (1987), historians, philologists, and literary scholars have been captivated by the different elements that circle the text in a book. As Genette proposes, it is the paratext that makes the text meaningful, contextualizes it, influences its reception, and allures and persuades readers to consume it.[50] Jonathan Gray added that even those who do not consume an entire text often form an opinion of it and its author based on its paratexts.[51] This helps to explain why paratexts are so important for my argument about different parts of books as a prime setting for constructing scholarly personae; they are a rostrum for individual and collective self-fashioning. Of course, readers do not necessarily react to paratexts in predicted or uniform ways, as their intertextual experience conditions the reading of paratexts. It is now agreed that paratexts tell more about the expectations which authors and publishers have about readers at a certain historical moment than about their actual reception. Yet readers recognize that texts and paratexts represent different aspects of modality.[52]

A large and growing body of literature has problematized the relationship between texts and paratexts, has questioned whether all the features that encircle a text are paratexts, or has focused on drawing taxonomies of the functions which paratexts serve.[53] Despite this sustained interest in paratexts, very little attention has been paid to paratexts in ***modern scholarly discourse***. Whereas book historians have explored early modern scholarly print culture, more recent paratexts have been discussed mostly by literary critics in fictional genres.[54] As it stands, Anthony Grafton's splendid history of the footnote remains the only book-long treatment of a scholarly paratext. This is unfortunate because, as Samuel Kinser argued already in 2004, "The assimilation [of fiction and non-fiction] is questionable because different genres of writing develop their text-paratext relationship differently."[55] It is not only the relationship between the text and its paratexts that make the sensitivity to the fiction/non-fiction distinction so critical. Scholarly paratexts have genre- and discipline-specific features and functions that stem from distinct epistemological, narrative, and aesthetic ideals and research practices, and more research is needed to enhance our understanding of the role which paratexts play in the presentation and construction of knowledge in scientific publications. Moreover, scholarly paratexts are easily exploited for non-scholarly, selfish, or ethically dubious purposes, and since they are by no means a neutral textual, visual, or physical space, the broader methodological claim I make is that it is vital that we start to pay adequate attention to paratexts in scholarly publications.

As this book investigates books as platforms for fashioning the scholarly persona, the paratext has proven to be a helpful concept in two specific ways. First, it neatly comprises all the elements that surround the texts in books and second, it highlights the functionality of these different components. Recently, some book historians have preferred "book parts" instead of "paratexts," arguing that the latter is too restrictive as Genette concentrated on textual elements in modern fiction.[56] However, Genette's choice of examples does not undermine the analytical usefulness of the concept. Since the publication of *Seuils* in 1987, scholars have fruitfully applied the concept to a broad spectrum of genres and components of manuscripts, books, and e-books. Hence, I adopt "paratext" in its most encompassing way and also distance myself from scholarship that denies paratextual status to bindings, ink, typography, or paper by appealing to the different semantic registers of material and aesthetic elements on the one hand and of textual ones on the other.[57] Authors, publishers, and readers, as the publishing of Victorian history books attests, engage with and interpret the paratextuality of all the material aspects of a book.

Furthermore, the idea of *function* that is embedded in the concept "paratext" is highly relevant for my argument. Dorothea Birke and Birthe Christ have further developed Genette's discussion about paratexts' functions and identify three major purposes for paratexts: interpretive, commercial, and navigational.[58] I wish to add to this list a fourth that is critical in scholarly publications: authoritative. As numerous examples in my book show, paratexts are pivotal for establishing credibility, authority, and value for historical texts and their authors as they supply the writers with a space for authorial and collective self-fashioning. As historians negotiated and performed their persona, they simultaneously emphasized the reliability of their narratives and shaped the readers' understanding of a historian, historical knowledge, and history as a discipline. Their message, then, was two-directional, as it contextualized the text within the book and the historian and the collective entity which they represented outside the book.

In practical terms, I trace the manifestations of the persona in various paratexts and investigate how historians utilized them for framing their persona and crafted paratexts that expressed that persona. I also examine how paratexts were commented on in letters, reviews, and other histories and how the connection between paratexts and personae were construed and established in these private and public accounts. This material reveals that historians were well versed about paratexts' interpretive, navigational, and commercial potential and recognized the rich opportunities which they presented for the persona project. Their paratextual practices were creative, as they—and their readers—appreciated paratexts' symbolical capacity. When John Wilson complained about a poor table of contents and a lacking index in the *Opere Inedite di Francesco Guicciardini*, he did not only defend the readers' right to a "rudder or compass" when "voyaging through" the bulky volumes but extended such "editorial delinquencies" to be indicative of the Italian national character.[59] Since paratexts were at the same time practical and symbolic gestures, it is not surprising that historians found them ideal for promoting their persona and appropriated a wide range of different paratextual configurations to confer authority and credibility to their texts and make the persona known to the widest possible audience.

To map the use of paratexts in history books, I have collected a comprehensive set of data that encompasses paratexts in 500 history books. This marks a significant departure from the traditional approaches which, apart from Andie Silva's *The Brand of Print: Marketing Paratexts in the Early English Book Trade*, rely on case studies of paratexts in a small number of texts. Since my study centers chronologically on the late-Victorian era, the database consists of 446 books that were published between 1860 and 1900, and as comparative material 54 books that were issued either before or after the temporal limits. As the database contains multi-volume histories and multiple editions, the total number of different titles is 265. The appendix holds a full list of the books in this database. Even more important than the size of this database is the diversity of formats, topics, and authors it comprises. This has been critical for my goal of capturing the richness and diversity of nineteenth-century history writing. In line with this aim, the database includes paratexts from the leading historians whose books form the Victorian historiographical "canon" as well as from those who were popular in their time but whose publications are now relegated to obscurity. The topics and formats are similarly diverse, and stretch from multi-volume histories of political events and institutions to children's histories, document editions, and lavishly illustrated histories of material culture. The size and heterogeneity of the material has obvious methodological benefits, as it has allowed me to identify trends and distinctive characteristics in paratexts in different types of histories and by historians who cultivated competing personae.

To collect this data, I took three consecutive steps. I began by systematically going through all the reviews of history books in two leading quarterlies, the *Edinburgh Review* and *Quarterly Review*, between 1860 and 1900, and selected from them all the histories that were reviewed in both journals. As a second step, I added any canonical works that were still missing by the scientific historians and their less familiar educational histories such as Seeley's *Short History of Napoleon the First*. Finally, I rounded up this process by including histories from women and Oxbridge tutors to enhance the gender and occupational diversity of the authors. Since drawing up this database, I have come across other histories that have proven to be significant for my argument. As I have consulted only some of their paratexts, they serve as sources but do not form part of the database. Furthermore, I have complemented the paratext data with other types of sources because historians' paratextual message was multiplied in reviews, excerpts, and advertisements as well as in gossip, letters, and other private documents. James Secord has fittingly observed how this kind of "literary replication" is an essential part of the distribution of a book.[60] Because of this, reviews and historians' correspondence with their colleagues and publishers at two notable publishing houses, Macmillan and Oxford University Press, constitute two additional sets of sources for my study. These letters reveal historians' astonishingly intense investment in paratexts and confirm the hypotheses of paratexts' importance in transmitting culturally and scholarly invested messages.

Chapter Outline

The book is divided into three sections according to historians' duties as scholars, educators, and entrepreneurs. The first part addresses historians as producers of knowledge and their paratextual attempts to set normative conditions for the persona, and to promote the virtues and methods which scientific history entailed and which marked it off from other approaches to the past. Chapter 2 discusses the authorial attributions on *title pages* as idealized projections of historians as well-educated and internationally connected men whose ties to academia intensified as the century drew to a close. This carefully curated image concealed the inadequate institutional development, resistance to the modernizing academia, and deep sense of insularity which all had an impact on the persona. Chapter 3 concentrates on proper names as *factual paratexts* that were used as a shorthand to denote specific sets of virtues, and argues that the heroization of a virtuous historian had epistemic consequences as it deterred critical assessment of the model historians. Chapter 4 explores the competing personae of historians and antiquaries. Historians maintained that their methods and aims differed from those of antiquaries, but the public debate about the *size* of history books undermined their efforts. As historians and antiquaries produced equally bulky volumes, readers continued to confuse history with antiquarianism.

The second part of the book considers historians as educators and paratexts as a pedagogical space. It tracks down historians' various educational duties and the possibilities and obstacles they posed for the promotion of the scholarly persona. Chapter 5 introduces historians as authors of small educational histories and examines how they used *prefaces* to present the writing of schoolbooks and manuals as a noble duty of a scientific historian. As an ideal historian produced unwieldy original studies, historians lobbied in the prefaces on the virtuosity of their books to convince readers that even small histories could be scientifically appropriate and grant acclaim to the persona. Chapter 6 turns attention to *footnotes* as an educational space where historians taught readers to apply the scientific method. While footnotes showed which skills and virtues a scientific historian needed to attain historical facts, they came to institutionalize the key components of the persona. Chapter 7 concentrates on radical historians who used *prefaces*, *running heads*, and *footnotes* to forge a persona of a politically inspired historian. The scholarly elite considered this as a severe threat to the dignity of the persona and to the reputation of the discipline and launched a campaign to refute the contending models.

The third part charts historians' entrepreneurial activities and the threats their commercial interests caused to the dignity of the persona. Chapter 8 investigates the conflict between scholarly altruism and the commercialization of historical knowledge and suggests that by demanding paratexts such as non-standardized *typography*, historians elevated themselves above the "trade," as they could present their books as dignified and prestigious commodities, not as equals to ordinary products such as food or hosiery. Chapter 9 explores historians as advertisers and argues that although they disassociated themselves from the perceived deceptiveness of professional marketing, their *prefaces* and *titles* betray variance in the grade of honesty in their promotional strategies. Chapter 10 examines history books as physical objects and shows how material and visual paratexts were interpreted as expressions of historians' inner qualities, persona, and social standing. Because of this, *design* and *quality* were important paratextual characteristics of history books.

Notes

1. Curthoys, "Jessopp," *ODNB*.
2. [Jessopp], "Brocas Book," 225.
3. "Death of the Bishop of Oxford," *Daily Telegraph*, April 23, 1901, 7.
4. [Hayward], "Lord Macaulay," 290.
5. Tollebeek and Porciani, "Institutions, Networks, and Communities," 7–12; Lambert, "Professionalization and Institutionalization of History," 42–43.
6. Goldstein, "Organizational Development," 181–188; Daunton, "Introduction," 19–20; Iggers, "Historiography," 4–5.
7. Daston and Galison, *Objectivity*, 217.
8. Howsam, *Past into Print*, 7; "Freeman's History of William Rufus," *Times*, April 17, 1882, 44.
9. [Warre-Cornish], "Gardiner's *Protectorate*," 446.
10. [Jessopp], "Brocas Book," 225.
11. Shapin, *Scientific Life*, 4–8.
12. Daston and Sibum, "Introduction: Scientific Personae," 2–6.
13. Paul, "What Is a Scholarly Persona," 358–365.
14. In addition to Daston and Sibum and Paul, essential studies include Bosch, "Scholarly Personae and Twentieth-Century Historians," *BMGN* (2016); Algazi, "Exemplum and Wundertier," BMGN (2016); Niskanen and Barany (eds.), *Gender, Embodiment, and the History of* the Scholarly Persona (Palgrave Macmillan, 2021); Paul (ed.), *How to Be a Historian?* (Manchester University Press, 2019); Niskanen, Bosch, and Wils (eds.), "Scientific Persona," special issue of *Persona Studies* (2018); Condren, Gaukroger, and Hunter (eds.), *Philosopher in Early Modern Europe* (Cambridge University Press, 2006).
15. Hakosalo, "Cut out for Medicine," 152–153; Paul, "Sources of the Self," 143.
16. See, for example, Bosch, "Scholarly Personae," 42; Paul, "Virtues of a Good Historian," 681–709; Cabanel, "Woman in a 'Man Made World'," 113–139; Hakosalo, "Cut out for Medicine," 154–164.

17. McKenzie, *Bibliography and the Sociology of Texts*, 4–5; Howsam, *Old Books*, 20–21, 32–35.
18. Lyons, *Readers and Society*, 157.
19. Feather, *History of British Publishing*, 130–134; Weedon, "Analysis of the Cost of Book Production," 234; Secord, "Progress in Print," 377–380.
20. Niskanen, Bosch, and Wils, "Scientific Persona," 2.
21. Gooday, "Ethnicity, Expertise and Authority," 15–29; Vandendriessche, Peeters, and Wils, "Introduction: Performing Expertise," 1–4; Myers, "Pedagogy and Performativity," 820.
22. Johns, *Nature of the Book*, 2.
23. Hilgartner, *Science on Stage*, 16–18.
24. Garritzen, "Revise, Edit, and Improve," 289–294.
25. Topham, "Scientific Publishing," 581–582; Secord, "Science," 444; Lightman, *Victorian Popularizers of Science*, 16; Ross, "Fixing Genius," 58–59; Jardine, "Books, Texts and the Making of Knowledge," 400.
26. Howsam, *Past into Print*, 1–2, 114.
27. Topham, "Scientific Publishing," 58; Ives, "Bibliographical Approach to Victorian Publishing," 273–274; Genette, *Paratexts*, 16; Arroyo Redondo, "Aproximaciones teóricas al prólogo," 61.
28. Hesketh, *Science of History* (Pickering & Chatto, 2011); Jann, *Art and Science* (Ohio State University Press, 1985); Bentley, *Modernizing England's Past* (Cambridge University Press, 2005); Lang, *Victorians and the Stuart Heritage* (Cambridge University Press, 1995); Melman, *Culture of* History (Oxford University Press, 2006); Kingstone, Victorian Narratives (Palgrave Macmillan, 2017); Maitzen, *Gender, Genre* (Garland, 1998); Spongberg, *Women and the Nation's Past* (Bloomsbury, 2018); Mitchell, "Busy Daughters of Clio," *Women's History Review* (1998).
29. Paul, "Scholarly Personae," 7–8.
30. Hesketh, *Science of History*, 86–93; Jann, *Art and Science*, 105–140.
31. Daunton, "Introduction," 10.
32. Ellis, *Masculinity and Science*, 26–27; Lightman, *Victorian Popularizers of Science*, 12–13; Collini, *Public Moralists*, 208–209.
33. Townsend, *History's Babel*, 3–4. There have been many attempts to define what a professional historian meant in the nineteenth-century context, but they share a tendency to isolate history from the wider social forces which influenced historians' developing sense of a profession. Pim den Boer means by professionalization the making of living with history, Rolf Torstendahl equates it with acceptance of shared norms and recognition as a professional by other historians, and for Stefan Berger it entails a development of the method and research infrastructure such as the opening of archives. Den Boer, *History as a* Profession, xiv–xv; Torstendahl, *Rise and Propagation of Historical* Professionalism, 16; Berger, "Professional and Popular," 14–15.

34. Fuchs, "Conceptions of Scientific History," 148.
35. Round, "Scientific History," 560.
36. Hesketh, *Science of History*, 13–33; Warren, "Rankean Tradition," 30.
37. Morus, "Worlds of Wonder," 808.
38. Goldgar, *Impolite Learning*, 4–7.
39. Goffman, *Presentation of Self*, 164, 207–222.
40. Collini, *Absent Minds*, 46–52.
41. Creyghton, "Generational Continuities and Composite Personae," 83–86. See also Creyghton, Huistra, Keymeulen, and Paul, "Virtue Language in Historical Scholarship," 925–926.
42. Bosch, "Scholarly Personae," 34–54; Niskanen, "Scholarly Persona Embodied," 316–319 and the essays in Niskanen and Barany (eds.), *Gender, Embodiment, and the History of the Scholarly Persona*.
43. Ellis, *Masculinity and Science*, 1–3, 7; see of this same Opitz, Lykknes, and Van Tiggelen, "Introduction," 1–4.
44. Freeman, "Review of My Opinions," 154.
45. "Last Words by Professor Freeman," *PMG*, April 8, 1892, 2.
46. Reader, *Professional Men*, 149–152; Crossick, "From Gentlemen," 161–165; For a recent re-evaluation of the British professional society and professions, see Heidi Egginton and Zoë Thomas (eds.), *Precarious Professionals: Gender, Identities and Social Change in Modern Britain* (University of London Press, 2021).
47. Wils and Huistra, "Scholarly Persona Formation," 85–86; Cabanel, "Woman in a 'Man Made World'," 114–117.
48. Hobbs, *Fleet Street in Every Town*, 6.
49. Freeman, *Methods*, 100–103.
50. Genette, *Paratexts*, 1–16.
51. Gray, *Show Sold Separately*, 23–26.
52. Marot, "Pour une poétique," 12–13; Armstrong, "Paratexts and Their Functions," 40–41; Gray, *Show Sold Separately*, 32.
53. Boundaries and definitions: Ruokkeinen and Liira, "Material Approaches," 110; Smith and Wilson, "Introduction," 1–14; Allen, "Perpetually Beginning," 181–183; Claes, "Supplements and Paratexts," 201; Mak, *How the Page Matters*, 9; Moner, "Introducción," xi; Rozzo, "Paratesto," 213–214, 223–224; Sabato, "Paratesto e narratività in antropologia," 283–285; Sedlmeier, "Paratext and Literary Narration," 70. Of functions: Birke and Christ, "Paratexts," 67–69; Armstrong, "Paratexts and Their Functions," 42; Marot, "Pour une poétique," 10–11; Masson, "Marginalité de la preface," 14–5; Santoro, "Ancora sull'indagine," 17; Ozment, "'She Writes Like a Woman'," 3; Murray, "Politics of the Preface," 50; Pearson, "Art of Self-Creation," 40–43.

54. Some exceptions include Anthony Grafton's *Footnote;* Tweed and Scott (eds.), Medical Paratexts (Palgrave Macmillan, 2018); Grimaldi, "Les prefaces des dictionnaires de langue," in Marot, *Textes liminaires*; Sabato, "Paratesto e narrativitå in antropologia," *Neohelicon* (2010); Skelton, "Paratext of Everything," *Book History* (2001).
55. Kinser, "Paratextual Paradise," 6.
56. Duncan and Smyth, "Introductions," 5–6.
57. Rozzo, "In paratesto," 213–214, 223–224.
58. Birke and Christ, "Paratexts," 67–68.
59. [Wilson], "Guicciardini," 439–440.
60. Secord, *Victorian Sensation*, 126.

References

Printed Primary Sources

[Anon.]. "Freeman's History of William Rufus." *The Times*, April 17, 1882, 4.

[Anon.]. "Last Words by Professor Freeman." *Pall Mall Gazette*, April 8, 1892, 2.

[Anon.]. "Death of the Bishop of Oxford. A Great Historian." *Daily Telegraph*, April 23, 1901, 7.

Freeman, Edward A. *The Methods of Historical Study*. London: Macmillan, 1886.

Freeman, Edward A. "A Review of My Opinions." *The Forum*, April 1892, 145–157.

[Jessopp, Augustus]. "The Brocas Book." *Edinburgh Review*, July 1887, 225–253.

[Hayward, Abraham]. "Lord Macaulay and His School." *Quarterly Review*, April 1868, 287–333.

Round, J. H. "Scientific History." *Athenaeum*, October 26, 1889, 560.

[Warre-Cornish, F. W.]. "Gardiner's *Protectorate*." *Quarterly Review*, April 1898, 446–470.

[Wilson, John]. "Guicciardini's *Personal and Political Records*." *Quarterly Review*, October 1871, 416–440.

Secondary Sources

Algazi, Gadi. "Exemplum and Wundertier: Three Concepts of the Scholarly Persona." *BMGN—Low Countries Historical Review* 131, no. 4 (2016): 8–32.

Allen, Robert. "Perpetually Beginning Until the End of the Fair: The Paratextual Poetics of Serialised Novels." *Neohelicon* 37, no. 1 (2010): 181–189.

Armstrong, Guyda. "Paratexts and Their Functions in Seventeenth-Century English *Decamerons*." *Modern Language Review* 102 (2007): 40–57.

Arroyo Redondo, Susana. "Aproximaciones teóricas al prólogo: su papel en la narrative española reciente." *Revista de Literatura* 76, no. 151 (2014): 57–77.

Bentley, Michael. *Modernizing England's Past: English Historiography in the Age of Modernism 1870–1970*. Cambridge: Cambridge University Press, 2005.

Berger, Stefan. "Professional and Popular Historians 1800 – 1900 – 2000." In *Popular History Now and Then: International Perspectives*, edited by Barbara Korte and Sylvia Paletschek, 13–29. Bielefeld: Transcript, 2012.

Birke, Dorothea and Birthe Christ. "Paratexts and Digitized Narrative: Mapping the Field." *Narrative* 1, no. 21 (2013): 65–87.

Bosch, Mineke. "Scholarly Personae and Twentieth-Century Historians." *BMGN—Low Countries Historical Review* 131, no. 4 (2016): 33–54.

Cabanel, Anna. "A Woman in a 'Man Made World': Erzsébet Kol (1897–1980)." In *Gender, Embodiment, and the History of the Scholarly Persona: Incarnations and Contestations*, edited by Kirsti Niskanen and Michael J. Barany, 113–146. Cham: Palgrave Macmillan, 2021.

Claes, Koenraad. "Supplements and Paratext: The Rhetoric of Space." *Victorian Periodicals Review* 43, no. 2 (2010): 196–210.

Collini, Stefan. *Public Moralists: Political Thought and Intellectual Life in Britain 1850–1930*. Oxford: Oxford University Press, 1991.

Collini, Stefan. *Absent Minds: Intellectuals in Britain*. Oxford: Oxford University Press, 2006.

Condren, Conal, Stephen Gaukroger, and Ian Hunter (eds.). *The Philosopher in Early Modern Europe: The Nature of a Contested Identity*. Cambridge: Cambridge University Press, 2006.

Creyghton, Camille. "Generational Continuities and Composite Personae: French Historiography from the 1870s to the 1950s." In *How to Be a Historian: Scholarly Personae in Historical Studies, 1800–2000*, edited by Herman Paul, 72–88. Manchester: Manchester University Press, 2019.

Creyghton, Camille, Pieter Huistra, Sarah Keymeulen, and Herman Paul. "Virtue Language in Historical Scholarship: The Cases of Georg Waitz, Gabriel Monod and Henri Pirenne." *History of European Ideas* 42, no. 7 (2016): 924–936.

Crossick, Geoffrey "From Gentlemen to the Residuum: Language of Social Description in Victorian Britain." In *Language, History and Class*, edited by Penelope J. Corfield, 150–178. Oxford: Blackwell, 1991.

Curthoys, M. C. "Jessopp, Augustus." *Oxford Dictionary of National Biography*. https://doi-org.libproxy.helsinki.fi/10.1093/ref:odnb/34187. Accessed July 13, 2022.

Daston, Lorraine and Peter Galison. *Objectivity*. New York: Zone Books, 2010.

Daston, Lorraine and H. Otto Sibum. "Introduction: Scientific Personae and Their Histories." *Science in Context* 16, no. 1–2 (2003): 1–8.

Daunton, Martin. "Introduction." In *The Organisation of Knowledge in Victorian Britain*, edited by Martin Daunton, 1–27. Oxford: Oxford University Press, 2005.

Den Boer, Pim. *History as a Profession: The Study of History in France, 1818–1914*. Princeton: Princeton University Press, 1998.

Duncan, Dennis and Adam Smyth. "Introductions." In *Book Parts*, edited by Dennis Duncan and Adam Smyth, 3–10. Oxford: Oxford University Press, 2019.

Egginton, Heidi and Zoë Thomas (eds.). *Precarious Professionals: Gender, Identities and Social Change in Modern Britain*. London: University of London Press, 2021.

Ellis, Heather. *Masculinity and Science in Britain, 1831–1918*. London: Palgrave Macmillan, 2017.

Feather, John. *A History of British Publishing*. London: Croom Helm, 1988.

Fuchs, Eckhard. "Conceptions of Scientific History in the Nineteenth-Century West." In *Turning Points in Historiography: A Cross Cultural Perspective*, edited by Q. Edward Wang and Georg G. Iggers, 148–161. Rochester: University of Rochester Press, 2002.

Garritzen, Elise. "Revise, Edit, and Improve: Writing and Publishing History as an Unending Process in Victorian Britain." *Clio: A Journal of Literature, History, and the Philosophy of History* 45, no. 3 (2016): 289–314.

Genette, Gerard. *Paratexts: Thresholds of Interpretation*. Cambridge: Cambridge University Press, 1997.

Goffman, Erving. *The Presentation of Self in Everyday Life*. London: Penguin Books, 1990.

Goldgar, Anne. *Impolite Learning: Conduct and Community in the Republic of Letters 1680–1750*. New Haven: Yale University Press, 1995.

Goldstein, Doris S. "The Organizational Development of the British Historical Profession, 1884–1921." *Bulletin of the Institute of Historical Research* 55 (1982): 180–193.

Gooday, Graeme. "Ethnicity, Expertise and Authority: The Cases of Lewis Howard Latimer, William Preece and John Tyndall." In *Scientists' Expertise as Performance*, edited by Joris Vandendriessche, Evert Peeters, and Kaat Wils, 15–29. London: Routledge, 2020.

Grafton, Anthony. *The Footnote: A Curious History*. Kent: Faber and Faber, 2003.

Gray, Jonathan. *Show Sold Separately: Promos, Spoilers, and Other Media Paratexts*. New York: NYU Press, 2010.

Grimaldi, Elisabeth. "Les prefaces des dictionnaires de langue du XIXe siècle, entre 'manifeste' et 'mode d'emploi'." In *Les textes liminaires*, edited by Patric Marot, 181–211. Toulouse: Presses Universitaires du Mirail, 2010.

Hakosalo, Heini. "Cut out for Medicine: Anatomical Studies and Medical Personae in Fin-de-Siècle Finland." In *Gender, Embodiment, and the History*

of the Scholarly Persona: Incarnations and Contestations, edited by Kirsti Niskanen and Michael J. Barany, 149–180. Cham: Palgrave Macmillan, 2021.
Hesketh, Ian. *The Science of History in Victorian Britain: Making the Past Speak*. London: Pickering & Chatto, 2011.
Hilgartner, Stephen. *Science on Stage: Expert Advice as Public Drama*. Stanford: Stanford University Press, 2000.
Hobbs, Andrew. *A Fleet Street in Every Town: The Provincial Press in England, 1855–1900*. Cambridge: Open Book Publishing, 2018.
Howsam, Leslie. *Old Books & New Histories: An Orientation to Studies in Book and Print Culture*. Toronto: University of Toronto Press, 2006.
Howsam, Leslie. *Past into Print: The Publishing of History in Britain 1850–1950*. London: British Library and University of Toronto Press, 2009.
Iggers, Georg G. "Historiography and Politics in the Twentieth Century." In *Societies Made Up of History: Essays in Historiography, Intellectual History, Professionalisation, Historical Social Theory & Proto-Industrialisation*, edited by Ragnar Björk and Karl Molin, 3–16. Stockholm: Molin, 1996.
Ives, Maura. "A Bibliographical Approach to Victorian Publishing." In *Literature in the Marketplace: Nineteenth-Century British Publishing and Reading Practices*, edited by John O. Jordan and Robert L. Patten, 269–288. Cambridge: Cambridge University Press, 1995.
Jann, Rosemary. *The Art and Science of Victorian History*. Columbus: Ohio State University Press, 1985.
Jardine, Nick. "Books, Texts and the Making of Knowledge." In *Books and the Sciences in History*, edited by Marina Frasca-Spada and Nick Jardine, 393–407. Cambridge: Cambridge University Press, 2000.
Johns, Adrian. *The Nature of the Book: Print and Knowledge in the Making*. Chicago: University of Chicago Press, 2009.
Kingstone, Helen. *Victorian Narratives of the Recent History: Memory, History, Fiction*. Cham: Palgrave Macmillan, 2017.
Kinser, Samuel. "Paratextual Paradise and the Devilish Arts of Printing." In *Paratext: The Fuzzy Edges of Literature*, edited by Carla Dauven, Daan den Hengst, Jelle Koopmans, and Lisa Kuitert, 5–38. Amsterdam: Institute of Culture and History, University of Amsterdam, 2004.
Lambert, Peter. "The Professionalization and Institutionalization of History." In *Writing History: Theory & Practice*, edited by Stefan Berger, Heiko Feldner, and Kevin Passmore, 42–60. London: Hodder Arnold, 2003.
Lang, Timothy. *The Victorians and the Stuart Heritage: Interpretations of a Discordant Past*. Cambridge: Cambridge University Press, 1995.
Lightman, Bernard. *Victorian Popularizers of Science: Designing Nature for New Audience*. Chicago: University of Chicago Press, 2007.
Lyons, Martyn. *Readers and Society in Nineteenth-Century France: Workers, Women, Peasants*. Basingstoke: Palgrave 2001.

Maitzen, Rohan Amanda. *Gender, Genre, and Victorian Historical Writing*. New York: Garland Publishing, 1998.

Mak, Bonnie. *How the Page Matters*. Toronto: University of Toronto Press, 2011.

Marot, Patrick. "Pour une poétique historique des textes luminaires." In *Les textes liminaires*, edited by Patrick Marot, 7–27. Toulouse: Presses Universitaires du Mirail, 2010.

Masson, Pierre. "Marginalité de la préface autoriale." In *L'Art de la préface*, edited by Philippe Forest, 11–23. Nantes: Éditions Cécile Defaut, 2006.

McKenzie, D. F. *Bibliography and the Sociology of Texts*. London: British Library, 1986.

Melman, Billie. *The Culture of History: English Uses of Past 1800–1953*. Oxford: Oxford University Press, 2006.

Mitchell, Rosemary Ann. "'The Busy Daughters of Clio': Women Writers of History from 1820 to 1880." *Women's History Review* 7, no. 1 (1998): 107–134.

Moner, Michel. "Introducción: el paratexto: ¿para qué?" In *Paratextos en la literatura española (siglos xv – xviii)*, edited by María Soledad Arredondo, Pierre Civil, and Michel Moner, xi–xviii. Madrid: Casa de Velázquez, 2007.

Morus, Iwan Rhys. "Worlds of Wonder: Sensation and the Victorian Scientific Performance." *Isis* 101, no. 4 (2010): 806–816.

Murray, Jessica. "The Politics of the Preface: Lady Anne Barnard's Gendered Negotiations in a Liminal Textual Space." *English Studies in Africa* 56, no. 2 (2013): 49–59.

Myers, Natasha. "Pedagogy and Performativity: Rendering Laboratory Lives in the Documentary *Naturally Obsessed: The Making of a Scientist*." *Isis* 101, no. 4 (2010): 817–828.

Niskanen, Kirsti. "The Scholarly Persona Embodied: Seclusion, Love, Academic Battles, and International Exchange in the Shaping of a Philosophy Career." In *Gender, Embodiment, and the History of the Scholarly Persona: Incarnations and Contestations*, edited by Kirsti Niskanen and Michael J. Barany, 315–348. Cham: Palgrave Macmillan, 2021.

Niskanen, Kirsti and Michel J. Barany (eds.). *Gender, Embodiment, and the History of the Scholarly Persona: Incarnations and Contestations*. Cham: Palgrave Macmillan, 2021.

Niskanen, Kirsti, Mineke Bosch, and Kaat Wils. "Scientific Persona," special issue of *Persona Studies* (2018).

Opitz, Donald L., Anette Lykknes, and Brigitte Van Tiggelen. "Introduction." In *For Better or For Worse? Collaborative Couples in the Sciences*, edited by Annette Lykknes, Donald L. Opitz, and Brigitte Van Tiggelen, 1–15. Basel: Birkhäuser, 2012.

Ozment, Kate. "'She Writes Like a Woman': Paratextual Marketing in Delarivier Manley's Early Career." *Authorship* 5, no. 1 (2016): 1–15.

Paul, Herman. "What Is a Scholarly Persona? Ten Theses on Virtues, Skills, and Desires." *History and Theory* 53 (2014): 348–371.

Paul, Herman. "Sources of the Self: Scholarly Personae as Repertoires of Scholarly Selfhood." *BMGN—Low Countries Historical Review* 131, no. 4 (2016): 135–154.

Paul, Herman. "The Virtues of a Good Historian in Early Imperial Germany: Georg Waitz's Contested Example." *Modern Intellectual History* 13, no. 3 (2018): 681–709.

Paul, Herman (ed.). *How to Be a Historian? Scholarly Personae in Historical Studies, 1800–2000*. Manchester: Manchester University Press, 2019.

Paul, Herman. "Scholarly Personae: What They Are and Why They Matter." In *How to Be a Historian: Scholarly Personae in Historical Studies, 1800–2000*, edited by Herman Paul, 1–14. Manchester: Manchester University Press, 2019.

Pearson, John H. "The Art of Self-Creation: Henry James in the New York Edition Prefaces." In *Marketing the Author: Authorial Personae, Narrative Selves and Self-Fashioning, 1880–1930*, edited by Marysa Demoor, 40–53. New York: Palgrave Macmillan, 2004.

Reader, W. J. *Professional Men: The Rise of the Professional Classes in Nineteenth-Century England*. London: Weidenfeld and Nicolson, 1966.

Rockenberger, Annika and Per Röcken. "Typograhpie als Paratext? Anmerkungen zu einer terminologischen Konfusion." *Poetik* 41, no. 3–4 (2009): 293–330.

Ross, Travis E. "Fixing Genius: The Romantic Man of Letters in the University Era." In *How to Be a historian: Scholarly Personae in Historical Studies, 1800–2000*, edited by Herman Paul, 53–71. Manchester: Manchester University Press, 2019.

Rozzo, Ugo. "Il paratesto e l'informazione bibliografica." *Paratesto: Rivista Internazionale* 3 (2006): 211–231.

Ruokkeinen, Sirkku and Aino Liira. "Material Approaches to Exploring the Borders of Paratext." *Textual Cultures* 11, no. 1–2 (2017): 106–129.

Sabato, Gaetano. "Paratesto e narratività in antropologia: il *prologue* a *Non-lieux* du Augé." *Neohelicon* 37, no. 1 (2010): 275–286.

Santoro, Marco. "Ancora sull'indagine paratestuale." *Paratesto: Rivista Internazionale* 4 (2007): 9–26.

Secord, James. "Progress in Print." In *Books and the Sciences in History*, edited by Marina Frasca-Spada and Nick Jardine, 369–389. Cambridge: Cambridge University Press, 2000.

Secord, James A. *Victorian Sensation: The Extraordinary Publication, Reception, and Secret Authorship of Vestiges of the Natural History of Creation*. Chicago: University of Chicago Press, 2000.

Secord, James A. "Science, Technology and Mathematics." In *The Cambridge History of the Book in Britain*, edited by David McKitterick, 443–474. Vol. 6 Cambridge: Cambridge University Press, 2009.

Sedlmeier, Florian. "The Paratext and Literary Narration: Authorship, Institutions, Historiographies." *Narrative* 26, no. 1 (2018): 63–80.

Shapin, Steven. *The Scientific Life: A Moral History of Late Modern Vocation*. Chicago: University of Chicago Press, 2010.

Skelton, Matthew. "The Paratext of Everything: Constructing and Marketing H. G. Wells's *The Outline of History*." *Book History* 4 (2001): 237–275.

Smith, Helen and Louise Wilson. "Introduction." In *Renaissance Paratexts*, edited by Helen Smith and Louise Wilson, 1–14. Cambridge: Cambridge University Press, 2011.

Spongberg, Mary. *Women and the Nation's Past: Empathetic Histories*. London: Bloomsbury, 2018.

Tollebeek, Jo and Ilaria Porciani. "Institutions, Networks and Communities in a European Perspective." In *Setting the Standards: Institutions, Networks and Communities of National Historiography*, edited by Ilaria Porciani and Jo Tollebeek, 3–26. New York: Palgrave Macmillan, 2012.

Topham, Jonathan R. "Scientific Publishing and the Reading of Science in Nineteenth-Century Britain: A Historiographical Survey and Guide to Source." *Studies in History and Philosophy of Science* 31, no. 4 (2000): 559–612.

Torstendahl, Rolf. *The Rise and Propagation of Historical Professionalism*. New York: Routledge, 2015.

Townsend, Robert B. *History's Babel: Scholarship, Professionalization, and the Historical Enterprise in the United States, 1880–1940*. Chicago: University of Chicago Press, 2013.

Tweed, Hannah C. and Diane G. Scott (eds.). *Medical Paratexts from Medieval to Modern: Dissecting the Page*. Cham: Palgrave Macmillan, 2018.

Vandendriessche, Joris, Evert Peeters, and Kaat Wils. "Introduction: Performing Expertise." In *Scientists' Expertise as Performance*, edited by Joris Vandendriessche, Evert Peeters, and Kaat Wils, 1–13. London: Routledge, 2020.

Warren, John. "The Rankean Tradition in British Historiography, 1840 to 1950." In *Writing History: Theory & Practice*, edited by Stefan Berger, Heiko Feldner, and Kevin Passmore, 23–41. London: Hodder Arnold, 2003.

Weedon, Alexis. "An Analysis of the Cost of Book Production in Nineteenth-Century Britain." In *Book Publishing*, edited by John Feather, 216–237. Vol. 2. New York: Routledge, 2011.

Wils, Kaat and Pieter Huistra. "Scholarly Persona Formation and Cultural Ambassadorship: Female Graduate Students Travelling Between Belgium and the United States." In *Gender, Embodiment, and the History of the Scholarly Persona: Incarnations and Contestations*, edited by Kirsti Niskanen and Michael J. Barany, 83–111. Cham: Palgrave Macmillan, 2021.

PART I

Historians as Scholars

The amateur historian and Conservative politician Henry Howorth concluded in 1892 that history had become "more difficult, more precise and more methodical" and, consequently, there was less room for "the untrained, untaught, and unscholarly amateur."[1] This was precisely what the scientific historians hoped to achieve when they began to forge themselves a scholarly persona that corresponded with the requirements of scientific history. The first part of this book spotlights historians' attempts to establish normative conditions for a historian, define the virtues which scientific historians should cultivate, and promote the methods and aims that separated scientific historians from antiquaries, genealogists, and all the others who explored past events and personalities. Through the persona, historians fixed epistemological, methodological, and ethical standards for the discipline and its representatives, but the persona they invented was characterized by a significant gap between their rhetorical moves and actual practices. This highlights the fact that personae are idealized models. The contradictions and paradoxes are also revealing about the growing pains of a modern discipline with long historical roots.

Note

1. Howorth quoted in Levine, *Amateur and the Professional*, 168.

CHAPTER 2

Educated and Well-Connected Oxbridge Men

When William Stubbs published in 1900 *Seventeen Lectures on the Study of Mediaeval and Modern History*, he invested its title page with an astounding number of seventeen attributes that promoted his merits. Readers learned that he was Bishop of Oxford, an honorary Student of Christ Church, the late Regius Professor of Modern History, a holder of five honorary doctorates, a member of international academies, and a recipient of the prestigious Knight of the Prussian Order's *Pour le Mérite*. This register of achievements invoked in the audience an impression of exceptionalism and created such a textually and visually arresting effect that it prompted one reader to scribble "wow!" on the title page.[1] As Whitney Trettien has suggested, a title page is an encoded bibliographical site that informs readers about the text and its author and generates confidence in both.[2] Historians' title pages set up interpretive possibilities and foregrounded authorial agency. As the attributions encompassed a range of titles and merits, the persona which they produced could be prismatic. This chapter takes its cue from this polychromatic character of the persona as an articulation of the institutional and normative conditions that underpinned it, and explores its relationship to various institutional, professional, and social settings in a manner that resembles the multidimensional nature of historians' scholarly persona (Fig. 2.1).

As title pages invite us to unearth meanings which historians assigned to memberships in learned societies, academic degrees, fellowships, and

E. Garritzen, *Reimagining the Historian in Victorian England*,
https://doi.org/10.1007/978-3-031-28461-8_2

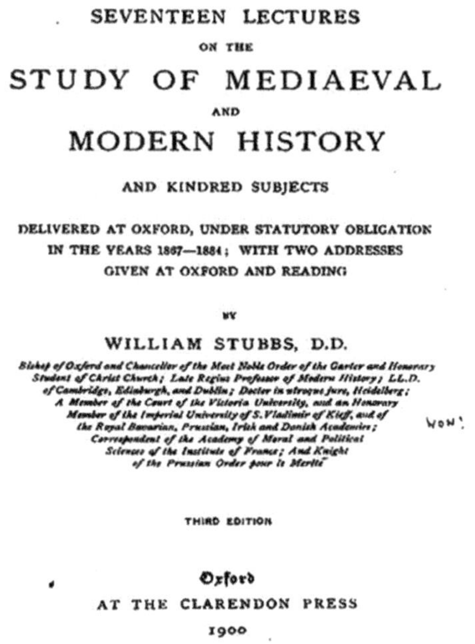

SEVENTEEN LECTURES

ON THE

STUDY OF MEDIAEVAL

AND

MODERN HISTORY

AND KINDRED SUBJECTS

DELIVERED AT OXFORD, UNDER STATUTORY OBLIGATION
IN THE YEARS 1867—1884; WITH TWO ADDRESSES
GIVEN AT OXFORD AND READING

BY

WILLIAM STUBBS, D.D.

Bishop of Oxford and Chancellor of the Most Noble Order of the Garter and Honorary Student of Christ Church; Late Regius Professor of Modern History; LL.D. of Cambridge, Edinburgh, and Dublin; Doctor in utroque Jure, Heidelberg; A Member of the Court of the Victoria University, and an Honorary Member of the Imperial University of S. Vladimir of Kieff, and of the Royal Bavarian, Prussian, Irish and Danish Academies; Correspondent of the Academy of Moral and Political Sciences of the Institute of France; And Knight of the Prussian Order pour le Merite

WOW!

THIRD EDITION

Oxford

AT THE CLARENDON PRESS

1900

Fig. 2.1 The title page of William Stubbs's *Seventeen Lectures* was a visually impressive testimony of his achievements and made a deep impact on its reader (*Source* Internet Archive, original in Harvard University Library)

professorships, they provide a unique opportunity to trace how history as a modern academic discipline and a profession began to take shape during the latter half of the century. The attributions on the title pages also document the broader developments in English academic culture during the later Victorian era. Oxford and Cambridge had faced growing external pressure in the 1840s and 1850s to prepare their students for a more complex and professional society. The reforms led to a gradual revision of the aims and methods of teaching and to the modernization of academic occupations. The traditions were not revised overnight, but by the end of the century, Oxford and Cambridge resembled modern, secular universities where the ambiance had shifted from the aristocratic and clerical to the bourgeoisie and where hard work, efficiency, and utility of knowledge were appreciated.[3] New schools and tripos and the reorganization of the roles and responsibilities of tutors and professors came to have an impact on historians' persona, attesting to its mutability. These alterations were also inscribed on the title pages. The repetition of the matrix of attributions from one title page to another promoted a scientific historian as an

Oxbridge educated middle-class Englishman with a university appointment, but a closer analysis of the meanings ascribed to the attributions suggests that the transition into a merit- and award-based professional academic culture was not that straightforward at all. Rather, historians fostered conflicting ideas about the role academic and professional institutions should have in sheltering their persona and stimulating original investigation. Owing to the dissonance between the vested symbolism of the attributions and actual reality, title pages offer a fascinating window into the restricted institutional conditions that made the stage for negotiating scholarly personae, and remind us once again why books were so important for promoting that persona.

The title pages, moreover, participated in scholarly boundary work. Stubbs's title page in *Seventeen Lectures* was an antithesis to the sparseness of title pages in early-Victorian history books or in late-Victorian amateur histories. Lingard, Hallam, and Macaulay had trusted almost solely on the power of their names.[4] Scientific historians, on the contrary, had no scruples about promoting their excellence on the title page. While listing virtues in prefaces was condemned as subjective and egomaniacal, the moralizing did not extend to the title pages, which recorded seemingly objective and verifiable facts about the writer. Professorships, fellowships, honorary degrees, and other academic honors could be granted on political and religious bases, but this widely known fact did not seem to diminish their symbolic value as markers of proven scientific excellence. This chapter holds that historians took advantage of the title page not just to establish normative standards for their persona but also to emphasize its difference to the personae cultivated by the men of letters or contemporary amateur historians. The appropriation of the title page as a discursive performance of academic excellence allowed historians to teach readers to identify the practitioners of scientific history.

Anatomy of a Title Page in a History Book

Title pages are gateways to books, and before dust jackets with author biographies became popular in the twentieth century, a title page was the first instance a reader could learn about the writer.[5] Title pages were crucial for enticing readers, and they have been critical entryway paratexts since the early days of printing. They have retained their core elements such as a title, author, publisher, and the time and place of publishing. Their architecture and design, though, have followed altering fashions.

The title pages in Victorian history books were a reversal of the richly decorated early modern title pages. The move from a pictorial to textual orientation began during the last decades of the eighteenth century, and decorativeness was reserved in late-Victorian histories for popular and amateur publications.[6] In historical studies, the restrained typography functioned as an outward display of historians' virtues such as sobriety, modesty, and gravity. Even vignettes were rare and epigraphs, which had been popular among the Romantic writers, were largely abandoned by the scientific historians.[7] The lack of ornamentation foregrounded both the author and the publishing house as sources of authority and information about the book. As publishers grew more specialized in the nineteenth century, they forged distinct identities which their imprints embodied. "Clarendon Press" on a title page awarded scholarly respectability but repelled those readers who sought amusement rather than instruction. The imprint simultaneously meant reliable scholarship and scholarly dullness.[8]

Although each component on a title page shaped readers' expectations, publishers were mainly concerned about the most commercially viable elements: the title, typography, and layout. They were strikingly indifferent about the authorial attributions, even if they comprehended that historians, too, could be fabricated into saleable brands. Their lack of interest in title page attributions is documented in historians' correspondence with publishing houses. The letters are packed with comments about titles and typography, and some even contain historians' sketches of the desirable layout, whereas remarks about the authorial attributes are almost non-existent.[9] A rare exception comes from C. E. Doble, who was an assistant secretary at Oxford University Press and whose special charge paratexts were. He once wrote to George Birkbeck Hill to explain how honorary doctorates added "to the interest" of a title page and therefore recommended Hill to add his doctorate to the title page of his *Johnsonian Miscellanies*.[10] Publishers' limited role in selecting authorial attributes suggests that historians had a rather free hand to make decisions about them. The attributes, then, represent historians' view of the components that were essential for their authority and persona. The drawback of the absence of such lively discussions is that the meanings which historians embedded into the attributions must be inferred from their general remarks about learned societies, academic degrees, or professions.

The following discussion of title page attributes builds on 349 title pages from 97 historians. To construct this title page data, I have used only the first volumes and later installments from multi-volume histories to avoid the multiplication of the very same attributes in my data. I have divided the historians in my sample into three categories—professors, professionals, and amateurs—to tease out the differences and similarities in their paratextual strategies. The first of these, "professors," includes 24 historians. They all held a professorship in history at some point in their careers. The second group, "professionals," includes 23 historians who earned a living in history-related occupations. They were lecturers, tutors, fellows, librarians, archivists, or museum curators. Historians who belong to these two groups were predominantly middle-class men with a university education. The third category, "amateurs," is the largest, with 50 historians. This is the most heterogeneous group in terms of class, gender, and occupation, comprising aristocrats, politicians, clergy, journalists, public servants, and gentle and middle-class women. I want to emphasize that the three categories do not consist of an equal number of historians or title pages. The differences derive both from my method of collecting the paratext data and from the divergences in the numbers of professors, professionals, and amateurs in the late-Victorian sphere of history writing. The professoriate was a small, yet highly productive community. The number of the other history-related occupations was similarly moderate. Amateur history was popular, but its representatives paled in productivity to the professors. Accordingly, my sample consists of 157 title pages from professors, 85 from professionals, and 107 from amateurs. Owing to the different sizes of the categories, the graphics in this chapter record the results in percentages per category. The numerical breakdown of the data is in Table 2.1.

As with all such categories, there are demarcation disputes at the margins, as the careers of some of the historians complicate the classification. For instance, James Anthony Froude is counted as a professor

Table 2.1 Numerical breakdown of the most common title page attributions

Category	*Amateurs*	*Professionals*	*Professors*
Learned societies	13	7	37
Honorary degree	9	16	65
Academic office	0	34	79
Academic degree	25	37	78
College affiliation	11	34	80

because he held a professorship in history, even though he dissociated himself from the scientific historians, and his rejection of history as a science and unorthodox views about the appropriate style led to his exclusion from the scholarly community. The small coterie of scientific historians relied largely on self-labeling, but a manual classification of the historians would have left too much room for interpretation to offer a viable alternative to a mechanical method. In the absence of other convenient formal markers of scholarly status such as a doctoral degree, which signified professionalism in many other countries, occupational status has offered enough practical utility to draw up the statistics.

The three groups—professors, professionals, and amateurs—adopted unique paratextual strategies. As Chart 2.1 shows, professors and amateurs had different ideas about the appropriate number of attributions and this contributed to historians' boundary work. The amateurs were prudent with attributions, as 66 percent of their title pages had either only one authorial marker or nothing, whereas the same group for the professors' title pages was 19 percent, and for the professionals 29 percent. The difference is even more evident with the title pages that were void of any authorial decorations: only 5 percent, that is eight out of 157 title pages of the professors, were left blank, while corresponding numbers for the amateurs were 20 percent, or 21 out of 107 title pages. Similar differences appear at the other end of the scale. The seventeen attributions in Stubbs's *Seventeen Lectures* were a statistical oddity, but there were also other professors who invested title pages with ten or more attributes. The amateurs abstained from such self-promotion altogether. The title pages of the professionals were pitched between those of the professors and the amateurs. They abstained from decorating their title pages as lavishly as the professors, but otherwise emulated them to show their alignment with the persona of a historian with scientific ambitions.

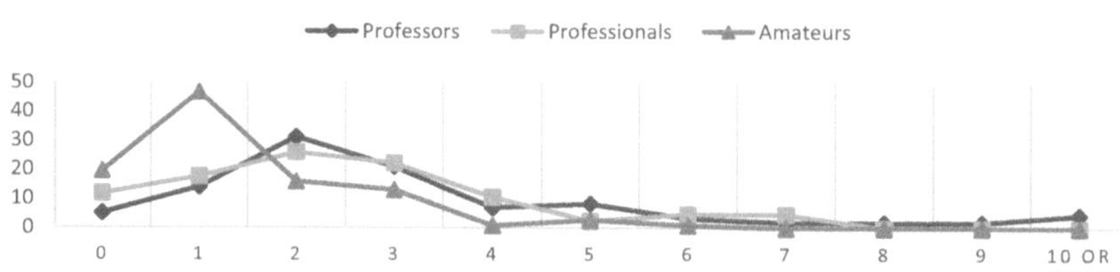

Chart 2.1 Percentage of the number of attributes on a title page

Can the perceivable differences in the title page strategies be attributed to the professors' persona project alone? Or could they rather be explained with the divergent types of histories in my sample? Title pages were adjusted according to the type of book: scholarly studies carried a high number of attributes while textbooks contained a minimal framing of the author. However, this does not offer a sufficient explanation for the contrast between the professors and amateurs, because most of the textbooks and primers in my sample were written by either professors or professionals. Another explanation could be that the historians who dedicated their life to research gained more merits to record on a title page than the amateurs cumulated. This is not an entirely viable explanation either, because many of the amateurs had typical title page merits but chose to suppress them. John P. Prendergast, acclaimed author of *The Cromwellian Settlement in Ireland* (1865), had received an academic education in England and Ireland and been commissioned to edit manuscripts for the Public Record Office alongside his legal work.[11] Nevertheless, his title page introduced him only as "Esq." In the second edition, he presented himself as a "Barrister at Law." A more likely explanation for the diverging paratextual strategies is the professors' and amateurs' different primary points of identification. The professors relished the opportunity to advertise their scientific status, but the amateurs did not have a similar need or impetus to attach themselves to the new kind of image of a historian. The vocabulary of scientification and professionalization had different currency among historians with diverse backgrounds and such an affiliation might have even been a reputational threat to those who strove to uphold the class distinctions between those who wrote history for leisure and those who considered history an occupation.

What, then, did historians include on their title pages? I have classified the attributes into four categories: (1) academic status, (2) scholarly honors, (3) non-academic occupations, and (4) social status including marital status and decorations. The first two categories were critical for the persona project and therefore my attention will be mostly on these. The first includes degrees, college fellowships, and academic offices, and the second honorary doctorates and memberships of learned societies. The value and prestige ascribed to the different fellowships, colleges, and memberships varied and the differences were evident to the scholarly elite. Some were willing to share their wisdom with the less-informed public. One writer instructed the middle-class readers of the *Englishwoman's Domestic Magazine* to choose "*one of the best colleges*" for their sons because each college set its unique stamp on young men and apparently "the stamp of the college remains for a life."[12] Largely, the non-academic audience might not have been well versed on the intricacies of academic institutions and perhaps the long lists of merits were a means to impress those who were unfamiliar with the intricate hierarchies between different colleges, universities, and societies. The profusion of attributes in Stubbs's *Seventeen Lectures* wowed at least one reader.

The memberships of learned societies offer a good example of these undisclosed hierarchies. These memberships had been crucial indications of scholarly authority on title pages for centuries but, as Chart 2.2 shows, in the late-Victorian period degrees and academic affiliations surpassed them in popularity. They were not entirely meaningless either and closer scrutiny shows how historians carefully curated the lists of memberships on their title pages to create distance from the culture of amateur history—at least in the paratextual sphere. They accomplished this by mentioning only the international societies of which they were members and excluding all references to their participation in national and regional societies, as those enjoyed an air of amateurism.

Learned societies were, for the Victorians, entanglements of knowledge and sociability. They brought together people with different social backgrounds and tickled the public imagination. Colonel Creighton, the amateur scholar in Kipling's *Kim*, had set his heart on one day affixing "F. R. S." to his name, believing that whereas other honors

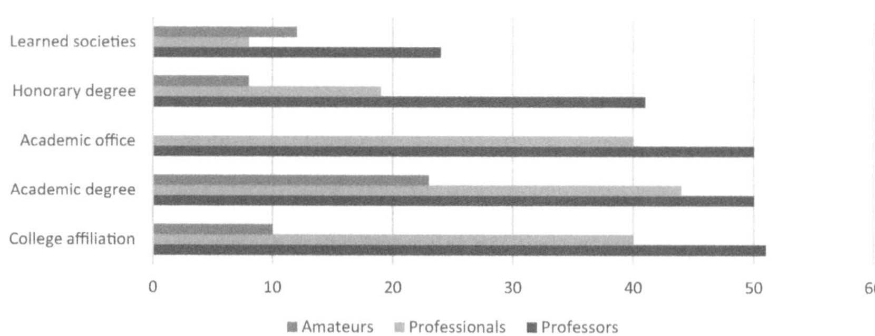

Chart 2.2 Percentage of academic attributions on title pages

were acquired with the help of friends in suitable offices, learned societies operated according to objective and merit-based standards. Only publications "on strange Asiatic cults and unknown customs" paved the way to a membership, he maintained.[13] The reality was different in many societies where the dissemination of knowledge was mixed with political, cultural, and social interests. The professionalizing scientists, wary of amateurism, either tried to reform the existing societies or founded new ones with strictly scientific aims. Yet throughout the century many of the societies remained contested spaces where amateurism and professionalism wrestled, and membership was a mix of leisured aristocrats, Anglican clergy, retired officials, civil servants, and full-time scientists. The societies fitted poorly to the new disciplinary organization of knowledge.[14] The Royal Historical Society, founded in 1868, embodied this tradition of amateurism. It was crowded by dilettanti for its first 30 years serving as a platform for social, cultural, and patriotic pursuits. Memberships were awarded to local and foreign eminences; *The Times* reported in June 1877 how the King of Sweden had accepted an Honorary Fellowship of the RHS. From the professors' point of view, the lack of a scientifically oriented national association was regrettable, as such an institution could have provided a space for cultivating their scientific ethos and nurturing young talent. The RHS was not such an institution until the turn of the century, when the more confidently professional historians seized control of it.[15]

The weak scholarly status of the RHS explains why it was invisible on the professors' title pages. Seeley's *Expansion of England* (1883) is the only exception to this. The same applies to the countless regional and local historical, archeological, and antiquarian societies to which historians belonged. The memberships which the professors listed were instead those granted by foreign societies. Names like *Società Romana di Storia Patria* or *Academy of Sciences of Saint Petersburg* indicated international peer approval and networks. Stubbs could barely hide his delight when he described in his last statutory lecture in Oxford in 1884 how it had been

> exceedingly pleasant to me to receive from the Academies of Germany their recognition that the labours of the Oxford School have not been thrown away. I am very proud to be the recipient of diplomas signed by Döllinger and Giesebrecht, by Curtius, Pauli, Ritschl, and Dove, and to be numbered among the members of the American Academy.[16]

The names of the foreign societies denoted a boundary between the professors and amateurs, as only three of the amateur historians in my sample mentioned membership of a foreign society.[17]

Memberships of foreign societies created an image of historians with wide international networks and of the desirability of such transnational contacts for their persona. However, memberships were not necessarily indications of personal contacts, as they were also granted by publications on their own. Indeed, English historians were insular, apart from the cosmopolitan Acton. They followed the international currents of research, read the *Historische Zeitschrift* and *Revue historique*, and Seeley even tried to acquaint himself with the fashionable German *Katheder* socialism, but mostly their contacts were limited to a few individuals abroad.[18] The meager connections, though, were precious to them, and their enthusiasm could amuse their foreign friends. When the German historian Reinhold Pauli visited England, attention was ensured. His *Lebenserinnerungen* depicts Freeman, and in a lesser degree also Stubbs, as eager hosts who showed him around like a trophy.[19] Seeley, too, capitalized on his acquaintance with Pauli by dedicating *Life and Times of Stein* (1878) to him. This was extraordinary, as dedications had gone out of fashion in scientific publications by this time. The altering economies of authorship and publishing had erased the financial role which laudatory epistles had taken in early modern books.[20] At Oxford University Press they were now considered unnecessary decorations and "out of place" in serious scholarly works.[21] Most historians seemed to agree with this. In my paratext data, dedications were used by amateurs and women who hoped that the names of eminent historians would lend distinction to their books.

However, the title pages should not be read as accurate records of historians' participation and non-participation in historical and antiquarian societies. Many of them were active in regional societies, but since they—or the RHS—conferred the wrong kind of image of their status, they expunged the memberships from their title pages. The international societies, on the contrary, enjoyed more prestige and their memberships created a desirable impression of international recognition and networks. The strategic use of memberships is a good indicator of how historians manipulated the title pages as they crafted their persona. The quantitative and qualitative use of the attributions on the title pages presented historians with a wide range of opportunities for depicting idealized portraits of themselves and the discipline they represented.

Academic Degree: Non-expert Skills and Male Sociability

Pedagogical reforms and the refashioning of a doctoral thesis into an exercise of original research contributed to the professionalization of history in nineteenth-century Europe. Leopold von Ranke's famous seminar equipped students with research skills as they learned how to apply the source-critical method, handle—literally—original sources, and conduct independent research. Seminars, moreover, facilitated students' acculturation to the historical community and molded their professional ethos. The thesis work and a PhD completed that process.[22] A university diploma came to mark integration into the scholarly community and a specific level of methodological aptitude, research skills, and historical knowledge. The inductive method identified scientific history in England, too, but its application was not taught systematically at Oxford or Cambridge and the first PhDs were not awarded until 1921. This did not stop the lower degrees from surfacing frequently on historians' title pages. As Chart 2.2 shows, a degree was among the three most common attributions among professors, professionals, and amateurs. Although a degree did not indicate mastery of the historian's craft, it had other important connotations: it symbolized mental habits and sociability appropriate for men serving the public good.

After the founding of the Honour School of Modern History at Oxford in 1871 and the Historical Tripos at Cambridge in 1873, history was no longer an auxiliary subject. This launched a debate about the content and aims of history teaching that lasted well into the twentieth century and had an impact on the symbolic meaning of a degree on a title page. The key question was whether teaching should adhere to the traditions of a liberal education and emphasize the cultivation of intellectual faculties or submit to the needs of the modernizing society and offer practical knowledge and training in skills. Similar questions were debated in other disciplines, and the resistance to change was significant. There was a consensus that universities should produce men of character with a sense of duty rather than experts with narrow specialization. As a contrast to the commercial middle class, such graduates would have epitomized the new "gentle man," a representative of the educated middle class who was located in the upper echelons of the middling orders.[23] Historians agreed with this, and the ethos was personified in Cambridge by Seeley and in Oxford by Freeman. While both were committed to scientific history, they argued that students should be educated for public service, not for historical research.[24] They had wide support for this, and the *Pall Mall Gazette* endorsed their pedagogical philosophy, professing how "Oxford and Cambridge are still the training ground of most of our politicians, and the presence of Mr. Freeman at the one university and Professor Seeley at the other ought to make itself felt on the next generation."[25]

The pressure to teach the scientific method began to grow in the 1880s when young men with degrees in history entered the ranks of tutors and lecturers. The more research-conscious generation pointed out in the 1890s that a student's awareness of the method was critical for a more pronounced demarcation between history and amateur endeavors such as antiquarianism. They did not question liberal education as such, but suggested revisions so that students would become acquainted with the method and original authorities. The response to this was lukewarm and many of the tutors, too, doubted the need for teaching the method, as only a fraction of the students aspired to careers in history. They believed that what most of them needed was general knowledge, not *Hilfswissenschaften*.[26]

It is not surprising that in this atmosphere the German style research seminar provoked little enthusiasm. Although some tutors integrated the method into their teaching and Seeley and Creighton held private seminars for the most promising students, these should not be confused with the thoroughness of the Rankean model. The historical "laboratory" was not properly embraced until the founding of history programs in Manchester and the other civic universities.[27] The more ambitious students sought opportunities to train their argumentative skills and to attempt original research outside the curriculum from the mid-1870s onwards. The Stubbs Society in Oxford and the Politics Society in Cambridge encouraged students, tutors, and professors to come together to discuss epistemologies, methods, source criticism, or hear papers that introduced different approaches to history. The societies embraced a conventional view of history as a storehouse of moral lessons and information about national virtues, vices, and character.[28] Although the societies could be formative for their members, they did not correspond in aims or depth with a seminar. The mixing of historical questions and practical political issues was more in line with the aims of liberal education.

It was indeed possible for history students to march through their Oxbridge studies without ever seeing or touching unprinted historical records or without any chance to test the inductive method in practice. The consequence was that the graduates were ill prepared for independent research and the degree on their title pages did not vouch for proficiency in research skills. This does beg the obvious question of how the aspiring historians were apprenticed to historical research. How were they received in the scholarly community? Where did they assume the persona? As these are highly relevant to the construction of a scholarly persona, a detour from the title pages is justified.

Many learned the historian's craft through trial and error and without proper mentoring from seasoned historians. As the more professionally conscious young historians' dissatisfaction with their inadequate education grew, they began to seek alternate venues for rehearsing research skills and molding their persona. Two such platforms emerged in the 1880s: the *Dictionary of National Biography* and the *English Historical Review* (*EHR*). T. F. Tout, a medievalist and Oxford graduate, later recollected how budding historians had been forced to approach "historical investigation without the least training or guidance in historical method." For him, the *Dictionary of National Biography* edited by Leslie Stephen served as an invaluable induction into research. Stephen was not a trained historian and admitted that his limited familiarity with history was a shortcoming. He certainly compensated for this with his commitment to editing. As Tout remarked, the stringent regulations and Stephen's manner of enforcing obedience to them "constituted for many of us our first training in anything like original investigation." No "École de Chartes in England" would have done the same to the medievalists than Stephen's good sense and "brusque but kindly" editing did, Tout added.[29] Historians recognized the role which the *Dictionary* had in nurturing new talent, and the English Historical Review observed in 1890 how the *Dictionary* "has become a training school of historical research, and we may hope for many fruits of the labours of those whom it is encouraging to devote their energies to [historical] studies."[30] Tout embraced his experiences and when he was appointed professor at Manchester in 1890, he made seminar and thesis work the pedagogical cornerstones of history.[31]

The *English Historical Review* was an even more important educational platform for young historians. The first issue of the *EHR* was released in 1886 after more than two decades of anticipation. Specialized scientific journals had been harbingers of professionalization since the first half of the century and historians had been discussing founding their own journal since the 1860s. Together with potential publishers they had been oscillating between popular and strictly scholarly models and debating the audiences and economic prospects of such journals. As historians gradually assumed a more professional and scientific persona, the younger fellows and professors came to prefer a journal that was aimed at a learned audience. This kind of journal, they emphasized, could compete with the distinguished German and French journals and increase the prestige of English historians abroad.[32] Moreover, it was important that the *EHR* provided space for mentoring new talent, and for that purpose historians needed a publication that both applied the inductive method and made visible in print their scientific procedures and standards.[33] Mandell Creighton, the first editor of the review and Dixie Professor of Ecclesiastical History at Cambridge, cherished his role as a mentor and trainer of upcoming historians. He was not ashamed to admit how the *EHR* became immediately an organ of the "clique," and not just Cambridge or Oxford, but of the entire "clique" of scientific historians.[34]

The *EHR* fulfilled its pedagogical purpose during Creighton's editorship in three important ways. It was a "training ground" that allowed inexperienced historians to refine their skills, a conduit for international dialog, and a manual of scholarly etiquette. Because of Creighton's intense engagement with the *EHR*, his input into setting the standards for historical research and appropriate scholarly conduct cannot be overstated. His main goal was to "raise the level of workmanship" of English historians and he wrote to Acton about how the *EHR* could function as *the* scholarly institution that brings together "a band of isolated & imperfectly trained but clever students."[35] He guided contributors to sharpen their methodological wit and style and quickly realized that he truly was shaping the development of historical research in England. After the first two issues of the *EHR*, he confidently wrote to the publisher Longman, explaining:

> Part of my policy is to be willing to give a start to a young man; a work in which I have really done good service to many. I have pointed out to many aspirants their shortcomings, have put them in the way of things they knew not, and have caused them to rewrite their articles to good purpose … the "English Historical Review" is really worked to some extent as a training ground.[36]

Although Creighton talked about "a young man," he also welcomed reviews and articles from women and, as Rosemary Mitchell has observed, "a substantial minority" of the articles were written by women during the early years of the *EHR*.[37]

Creighton hoped that the *EHR* would challenge the insularity of English historians. He invited non-English topics, included reviews of foreign books, and recruited reviewers from abroad. His mission, however, was hampered by the realities of the publishing industry. Foreign publishers were reluctant to send review copies and he cursed the German publishers whose indifference suggested that "Germans cared little for our opinion." English publishers, too, were at first disinclined to give copies of their books for the *Review*. This helps to explain the somewhat eclectic quality of the list of books that were reviewed during Creighton's editorship. As he explained to Acton, an editor had to be also "a man of business" and accept less valuable books from the publishers in a hope of receiving more valuable ones later.[38]

Finally, the *Review* shaped the historians' persona by defining the boundaries of acceptable scholarly conduct. Creighton was articulate about this, making evident how correct application of the method or the appropriate style were not enough for scientific legitimacy. Conduct and moral habits were also conceived of as essential conditions for reliable scientific knowledge in the Victorian world of science.[39] Creighton hence assumed the role of moderator in controversies; he set guidelines for polemical reviews and promoted scholarly politeness. To Acton he revealed how he turned down entries with abusive language and made decisions about what was appropriate to discuss in public and what suited private correspondence better.[40] This mediation between public and private spheres in controversies set a model for fair criticism and scholarly disputes. Yet scholarly altercations tended to leak from the private into the public. Nor should Creighton's lofty ambitions mislead modern readers to assume marked politeness from the reviews in the *EHR*.

Despite Creighton's editorial interventions, he was disappointed with the quality of the contributions. They were, according to him, mostly "indifferent" and lacked originality, methodological vigor, and organization. Even after his heavy editing, they remained mediocre at best. He vented his frustration in letters to Acton and Browning by ungraciously dissecting his contributors' faults. At least Acton eagerly participated in this backstage screening.[41] Their correspondence bears witness first to their persistent anxieties about the backwardness of English historical scholarship. Second, the letters show how they believed that the poor quality stemmed not just from the writers' methodological weaknesses, but from a lack of character traits and manners which they considered critical for a historian. For example, the first issue of 1889 stirred Creighton to scrutinize submissions from his two Cambridge colleagues. H. M. Gwatking, a medievalist who succeeded Creighton in the Dixie professorship, was "hopeless in the fay of history; he can only follow the line of his own thought & has no possible reader in view." He was even harsher about James Bass Mullinger, a librarian of St John's College. Apparently, Mullinger was "a man of such ruinous temper that he cannot be interfered with; his work is sound but it is never independent." He also wished that Mary Robinson would just stop writing history, for "her knowledge is great; but she is untrained and untrainable & does not know what she wants to be at."[42]

Much of Creighton's criticism can be attributed to the growing pains of the *EHR* as the writers tried to get grips with the novel mode of writing research-based articles and short book reviews. Then again his disappointment derived also from his frustration at the inadequacy of English historians' education and the insufficiency of the scholarly community in nurturing new talent. His reference to the band of isolated students is telling in this sense. Further, the writers' inability to perform according to the disciplinary standards was an embarrassment for the historical community, as the *EHR* was one of the most important arenas in which they promoted their persona. In Goffman's terms, they should have kept the backstage closed from the audience and ensured that the contributors acted according to the script.[43] The weak entries, however, uncovered the shameful state of history teaching and historical research to both the national and international audience. Despite Creighton's harsh verdict, the *EHR* was nonetheless an important innovation and provided historians with their own institutional space where they could demonstrate how to apply the inductive method, how to write scientific texts, and how to behave as dignified historians.

As the need for alternate venues for developing and demonstrating methodological expertise indicates, an academic degree on a title page was not an indicator of scholarly competence, but of a more abstract and general ability for intellectual and literary work. The degree had another important meaning as well, as it signified the author's belonging to an academic community. More than half of the professors' and 40 percent of professionals' title pages specified these loyalties as they announced the writers' links to the Oxbridge colleges, as Chart 2.2 shows. Sociability and conviviality were part of the manly middle-class culture in Victorian society and college life allowed undergraduates to rehearse their social skills. The same writer who advised mothers in the *Englishwoman's Domestic Magazine* to choose a good college for their sons also shared his insights about the impact which universities had on forming "a correct taste of the proprieties of social intercourse" and providing the attendee with an extensive circle of "college contemporaries" in the highest positions in society. Professors and lecturers, the same writer continued, could make a scholar but not so much a proper "Christian gentleman." The maturation of the sociable character took place outside the lecture halls at least as much as it did inside.[44] This was a popular view, and it was this "social character," another article in the same magazine purported, that explained the "*differentia*" of the English universities.[45] This statement is revealing, as it shows both how tight the connection was between liberal education and Englishness and how ignorant some commentators could be about foreign universities and their incentives for students' character formation.

Colleges were, as Julia Stapleton has explained, central in creating communities that glued intellectual circles together before specialization undermined their force.[46] Instead of an influential *doktorvater* who transmitted the scientific ethos to students in Germany, the bonds between students were formative and Browning maintained that the "clash of mind upon mind" and the free intercourse of undergraduates, not the books and lectures, were what gave "real value to an academic training."[47] These communities were, moreover, essential for making a name and reputation at an early stage. Freeman thought that it had been foolish of John Richard Green not to read for honors as he had lost a valuable opportunity to become more widely known in the right circles. Now, Freeman claimed, *he* had to "explain to people who Green is."[48]

The idea that undergraduate sociability shaped students' minds was widely accepted. The unusual academic career of Montagu Burrows provided an ideal backdrop for historians to assess how early social contacts affected the formation of the persona and scholarly aptitude since Burrows had missed the opportunity for such comradeship. Burrows was appointed the first Chichele Professor of Modern History at Oxford in 1862. He entered Oxford at the relatively mature age of 34, serving in the navy before that. He had earned a Double First in 1857 and stayed in Oxford, becoming a private tutor. His nomination to the professorship took everyone by surprise and many wondered how the chair had been given to someone who was so "young in University standing" when the list of candidates had included such names as Freeman and Stubbs. The observers were also shocked that the electors had chosen someone who had not published anything about history when the chair had been established precisely to strengthen the cultivation of research at Oxford. Instead of a serious scholar, they had elected a tutor.[49] Burrows was reminded about his unusual career trajectory throughout his life. As late as 1888, the *Athenaeum* speculated whether he should be called professor or captain and settled tellingly on "Capt. Burrows."[50] More serious for his scholarly credibility were the persistent doubts about his suitability for the professorship.

Jessopp was puzzled by Burrows's career. It was his review of Burrows's *The Family of Brocas* (1886) which he prefaced with the Jubilee eulogy, and his treatment of Burrows's deficient intellectual aptitude was inspired by his tribute to scientific history. Jessopp went so far as to establish a link between Burrows's lack of Oxford education at an early stage and the mediocrity of his academic performance, arguing that an Oxbridge undergraduate experience in youth was crucial for internalizing scholarly standards. He listed Freeman, Stubbs, and Gardiner as paragons of "the new order" of historical discipline and compared Burrows's career instead to a "curious romance" that no "writer of fiction would have dared to invent." His formative years at sea had excluded him from the company of his extraordinarily talented near-contemporaries whose intellectual maturation owed much to their early years in Oxford. "[W]ho is not influenced by the companionships and the rivalries of his early manhood; by the agreements and discussions and the very quarrels, which all contribute to the making of us?" he wondered, and went on, noting that because Burrows had been deprived of such experiences, his studies missed "the same complexion as that which some of his colleagues have been producing in the meantime."[51] The popularity of an academic degree on historians' title pages shows how important university education was for their credibility, but the educational experience that contributed to their authority and collective scholarly persona was narrowly defined. Burrows was one of the few who deviated from the model, and it exposed him to prejudice and marginalization.

An academic degree on a title page conveyed a general level of learning, intellectual habits, internalization of a set of cultural values, sociability, and male camaraderie. It did not, then, denote specialist skills, and Stefan Collini has suggested that as the demand for expertise grew during the second half of the century, a degree no longer sufficed as proof of competence.[52] It is unclear how widely this kind of thinking spread before the turn of the century, when a more explicit link between university education and professional credentials emerged. My statistics, nonetheless, indicate that the professors and professionals were rarely satisfied with only an academic degree on a title page and complemented it with merits that were more manifestly about their expertise.

Historians' Careers and Three Possible Personae

The university reforms altered historians' career prospects and gave an impetus to academic professionalization in England. Charles Oman captured the change in his inaugural lecture in 1906 as he acknowledged the novelty of his career trajectory by admitting how

> I have come before you, not like so many of my predecessors with all the prestige of a reputation gained outside Oxford, not with the glamour of the unknown about me, but simply as a veteran college tutor with twenty-one years of practical teachings interspersed with such research as my leisure would allow.[53]

An academic affiliation was a popular attribution among the professors and professionals as Chart 2.2 indicates. Their title pages record the gradual shift from history as a domestic pursuit into history as a career-oriented profession and introduce historians' three alternate career options and their corresponding personae. Each of them placed different demands on historians but found common ground in scholarly inquiry and in the principles of scientific history regarding methods, epistemologies, virtues, and research ethics. First, there was the early scientific historian who was, in Oman's words, the historian who returned to Oxford after first earning fame elsewhere. Then there were the tutors and professors whose roles and duties the mid-century reformers had wanted to redefine to create two distinct academic occupations, as the modernizing society needed professors who produced fresh knowledge and teachers who were specialists in their subjects. Out of these reforms emerged the career paths which Oman's title pages document exceptionally well, showing how he launched his career in the 1880s as a "B.A."

and a "Fellow of All Souls College," then in the 1890s received an "M.A." and a lectureship at New College. After this, he became a deputy professor of modern history before his appointment as Chichele Professor of Modern History at Oxford in 1905.[54]

A turning point in historians' career prospects was the relaxation of the colleges' celibacy regulations and the election of married fellows at Oxford in the 1870s and Cambridge in the 1880s. Before that, fellows had been compelled to abandon college life upon marriage. This had rendered the domestic sphere at least equally important as the college rooms for historians to formulate the principles of scientific history.[55] Jessopp wrote in the Jubilee eulogy with great insight about the peculiarly domestic state of historical research in England as "no great historical work has ever been written *at* Oxford." The giants like Freeman, Stubbs, and Gardiner had left the university soon after receiving their degrees and had "won their spurs in a larger arena."[56] The culture of domestic history writing was inscribed on the title pages in attributes such as "late" student or "late" fellow of this or that college, presenting historians as approved members of a loose scholarly network. The frequent references to former college memberships suggest that an institutional affiliation was not an insignificant qualifier for the early scientific historians, but it was not decisive for their success or influence in the scholarly community either. While this was not considered extraordinary in England, Freeman discovered that foreigners were unable to comprehend how prominent English historians were not professors.[57]

Freeman, who found himself correcting people about his non-professorial status, exemplifies a historian who built a great public profile as a scientific historian while working in the seclusion of his countryside house. This was a firmly middle-class arrangement and Freeman should not be confused with the aristocratic gentleman-scientists who toiled in their private laboratories.[58] His overlapping roles as a historian, public intellectual, fervent defender of Gladstonian liberalism, and prolific writer, together with his formidable personality, explosive temper, and, what Stubbs called, "vivacity of combative energy," captivated his contemporaries and have fascinated historians ever since.[59] After applying unsuccessfully for a professorship twice in the 1860s, he retreated to Somerleaze and returned to Oxford only in 1884 when he was appointed Regius Professor of Modern History. Despite his remoteness, he established himself as a proponent of the inductive method and as someone who was on a constant crusade against Froude and all the others who compromised historical truth for dramatic effect.[60] He was best known for his five-volume *History of the Norman Conquest of England* (1867–1876), in which he traced the origins of the English Parliament to the Anglo-Saxon era, but his production knew no bounds in variety or vastness. His *oeuvre* is indeed a testimony to the diversity of late-Victorian historians' literary engagements and of the multiplicity of audiences they addressed. He authored textbooks, popular histories, and historical essays. He wrote reviews and political pamphlets and even composed an essay about English architecture for the Baedeker's handbook on Great Britain.[61] His journalistic input was equally impressive; he contributed to the *Saturday Review* alone more than 700 articles between 1860 and 1869.[62]

The rural life suited Freeman's intellectual temper exceptionally well. While Creighton longed for libraries and archives when he resided in Embleton, Freeman's research was not obstructed by the lack of such facilities. As his contemporaries marked with astonishment, the advocate of original authorities abhorred archives and libraries. He rarely consulted anything else but the selection of printed sources which he had in his study, and he was indifferent about paleography and the other hands-on skills that manuscript scholars needed.[63] He also had his wife and daughters on standby waiting for his orders to transcribe notes and prepare indexes. Historical research was, as Bonnie Smith has proposed, a family business.[64] Freeman reflected the benefits of writing history "far from either the advantages or the distractions of a capital or an [sic] University" in the final fifth volume of *Norman Conquest*. Romantically, he narrated how he lived, breathed, and thought about history on the very soil where the events which he recounted had happened. This had made the subject "more of a living thing" to him.[65]

Freeman's bucolic lifestyle became part of his public image. The obituarists in *The Times* and *Daily Telegraph* bracketed his "literary solitude in the country" as his peak period of scholarly creativity. He had impressed upon the public his views on "how history should be studied" through the numerous books, essays, and reviews which he had produced. The only drawback of the isolation was, according to the writers, that when he had returned to Oxford, he had been "past his prime" and "somewhat out of touch" with the new ways of the university.[66] As he was a staunch defender of liberal education and the traditional Oxford culture, it is unlikely that he would have been any more tolerant about football matches, boat races, theatricals, and other new fashions had he returned there at a younger age.

The generations who came in the wake of Freeman had novel possibilities for forming an academic career. The university reforms established separate models for professors and tutors and, with the cultural and social changes in the sciences, endorsed a re-evaluation of the life of a scientist. As James Secord has explained, by the mid-nineteenth century, a generation of scientific careerists had emerged; these young men wanted to settle down, start families, and pursue science as a paid profession without the constant need to submit to journalism, lecture tours, or showmanship.[67] Once the number of available fellowships for historians began to slowly grow in the 1870s and 1880s, historians too could envision history as a possible and respectable occupation for someone with a family. The growth should not be exaggerated, though, as the number of history professors in each of the English universities could be counted on one hand and the expansion of fellowships was moderate.[68] At least equally important as the new teaching positions was the change in the attitudes toward tutors. Oxbridge dons were no longer perceived as failures whose scholarly aspirations had made them appear unmanly. The early-Victorian curricula had not encouraged ambitiousness, and most dons had lacked a research ethos and settled for passing on what was already known.[69] The new tutor was a specialist who both taught and wrote history. Scientific history was now increasingly advanced from the campuses by men who had earned a degree in history. The persona of the domesticated historian, however, did not disappear overnight. Academic historians did not abruptly start to reject independent scholars to protect their authority.[70] They appreciated those non-affiliated historians who followed the standards of scientific history and emulated the appropriate persona.

The university reforms had a direct bearing on the historians' persona project and they materialized in the formation of the persona of a professor and that of a tutor. The new models highlighted the composite nature of the persona, as both consisted of research, teaching, and committee work, although emphasizing them differently. The former prioritized original research and the latter pedagogical skills and the authoring of educational histories. The redefinition of the boundaries and educational responsibilities caused a lasting struggle over institutional and pedagogical power between the professors and tutors.[71] The situation was especially delicate in Oxford where the tutors saw the professors as a threat to their position and were determined to retain control over the School of Modern History and its curricula.[72] The bitterness was inscribed in their rhetoric. Freeman talked about "cock-a-hoop dons" and Oman spilled out his resentment in his inaugural by accusing his predecessors of lecturing on abstruse topics at inconvenient hours, of being "out of date," and of glorifying their indifference or incompetence as teachers.[73]

The main duty of the new kind of professor was to conduct research and publish studies that enabled him to lead and set an example for the disciplinary community. Professors were also assigned annual teaching obligations and encouraged to teach according to their own expertise. Both Freeman and Creighton envisioned themselves in their inaugural lectures as teachers who would inspire a small group of dedicated students to take up original research and form a small "school" of prospective scholars.[74] They quickly discovered that this was naïve optimism. Especially in Oxford, professorial lectures were poorly attended, and professors were isolated from the students because the embittered tutors discouraged their students from attending professors' lectures, presenting them as worthless for the examinations. If history students could pass their studies without gaining any experience with the original authorities, they could also do so without any contact with the professors.

The third constituent in the professors' vocational persona was administrative work. Hardly anyone admitted to enjoying the meetings or the maneuvering and scheming they involved. Perhaps an exception to this were the posts as Oxford Delegates or Cambridge Syndicates, which gave professors a unique opportunity to steer the course of academic publishing at a moment when even the university presses had to revise their age-old publishing practices to meet the challenges of the modern and highly competitive literary marketplace. As the press secretaries relied on historians' expertise when they made publishing decisions or needed an opinion about paratextual design, Stubbs and Frederick York Powell at Oxford and Seeley and Frederic William Maitland at Cambridge left an indirect imprint on numerous history books. Maitland summed this up when he reminisced about Powell and how "a good deal of him" continued to live "in other peoples' books."[75]

The persona of a teaching fellow was composed of the same elements as the professorial persona, but the educational dimension took priority, as tutors bore the responsibility of teaching the courses that prepared students for the examinations. This gave the professors an excuse to disparage them as educational machines who only taught "subjects" and "periods" and crammed the students for the exams. This was an unfair observation. By interacting with the undergraduates and imposing standards for the exams, the tutors shaped the students' historical thinking more profoundly than the professors were willing to admit.[76] Moreover, pedagogical skills and virtues gained prominence in a tutor's persona, and not only because they were considered merits for permanent teaching positions. The tutors were in general keen to improve their teaching and to develop pedagogical skills, as many were painfully aware of their inadequacies as teachers and examiners.[77]

The professors' deliberate misrepresentation of the tutors created a persistent image of them as mere teaching drudges, and even later historians have projected them as teachers with little inclination for research.[78] Yet the tutor's persona and job description included research as well. The pressure to teach, supervise students, and conduct research made the life of young tutors intense and stressful, as they knew that both educational and scholarly output had an impact on the continuation of their fellowship. Fellowships were usually granted for seven years and could then be renewed for life after a vetting that betrayed how middle-class efficiency had reached academia, as the colleges wanted to know how many students with a First the fellow had turned out and how many books he had published.[79] The true nature of tutors' demanding workload unraveled gradually after the implementation of the university reforms. When Creighton chose the life of an Oxford don in 1866, he expected to be resigning himself "to a bookworm's fate" with ample time for study. His illusions were quickly shattered, and eight years later, he exchanged Oxford for a parish life that promised more time for learning and less stress about teaching or the future of his fellowship.[80] Many who came after Creighton were better prepared but nonetheless longed for quiet study, regretting that teaching assumed too much time to permit original research.

The tutors adapted their research to these realities and instead of launching time-consuming projects and comprehensive archive tours around the Continent, they wrote educational histories. Since their historical pursuits differed from the research ideal, their publications have evaded the historiographical gaze and the character and comprehensiveness of their work has not been recognized. They wrote textbooks, manuals, and short popular histories and participated in debates about the state of historical research through reviews and essays in the *English*

Historical Review and other periodicals. These were short enough texts to compose during the odd hours between the terms and summer vacations in Switzerland where many a tutor journeyed to breathe "the mountain air."[81] William Holden Hutton, Fellow of St John's College, wrote a compact history of Hampton Court and provided a vivid portrait of a tutor's life in its preface:

> It [*Hampton Court*] originates in what schoolmasters used to call a "holiday task." College Dons, unlike the popular idea of them, are busy folk, who are constantly at work, teaching or preparing to teach, examining or writing, and each of these things as part of their professional obligations.[82]

Textbooks were an appealing format for the tutors, as they allowed them to combine their pedagogical and historical insights, and therefore the words "Fellow," "Tutor," or "Lecturer" had high currency on the title pages of textbooks and manuals. Their engagement with educational histories had significance for the entire discipline too, as the scientific historians strove to seize control of the educational histories from the amateur historians and incompetent hacks, as Chapter 5 shows. The names and academic credentials of the tutors on the title pages of dozens of textbooks lent support to the attempt to claim that even the smallest histories were an entitlement of qualified historians alone.

But the educational histories also held risks for the tutors, as the small histories fit poorly to the widely accepted idea that only the writing of a "big" history qualified one as a proper historian. Contesting this cult of original research was practically impossible, and similar tensions between teaching, research, and textbooks were also apparent in other disciplines where the pressures of time compelled tutors to scale down their scientific ambitions.[83] Whereas the cult of big history was a cultural template, the format of small histories embodied its own perils as well, as it seemed to encourage intellectual laziness. It was too easy to fall back on the satisfaction which the steady production of textbooks caused when a commitment to patient research yielded scholarly rewards only slowly. The fictional don Edward Slade in W. Warde Fowler's *Oxford Correspondence of 1903* compared himself as a scholar to a kettle that never reached boiling point. Unambitiously, he submitted himself to the wishes of the publishers, who persuaded him to write schoolbooks or books that communicated third- and fourth-hand knowledge: in short, "work that may be done without boiling over." These books were weak tea, made with water from a kettle that "never got beyond a gentle singing on the hob."[84] Fowler, himself a historian and tutor at Lincoln College, probably wrote from experience when he made Slade despondently concede how he would never be able to obtain enough scholarliness to accomplish a real "first-hand work."

The title pages in history books were reflections of historians' changing career structures and of their three alternate personae that derived from the different vocational positions. The title pages introduced both historians who withdrew to their domestic studies and were "late" fellows, as well as the newly fashioned professors, fellows, tutors, and lecturers. By the end of the century, a university appointment had become a constitutive component on the title pages and in the historians' scholarly persona. While each of the persona models adhered to the principles of scientific history, the cult of original research ensured that forging especially the persona of a historian-cum-tutor was anything but straightforward.

Questioning the Alternate Paths to Expertise in History

As technical expertise gained a stronger footing during the last quarter of the century, the relationship between non-historical technical knowledge and history writing drew historians' attention. During the first half of the century, historians had been, among other things, bankers, lawyers, priests, or journalists, but had not seen any use for advertising their occupational status on their title pages. Their reputation as a man with a vision had vouched for their authority. The late-Victorian amateur historians, however, responded to the increasing demand for specialist knowledge by promoting their vocation or profession on a title page. More than 10 percent of their title pages contained a reference to non-academic occupations in the church, law, military, politics, civil engineering, and miscellaneous authorship. While these attributions were calculated to generate confidence, they prompted historians and reviewers to ask whether such professional experience supplied insights that could cumulate reliable and useful historical knowledge. The question split opinion. Freeman was known for his dislike of lawyers who claimed competence in history. He proclaimed in *The Growth of the English Constitution* (1872) that constitutional history had been "perverted at the hands of lawyers," and entertained in 1885 an Oxford audience who, according to a "Lady commentator," was provoked into "repeated bursts of laughter" by his remarks about lawyers' "invincible ignorance" of history.[85] Yet many were less zealous about the boundaries, allowing that some professional experience might prepare men for historical research. Military and naval history were mentioned as examples of such fields, and it was even asked whether it was possible at all to write about historical battles without military training.

Thomas Kebbel, when reviewing Frank S. Russell's *Earl of Peterborough and Monmouth* (1887), endorsed officials' suitability for historical work as their education developed the very same virtues which a historian needed. Russell's title page introduced him as a colonel of the Royal Dragoons, and for Kebbel this was "a guarantee of his accuracy." As a soldier he was disposed to eschew "rhetorical embellishment or superfluous digression."[86] Those who conceived military officials as capable historians stressed their specialist knowledge: they offered practical estimates and made battles and tactics comprehensible to civilians.[87] Yet reviewers found fault in their research skills and critical judgment. Too often, their histories were plagued by morbid hero worship, anachronisms, a lack of generalization, and an undue use of dramatization and imagination.[88] When the scientific method was accepted as a precondition for reliable knowledge, non-historical expertise no longer compensated methodological weaknesses. If a military background did not qualify for historical research, did specialization in history prepare someone for writing military history? This troubled some of the scientific historians who wrote about battle scenes.

Samuel Rawson Gardiner, Professor of Modern History at King's College and an undisputed authority on the Civil War era, raised this question in the preface to the first volume of *History of the Great Civil War* (1886). He feared that it may appear "an impertinence for one who is not only not a soldier, but who knows nothing of the military art, to write about war." He nonetheless defended historians' right and duty to address military history, arguing that war was always more than mere tactics and battles. Whereas soldiers excelled at recounting campaigns, historians mastered what truly mattered for historical comprehension: the "general considerations" and the "true causes."[89] Gardiner's prefatorial reflections caught Adolphus Ward's attention. Ward, professor at Owens College in Manchester, agreed with Gardiner and maintained in the English Historical Review that because "the results of the series of campaigns are not solely, or even mainly, dependent [on] military considerations," the staff officers were not disposed to explain events to the readers. They lacked the competence "to bring home to untrained readers a notion of how, in Ranke's phrase, the thing really happened."[90] The discussion was picked up several years later by Hereford B. George, a history tutor at New College, Oxford, in the preface to his *Battles of English History* (1895). He justified his intervention in military history with the fact that the battles which he addressed had happened in the past. The training of modern soldiers did not prepare men in any way to write about them, he contended.[91] Gardiner, Ward, and George rationalized the scientific historians' claim for military history with a need for a comprehensive contextualization of battles and campaigns and the necessity of a historical perspective which only qualified historians had.

While the scientific historians insisted that military history was an appendix to general history and as such demanded a command of historical methods and a thorough understanding of historical forces, a small group of historians objected to the relegation of military history to a "mere episode of general history." They proposed a persona of their own, requesting both specialist knowledge about military affairs and a rigorous application of the inductive method. It was not a persona for unqualified dramatists or non-experts in military matters.[92] John Laughton was the leading proponent of this approach. Laughton, now a forgotten Professor of Modern History at King's College, embodied expert naval knowledge and proficiency in historical research. He set the model for the others to emulate, but naval and military history were at that moment too niche to find much of a serious following.

After studying mathematics at Cambridge, Laughton worked as an instructor in the navy and the Royal Naval College, where he realized how the past formed the grounds for developing contemporary naval doctrine. He was disillusioned about the current state of naval history and obliged himself to write a more accurate account of naval events in British history. He was instructed in his pursuits by Gardiner and his career took a significant turn when he was nominated for the professorship in 1885. He founded the Navy Record Society in 1893 to advance research on Britain's naval past and was pleased that the society attracted approximately 500 members annually and published three volumes of records each year. Using his institutional status, he advocated naval history as an independent academic topic with high relevance for naval education and doctrine planning.[93] While he was convinced about the usefulness of historical knowledge for officers, he was confident that it was best to leave the writing of history to competent historians. Indeed, he had little patience with poorly executed naval histories.[94]

Although the boundaries between scientific and amateur history were somewhat porous and the paths to historical expertise were not as narrowly restricted to the academic career trajectories as the title pages of the professors and professionals propagated, the debates about military and naval history indicate that the scientific historians strove to marginalize those who justified their competence in history writing with non-academic professional experience. The references to such occupations on title pages indicated an ownership of technical knowledge, but the established historians assigned an auxiliary role to this kind of expertise. It could not replace or make up for inadequate skills in methods which were now considered as the assurance of reliable and truthful knowledge.

Gendered Personae?

Title pages evoked associations of gender, class, and ethnicity and projected the scientific historian as a middle-class Englishman. The vocabulary of expertise was markedly masculinist because women were excluded from universities until the last quarter of the century. Degrees, fellowships, and academic appointments were a manly entitlement and therefore women had limited opportunities for claiming scholarly authority on title pages. Listing previous publications or referring to connections to male authorities whose names could lend prestige to them were the only chances women had for promoting their competence on title pages. Alice Stopford Green capitalized on the reputation of her late husband, the best-selling historian John Richard Green, by introducing herself as "Mrs. J. R. Green" in her first histories, *Henry the Second* (1888) and *Town Life in the Fifteenth Century* (1894). This had some evident risks, as such close affinity was interpreted as a sign of women historians' reduced autonomy and submission to male supervision.[95] If the title pages make visible the gendered and structural obstacles which women historians encountered, they also illustrate how their scholarly opportunities expanded as the century drew to close, and how they negotiated the sociocultural hierarchies to carve themselves space as respected historians.

An academic degree was a symbol of the gendering of expertise and the gender hierarchies in Victorian society. Y. Y. declared in the *Bookman* in 1894 that women historians did not need any formal education. Apparently, "The needle is a good school for patient diligence and methodical nicety," and so needlework was all the training women needed for writing the kind of histories they preferred.[96] Such remarks dismissed women's educational ambitions and disposed their histories as feminine and, obviously, worthless. At the same time, women such as Kate Norgate were writing scientifically acclaimed histories and Oxbridge women were campaigning for educational equality and women's right to earn university degrees.[97] The campaigns disclosed both the gender inequalities in the ancient universities and the symbolic, cultural, social, and economic value of an academic degree. Although a degree on a title page did not denote specialist skills, its meaning became evident when the right to it was denied.

The University of London and University College London were the first to admit degrees to women in 1878 and the new civic universities followed the precedent. The situation was different at the ancient universities. The founding of Girton College in 1869 and Newnham College in 1871 at Cambridge and Lady Margaret Hall in 1878 and Somerville College in 1879 at Oxford brought women to the universities. They were allowed to enter the examinations in the 1880s, but instead of a degree they received a diploma marking the completion of their studies.[98] The problem was that, as Alice Gardner protested, "the world" was unable "to understand that a certificate stating that a woman had attained the standard required for a degree in honours is really as good a guarantee of attainments as the letters B.A. to which every poll man is entitled."[99] Gardner, a history lecturer at Newnham, followed anxiously the debates about women's degrees at Cambridge and Oxford in the 1890s. Men opposed the motion vehemently—and successfully. Even Gardner's archeologist brother Percy objected to it and refused to believe that the lack of a degree truly hindered women's careers.[100] His sister knew better. When she wrote about those events later, she stressed how women had not sought the right for "the letters" out of vanity but because they were perceived as "a certain standard of mental equipment" and therefore had "market value to educated women."[101]

Although the absence of degrees characterizes women's title pages and underlines the inequalities, marks of women's expanding academic opportunities are also imprinted on them. Gardner's title pages encapsulate this, as she appeared in *Synesius of Cyrene. Philosopher and Bishop* (1886) as a "Resident Lecturer, Newnham College, Cambridge" and seven years later identified herself on the title page of *Julian. Philosopher and Emperor* as a "Lecturer of Newnham College, Cambridge, Associate of Newnham

College, Cambridge, Author of 'Synesius of Cyrene'." These affiliations show how the women's colleges offered academic positions for some women historians before the First World War. The symbolic value of the appointments as indicators of expertise and scholarliness was enhanced in the 1890s when the colleges started to demand a university education from their teachers, making the importance of women's education palpable. Moreover, the colleges endorsed research and established fellowships to support young ambitious women.[102] Gardner recollected once how important it had been for her that, when she had been torn between her teaching duties and writing, Miss Clough, the Principal of Newnham, had encouraged her intellectual work by telling her that "it is good for the College that our Lecturers should write books."[103] This legitimized research for her, as it was something that not only fulfilled her selfish curiosity, but had wider implications for women's academic endeavors.

Title pages uncover on the one hand prejudices and structural obstacles, and on the other the expanding opportunities and women's chances to participate in history writing regardless of the prevailing gender hierarchies. The community of historians was vested in the Victorian notions of manliness, but the approaches to women historians varied and were often ambivalent and contextual. Powell and John Horace Round argued that women were capable only of simple plodding work, not independent research.[104] Jessopp disclosed such prejudices in a two-part essay about women historians in 1899. Despite all the "emancipation of the sex," he considered women as historians to be inferior and confidently declared that the future of history belonged to men: "the great builders-up, the great discoverers, the great thinkers, the great historians will not be women."[105] Jessopp's publicly pronounced prejudices led him to exaggerate women's incompetence and invisibility as historians. Women, quite the contrary, were active producers of historical knowledge and engaged in a great variety of topics from the history of gardening and lace to the history of philosophical ideas and conventional political history. Moreover, many of the leading historians supported women and encouraged them to adopt the scientific method. Women's histories were reviewed in leading periodicals and referred to as reliable sources in the footnotes of historical studies. Collaboration and networking across the gender line was also common; Freeman teamed up with Edith Thompson, Creighton with Stopford Green, Gardiner with Mary Dormer Harris, and Maitland with Mary Bateson. Powell, who otherwise loathed intellectual women, found in Stopford Green an important scholarly ally, and even Jessopp had to privately acknowledge women's achievements. He wrote in 1898 to Dormer Harris, a medievalist with Oxford education, how she and the other younger historians were proving to be worthy of the new advantages that arose in history as they used them "so luminously and intelligently."[106]

These gestures of appreciation contest the conventional narrative of the professionalization of sciences which argues that it was a masculine project that validated men's expertise and restricted women's chances for engaging in scientific research. Accordingly, women were assigned a role as amateurs or assistants of their husbands, fathers, and other male relatives, or as popularizers of scientific discoveries. According to the new biological theories, this was only natural, as women were perceived to lack the intellectual power and bodily physique necessary for generating original knowledge.[107] Although this account holds some truth and the disciplinary contours of history were largely shaped by university-educated men, it nonetheless simplifies the complex nature of late-Victorian women's history writing and the fluidity of the boundaries in the scholarly community. The indeterminacy of professionalization and the institutionalization of expertise allowed women to expand the limits of what was possible for them. They gained novel opportunities for engaging with the past and earned, if not full membership, at least partial membership in the historians' community. This is in accordance with the "Thirsk Law" which Joan Thirsk has formulated: whenever there have been new openings in social, economic, cultural, or scientific life, "women have usually been prominent alongside the men" until the institutionalization and formalization hardens the formal structures and "the direction ... always fall[s] under the control of men."[108] The late-Victorian women were experiencing such a transitional period before professionalization and institutionalization made the advancement more challenging for those who aspired to an academic career. But even then women found alternate avenues to practice historical research.[109]

As the late-Victorian women were pitched between the marked amateurism of their predecessors and the academically oriented early twentieth-century historians such as Eileen Power, their historical endeavors have been hard to pinpoint. Perhaps because of this, the focus has been on the early-Victorian women who appropriated a range of genres from novels to biographies and memoirs of feminine worthies. They explored the history of women, social history of families and households, and material culture of daily life.[110] The late-Victorian women continued these traditions but also created new ones as they joined their male colleagues in debates about the nature of historical research and historians' scholarly persona. Their endeavors were not isolated from the scientification of history, and it would be misleading to ascribe a specifically feminine or amateur persona to them. It is therefore critical to give voice to Bateson, Dormer Harris, Gardner, Stopford Green, Thompson, and many other women on the pages of this book.

*

A title page enabled historians to modify and perform in print the normative elements that were considered constitutive for the persona. The attributions were a communal act of institutionalizing historical expertise and curating an authoritative scholarly persona. Scholarly authority no longer derived only from an individual historian's reputation, but also from the collectively established markers of verifiable qualifications that reflected developments in academia and the world of science and learning. The repetition of the matrix of approved qualifications on the title pages introduced several alternate personae which historians could adopt. There were the personae of a domesticated historian, a professor, and a tutor. Despite this stock of available models for a scientific historian, the templates had overarching qualities, as all were grounded on the ideals of scientific history and Victorian middle-class manliness. A scientific historian was portrayed as a university-educated Englishman with wide networks home and abroad and, during the latter quarter of the century, held an academic position, but the reality was messier than this clean-cut image permitted. Title pages established the normative frame for the persona, but historians needed other paratexts to ratify the epistemic and moral virtues which their persona entailed. Turning proper names into paratexts of virtuosity proved to be a successful recipe for this.

Notes

1. This specific copy belongs to the Harvard Library: https://archive.org/details/seventeenlectur00stubgoog/page/n9.
2. Trettien, "Title Pages," 41.
3. Heyck, *Transformation of Intellectual*, 155; Stone, "Size and Composition," 65–66.
4. John Kemble's *The Saxons in England* (1849) breaks the pattern by introducing its author's degrees and memberships in learned societies.
5. Tanselle, *Book-Jackets*, 8–13, 64.

6. Pettigree, *Book in the Renaissance*, 35–36, 99–100; Maclean, *Scholarship*, 125–129; Blair, *Too Much to Know*, 53; Shevlin, "To Reconcile *Book*," 49–52.
7. Gratman, "How to Do Things with Mottoes," 140–141; Scott, "Signs in the Text," 28–29; Buurma, "Epigraphs," 167–168.
8. Macmillan to Price, October 25, 1893, Add. MSS 55394 (2), BL; Hammond, *Reading*, 97; Garritzen, "Paratexts and Footnotes," 2–3, 8; Mason, *Literary Advertising*, 75.
9. Freeman to Macmillan, January 1, 1871, Add MSS 55049, BL; Freeman to Macmillan, March 25 and 27, 1872, Add MSS 55050, BL; Freeman to Macmillan, February 11, 1876, Add MSS 55051, BL; Freeman to Macmillan, July 6, 1879 and April 12, 1880, Add MSS 55052, BL; Gabbet to Stubbs, February 16, 1880, Letter Books 22, OUP; Price to Freeman, June 8, 1880, Letter Books 23, OUP; Freeman to Macmillan, March 25, 1886 and January 29, 1892, Add MSS 55053, BL; Stopford Green to Macmillan, February 9, 1926, Add MSS 55061, BL.
10. Doble to Hill, December 10, 1896, Letter Books 67, OUP; Eliot, "Evolution of a Printer," 106–107.
11. Barnett, "Prendergast," *ODNB*.
12. "Classical School," *Englishwoman's Domestic Magazine*, 1864, 79–80.
13. Kipling, *Kim*, 170–171.
14. Turner, *Contesting Cultural Authority*, 179–180; Lubenow, '*Only Connect*', 27.
15. "Dr. Schliemann," *Times*, June 16, 1877, 9; Prothero, "Historical Societies," 232–234, 241–242; Goldstein, "Organizational Development," 184–186; Lubenow, *'Only Connect,'* 111.
16. Stubbs, *Seventeen Lectures*, 381. All the following references are to the first edition of Stubbs's *Seventeen Lectures* (1886) unless otherwise noted.
17. Henry Howorth, the author of *History of the Mongols*, was a corresponding member of the Royal Academy in Lisbon. Earl Stanhope mentioned on the title page of *History of England Comprising the Reign of Queen Anne* that he was a corresponding member of the Institute of France. Henry Yule was an honorary member of the Geographical Society of Italy, Geographical Society of Berlin, and a corresponding member of the Geographical Society of Paris according to the title page in *The Book of Ser Marco Polo.*
18. Seeley to Browning, April 28, [1877], GBR/0272/OB/1/1455/A, King's Cam.
19. Pauli, *Reinhold Pauli*, 291–292, 297–298.
20. Genette, *Paratexts*, 123–124.
21. Gell to C. Plummer, June 23, 1885, Letter Books 37, OUP; Wheatley, *Dedications of Books*, 1.

22. Eskildsen, "Leopold von Ranke," 463–468.
23. Stapleton, *Political Intellectuals*, 17–20; Perkin, *Rise of Professional Society*, 83–84, 119–120; Reader, *Professional Men*, 158.
24. Hesketh, *Science of History*, 230–231; Jann, *Art and Science*, 226–227.
25. "Additional Notes," *PMG*, March 17, 1884.
26. Oman, *Inaugural*, 20–23; Slee, *Learning and a Liberal Education*, 78–79, 83–93; McLachlan, "Origin and Early Development," 86–91.
27. Creighton, *Life and Letters*, 288–290; Slee, *Learning and a Liberal Education*, 153–160.
28. Soffer, *Discipline and Power*, 168–171.
29. Stephen to Jessopp, November 5, 1886, 1–10 c.1, RUB Bay 0039:07, RUB; Maitland, *Life and Letters*, 370–371.
30. "Review of *The Dictionary*," *EHR*, 1890, 788.
31. Lambert, "Professionalization and Institutionalizing of History," 50.
32. Howsam, "Academic Discipline," 530–534.
33. Kadish, "Scholarly Exclusiveness," 196.
34. Creighton to Acton, September 14, 1885, Add MSS 8119/1/C249, CUL; Creighton to Freeman, November 20, 1889, FA1/7/127a–127b, JRL.
35. Creighton to Acton, January 26, 1886, Add MSS 8119/1/C253, CUL.
36. Creighton to Longman, May 8, 1886, in Creighton, *Life and Letters*, 341.
37. Mitchell, "Women," 2–3.
38. Creighton to Acton, August 6, 1885, in Creighton, *Life and Letters*, 336; Creighton to Acton, April 29, 1886, Add MSS 8119/1/C254, CUL; Creighton to Acton, December 13, 1888, Add MSS 8119/1/, CUL.
39. Bellon, "Moral Dignity," 937–938.
40. Creighton to Acton, July 28, 1885, in Creighton, *Life and Letters*, 334–335; Creighton to Browning, May 1, 1888, OB/1/423/A, King's Cam; Creighton to Acton, January 23, 1889, Add MSS 8119/1/C272, CUL; Creighton to Freeman, November 20, 1889, FA1/7/127a–127b, JRL.
41. Acton to Creighton, January 19, 1886, Add MSS 6871/35–38, CUL; Acton to Creighton, April 8, 1886, Add MSS 6871/39–40, CUL; Creighton to Browning, April 20, 1886, OB/1/423/A, King's Cam; Creighton to Acton, April 29, 1886, Add MSS 8119/1/C254, CUL; Creighton to Acton, July 29, 1888, Add MSS 8119/1/C264, CUL; Acton to Creighton, January 20, 1889, Add MSS 6871/84–85, CUL.
42. Creighton to Acton, January 23, 1889, Add MSS 8119/1/C272, CUL.
43. Goffman, *Presentation of Self*, 116.
44. "Classical School," *Englishwoman's Domestic Magazine*, 1864, 81–82.
45. "Oxford," *Englishwoman's Domestic Magazine*, 1870, 147.
46. Stapleton, *Political Intellectuals*, 22.

47. Browning, *Memories*, 38–39. See Deslandes, *Oxbridge Men*, 57–62 for why the dons could not assume a role of a doktorvater.
48. McDowell, *Alice Stopford Green*, 39.
49. "The Oxford Professoriate," *Saturday Review*, July 12, 1862, 46; Garritzen, "Montagu Burrows," 1–5.
50. "Historic Towns—Cinque Ports," *Athenaeum*, November 3, 1888, 586–587.
51. [Jessopp], "Brocas Book," 232–234.
52. Collini, *Public Moralists*, 235–236.
53. Oman, *Inaugural*, 3.
54. For Oman's career path, see the title pages of *Art of War* (1885), *History of Greece* (1890), *Europe 476–918* (1895), and *History of the Peninsular War* (1903).
55. Deslandes, *Oxbridge Men*, 58–59.
56. Jessopp, "Brocas Book," 232.
57. Freeman to Macmillan, June 3, 1868, Add MSS 55049, BL.
58. Opitz, "'Not Merely Wifely Devotion'," 33–43.
59. Stubbs, *Seventeen Lectures*, 3rd. ed., x. There are no biographies of Freeman, but a good overview of the recent scholarship is G. A. Bremner and Jonathan Conlin (eds.), *Making History: Edward Augustus Freeman and Victorian Cultural Politics* (British Academy, 2015).
60. Hesketh, *Science of History*, 81–93; Jann, *Art and Science*, 173–183; Bratchel, *Edward Augustus Freeman*, 1969, 32–33.
61. Freeman, "Historical Sketch," xxxiii–lxii.
62. Hesketh, *Science of History*, 134.
63. Creighton to Acton, December 21, 1882, in Figgis and Laurence, *Selections*, 306; Stephens, *Life and Letters*, 2: 471–472.
64. Smith, *Gender of History*, 83–86; Opitz, "Not Merely Wifely," 34–35.
65. Freeman, *NC*, 5: ix–x.
66. "Death of Professor E. A. Freeman," *Times*, March 17, 1892, 5; "Death of Professor E. A. Freeman," *Daily Telegraph*, March 18, 1892, 4–5.
67. Secord, *Victorian Sensation*, 478.
68. Slee, *Learning and a Liberal Education*, 70–74; Levine, *Amateur and the Professional*, 156–158.
69. Daunton, "Introduction," 14.
70. Cf. Soffer, *Discipline and Power*, 41.
71. Heyck, *Transformation of Intellectual*, 168.
72. Soffer, *Discipline and Power*, 80–81.
73. Freeman to Thompson, March 15, 1889, U DX9/169, HHC; Oman, *Inaugural*, 5–6, 16.
74. Creighton, *Life and Letters*, 281–282; Freeman, *Methods*, 39–40.
75. Maitland to Poole, May 19, 1904, GBR/0012/MS Add. 7474/45, CUL.

76. Soffer, *Discipline and Power*, 135.
77. James Frank Bright to Browning, November 6, 1881, GBR/0272/OB/1/216/A, King's Cam; Gibbins, "'Old Studies'," 235–238.
78. Annan, *Dons*, 5; Slee, *Learning and a Liberal Education*, 101–102.
79. Covert, *Victorian Marriage*, 57; Heyck, *Transformation of Intellectual*, 226–227.
80. Creighton to C. A. Potter, October 1, 1866, and to Louise Creighton, September 23 and 24, 1874, in Creighton, *Life and Letters*, 40–41, 140–141.
81. James Bass Mullinger to Browning, August 23, 1881, February 6, 1884, June 12, 1884, and September 26, 1893, GBR/0272/OB/1/1165/A, King's Cam.
82. Hutton, *Hampton Court*, v.
83. Porter, "Gentlemen and Geology," 832–835.
84. Fowler, *Oxford Correspondence*, 58–60.
85. Freeman, *Growth*, vii; "Ladies Column," *Evening Telegraph*, December 25, 1885, 5.
86. [Kebbel], "Earl of Peterborough," 187.
87. [Kebbel], "Earl of Peterborough," 187; Lane-Poole, "Review of Jurien de la Graviére, 784.
88. [Henderson], "Lord Wolseley's *Life of Marlborough*," 288; [Laughton], "Lindsay's *Merchant Shipping*," 420–422; [Laughton], "Low's *History*," 344.
89. Gardiner, *History of the Great Civil War*, 1:x.
90. Ward, "Review of S. R. Gardiner," 381.
91. George, *Battles*, viii.
92. [Thursfield], "Armada," 3–4.
93. Laughton to Browning, November 16, 1895 and January 31, 1897, GBR/0272/OB/1/939/C, King's Cam; Lambert, "Laughton's Legacy," 274–280.
94. [Laughton], "Lindsay's *Merchant Shipping*," 420–422.
95. Garritzen, "Women Historians," 655–658.
96. Y.Y., "Town Life," 113.
97. About Norgate, see Garritzen, "Pasha and His Historic Harem," 99–100.
98. Dyhouse, *No Distinction of Sex?*, 12–13; Perronne, "Women Academics," 340–341; Deslandes, *Oxbridge Men*, 185–189.
99. Gardner, *Short History of Newnham*, 100.
100. Gardner, Percy, "The Proposed Degree," *Times*, January 31, 1896, 10; Gardner, Percy, "The Proposed Degree," *Times*, February 15, 1896, 12.
101. Gardner, *Short History of Newnham*, 44–46.
102. Gardner, *Short History of Newnham*, 92–94; Dyhouse, *No Distinction of Sex?*, 65.

103. Gardner to B. A. Clough, September 8, 1896, Add MSS 72824 A, ff. 128–137, BL.
104. Powell, *John Horace Round*, 160–161, 186; Elton, *Frederick York Powell*, 1: 74–75, 1: 119–120.
105. Jessopp, "Women as Historians (II)," 67–68.
106. Jessopp to Dormer Harris, May 22, 1898, CR3874/1/8, WCA.
107. Looser, *British Women Writers*, 8–9; Laurence, "Women Historians," 127–128; Shteir, "Elegant Recreations," 236.
108. Thirsk, "History Women," 1–2.
109. Carter, "Women Historians," 263–285.
110. Spongberg, *Women and the Nation's Past*, 1–17; Maitzen, *Gender, Genre*; Mitchell, "Busy Daughters," 107–134; Melman, "Gender, History," 5–41. Of the early academic professionals see Berg, *Woman in History*.

References

Unpublished Primary Sources

British Library: B. A. Clough Papers.
British Library: The Macmillan Papers.
Cambridge University Library: Acton Papers.
Cambridge University Library: Reginald Lane Poole Papers.
Hull History Centre: Letters from Edward Augustus Freeman to Edith Thompson.
John Rylands Library, Manchester: E. A. Freeman Archive.
King's College Archive Centre, Cambridge: Oscar Browning Papers.
Oxford University Press Archive: Letter Books.
Rubinstein Library, Duke University: Augustus Jessopp Papers.
Warwickshire County archive: Mary Dormer Harris Collection.

Printed Primary Sources

[Anon.]. "The Oxford Professoriate." *Saturday Review*, July 12, 1862, 46–47.
[Anon.]. "A Classical School and College Education." *Englishwoman's Domestic Magazine* 9, no. 50 (1864): 77–82.
[Anon.]. "Oxford & the Commemoration." *Englishwoman's Domestic Magazine* 8, no. 119 (1870): 147–150.
[Anon.]. "Dr. Schliemann and the Royal Historical Society." *The Times*, June 16, 1877, 9.
[Anon.]. "Additional Notes." *Pall Mall Gazette*, March 17, 1884.
[Anon.]. "Ladies Column. Notes in Passing." *Evening Telegraph*, December 25, 1885, 5.

[Anon.]. "Historic Towns—Cinque Ports." *Athenaeum*, November 3, 1888, 586–587.

[Anon]. "Review of *The Dictionary of National Biography*." *English Historical Review* 5, no. 20 (1890): 783–788.

[Anon.]. "Death of Professor Freeman." *The Times*, March 17, 1892, 5.

[Anon.]. "The Death of Professor E. A. Freeman." *Daily Telegraph*, March 18, 1892, 4–5.

Browning, Oscar. *Memories of Sixty Years at Eton, Cambridge and Elsewhere.* London: John Lane the Bodley Head, 1910.

Creighton, Louise. *Life and Letters of Mandell Creighton, Sometime Bishop of London.* Vol. 1. London: Longman, 1904.

Elton, Oliver. *Frederick York Powell. A Life and a Selection from His Letters and Occasional Writings.* Vol. 1. Oxford: Clarendon Press, 1906.

Figgis, John Neville and Reginald Vere Laurence. *Selections from the Correspondence of the First Lord Acton.* Vol. 1. London: Longman, 1917.

Fowler, W. Warde. *An Oxford Correspondence of 1903.* Oxford: Blackwell, 1904.

Freeman, Edward A. *The Growth of the English Constitution from the Earliest Times.* London: Macmillan, 1872.

Freeman, Edward A. *The History of the Norman Conquest of England.* Vol. 5. Oxford: Clarendon Press, 1876.

Freeman, Edward A. *The Methods of Historical Study.* London: Macmillan, 1886.

Freeman, Edward A. "Historical Sketch of Architecture in England." In *Great Britain*, xxxiii–lxii. Leipzig: Karl Baedeker, 1887.

Gardner, Alice. *Synesius of Cyrene. Philosopher and Bishop.* London: Society for Promoting Christian Knowledge, 1886.

Gardner, Alice. *Julian. Philosopher and Emperor and the Last Struggle of Paganism and Christianity.* London: G. P. Putnam, 1895.

Gardner, Alice. *A Short History of Newnham College Cambridge.* Cambridge: Bowes & Bowes, 1921.

Gardner, Percy. "The Proposed Degree for Women." *The Times*, January 31, 1896, 10.

Gardner, Percy. "The Proposed Degree for Women." *The Times*, February 15, 1896, 12.

Gardiner, Samuel Rawson. *History of the Great Civil War. 1642–1649.* Vol. 1. London: Longman, 1886.

George, Hereford B. *Battles of English History.* London: Methuen, 1895.

Green, Mrs. J. R. *Henry the Second.* London: Macmillan, 1888.

Green, Mrs. J. R. *Town Life in the Fifteenth Century.* Vol. 1. London: Macmillan, 1894.

[Henderson, George Francis Robert]. "Lord Wolseley's *Life of Marlborough*." *Edinburgh Review*, October 1894, 259–288.

Howorth, Henry H. *History of the Mongols from the 9th to the 19th Century*. Vol. 3. London: Longman, 1888.

Hutton, William Holden. *Hampton Court*. London: John C. Nimmo, 1897.

[Jessopp, Augustus]. "The Brocas Book." *Edinburgh Review*, July 1887, 225–253.

Jessopp, Augustus. "Women as Historians (II)." *Literature*, January 21, 1899, 67–68.

[Kebbel, T. E.]. "The Earl of Peterborough." *Quarterly Review*, July 1887, 185–217.

Kemble, John Mitchell. *The Saxons in England. A History of the English Commonwealth till the Period of the Norman Conquest*. Vol. 1. London: Longman, 1849.

Kipling, Rudyard. *Kim*. New York: Barnes & Nobles, 2003.

Lane-Poole, S. "Review of Vice-Amiral Jurien de la Graviére's *Doria et Barberousse*." *English Historical Review* 2, no. 8 (1887): 784–786.

[Laughton, John Knox]. "Lindsay's *Merchant Shipping and Ancient Commerce*." *Edinburgh Review*, April 1876, 420–455.

[Laughton, Jhon Knox]. "Low's *History of the Indian Navy*." *Edinburgh Review*, October 1878, 343–379.

Maitland, Frederic William. *The Life and Letters of Leslie Stephen*. London: Duckworth, 1906.

Oman, C. W. C. *The Art of War in the Middle Ages A.D. 378–1515*. Oxford: Blackwell, 1885.

Oman, C. W. C. *History of Greece from the Earliest Times to the Mediterranean Conquest*. London: Rivington, 1890.

Oman, Charles. *Europe 476–918*. London: Rivington, 1895.

Oman, Charles. *A History of the Peninsular War*. Oxford: Clarendon Press, 1903.

Oman, Charles. *Inaugural Lecture on the Study of History*. Oxford: Clarendon Press, 1906.

Pauli, Elisabet. *Reinhold Pauli: Lebenserinnerungen nach Briefen und Tagebücher zusammengestellt*. Halle: Erhardt Karras, 1895.

Prothero, George W. "Historical Societies in Great Britain." *Annual Report of the American Historical Society for the Year 1909*. Washington, D.C. (1911): 231–242.

Seeley, J. R. *Life and Times of Stein, or Germany and Prussia in the Napoleonic Age*. Vol. 1. Cambridge: Cambridge University Press, 1878.

Stanhope, Earl. *History of England Comprising the Reign of Queen Anne until the Peace of Utrecht*. London: John Murray, 1870.

Stephens, W. R. W. *The Life and Letters of Edward A. Freeman D.C.L., LL.D.* Vol. 2. London: Macmillan, 1895.

Stubbs, William. *Seventeen Lectures on the Study of Medieval and Modern History and Kindred Subjects*. Oxford: Clarendon Press, 1886.

Stubbs, William. *Seventeen Lectures on the Study of Medieval and Modern History and Kindred Subjects*. 3rd ed. Oxford: Clarendon Press, 1900.
[Thursfield, James]. "The Armada." *Quarterly Review*, July 1895, 1–30.
Ward, A. W. "Review of S. R. Gardiner's *History of the Great Civil War*." *English Historical Review* 2, no. 6 (1887): 381–385.
Wheatley, Henry B. *The Dedications of Books to Patron and Friend. A Chapter in Literary History*. New York: A. C. Armstrong, 1887.
Yule, Henry. *The Book of Ser Marco Polo, the Venetian*. 2nd ed. Vol. 1. London: John Murray, 1875.
Y.Y. "Town Life in the Fifteenth Century." *Bookman*, July 1894, 113–114.

Secondary Sources

Annan, Noel. *The Dons: Mentors, Eccentrics and Geniuses*. Chicago: University of Chicago Press, 2001.
Barnett, Thomas. "Prendergast, John Patrick." *Oxford Dictionary of National Biography*. https://doi-org.libproxy.helsinki.fi/10.1093/ref:odnb/22714. Accessed September 17, 2020.
Bellon, Richard. "The Moral Dignity of Inductive Method and the Reconciliation of Science and Faith in Adam Sedgwick's *Discourse*." *Science & Education* 21 (2012): 937–958.
Berg, Maxine. *A Woman in History: Eileen Power, 1889–1940*. Cambridge: Cambridge University Press, 1996.
Blair, Ann M. *Too Much to Know: Managing Scholarly Information before the Modern Age*. New Haven: Yale University Press, 2010.
Bratchel, M. E. *Edward Augustus Freeman and the Victorian Interpretation of the Norman Conquest*. Devon: Arthur H. Stockwell, 1969.
Bremner, G. A. and Jonathan Conlin (eds.). *Making History: Edward Augustus Freeman and Victorian Cultural Politics*. London: British Academy, 2015.
Buurma, Rachel Sagner. "Epigraphs." In *Book Parts*, edited by Dennis Duncan and Adam Smyth, 167–175. Oxford: Oxford University Press, 2019.
Carter, Laura. "Women Historians in the Twentieth Century." In *Precarious Professionals: Gender, Identities and Social Change in Modern Britain*, edited by Heidi Egginton and Zoë Thomas, 263–285. London: University of London Press, 2021.
Collini, Stefan. *Public Moralists: Political Thought and Intellectual Life in Britain 1850–1930*. Oxford: Oxford University Press, 1991.
Covert, James. *A Victorian Marriage: Mandell and Louise Creighton*. London: Hambledon and London, 2000.
Daunton, Martin. "Introduction." *The Organisation of Knowledge in Victorian Britain*, edited by Martin Daunton, 1–27. Oxford: Oxford University Press, 2005.

Deslandes, Paul R. *Oxbridge Men: British Masculinity and the Undergraduate Experience, 1850–1920*. Bloomington: Indiana University Press, 2005.

Dyhouse, Carol. *No Distinction of Sex? Women in British Universities, 1870–1939*. London: Routledge, 1995.

Eliot, Simon. "The Evolution of a Printer and Publisher." In *The History of Oxford University Press*, edited by Simon Eliot, 77–112. Vol. 2. Oxford: Oxford University Press, 2013.

Eskildsen, Kasper Risbjerg. "Leopold von Ranke, la passion de la critique et le séminaire d'histoire." In *Lieux de savoir: espaces et communautés*, edited by Christian Jacob, 462–482. Paris: Albin Michel, 2007.

Garritzen, Elise. "Paratexts and Footnotes in Historical Narrative: Henry Biaudet and the Scholarly and Nationalistic Ambitions of Historical Research, 1902–1915." *Scandinavian Journal of History* 37, no. 4 (2012): 407–429.

Garritzen, Elise. "Montagu Burrows and Generational Anxieties of a Victorian Historian." *Journal of Victorian Culture* 24, no. 3 (2019): 1–7.

Garritzen, Elise. "Pasha and His Historic Harem: Edward A. Freeman, Edith Thompson, and the Gendered Personae of late-Victorian Historians." In *How to Be a Historian: Scholarly Personae in Historical Studies, 1800–2000*, edited by Herman Paul, 89–106. Manchester: Manchester University Press, 2019.

Garritzen, Elise. "Women Historians, Gender, and Fashioning the Authoritative Self in Paratexts in Late-Victorian Britain." *Women's History Review* 30, no. 4 (2021): 651–668.

Genette, Gérard. *Paratexts: Thresholds of Interpretation*. Cambridge: Cambridge University Press, 1997.

Gibbins, John R. "'Old Studies and New': The Organisation of Knowledge in University Curriculum." In *The Organisation of Knowledge in Victorian Britain*, edited by Martin Daunton, 235–261. Oxford: Oxford University Press, 2005.

Goffman, Erving. *The Presentation of Self in Everyday Life*. London: Penguin Books, 1990.

Goldstein, Doris S. "The Organizational Development of the British Historical Profession, 1884–1921." *Bulletin of the Institute of Historical Research* 55 (1982): 180–193.

Gratman, Rainier. "How to Do Things with Mottoes: Recipes from the Romantic Era (with Special Reference to Stendahl)." *Neohelicon* 37, no. 1 (2010): 139–153.

Hammond, Mary. *Reading, Publishing, and the Formation of Literary Taste in England, 1880–1914*. London: Routledge, 2019.

Hesketh, Ian. *The Science of History in Victorian Britain: Making the Past Speak*. London: Pickering & Chatto, 2011.

Heyck, T. W. *The Transformation of Intellectual Life in Victorian England*. London: Groom Helm, 1982.

Howsam, Leslie. "Academic Discipline or Literary Genre?: The Establishment of Boundaries in Historical Writing." *Victorian Literature and Culture* 32, no. 2 (2004): 525–545.

Jann, Rosemary. *The Art and Science of Victorian History*. Columbus: Ohio State University Press, 1985.

Kadish, Alon. "Scholarly Exclusiveness and the Foundation of the *English Historical Review*." *Historical Research* 61, no. 145 (1988): 183–198.

Lambert, Andrew. "Laughton's Legacy: Naval History at King's College London." *Historical Research* 77, no. 196 (2004): 274–288.

Lambert, Peter. "The Professionalization and Institutionalization of History." In *Writing History: Theory & Practice*, edited by Stefan Berger, Heiko Feldner, and Kevin Passmore, 42–60. London: Hodder Arnold, 2003.

Laurence, Anne. "Women Historians and Documentary Research: Lucy Aikin, Agnes Strickland, Mary Anne Everett Green and Lucy Toulmin Smith." In *Women, Scholarship and Criticism: Gender and Knowledge c. 1790–1900*, edited by Joan Bellamy, Anne Laurence, Gill Perry, 125–141. Manchester: Manchester University Press, 2000.

Levine, Philippa. *The Amateur and the Professional: Antiquarians, Historians and Archaeologists in Victorian England, 1838–1886*. Cambridge: Cambridge University Press, 1986.

Looser, Devoney. *British Women Writers and the Writing of History, 1670–1820*. Baltimore: The Johns Hopkins University Press, 2000.

Lubenow, William C. *'Only Connect': Learned Societies in Nineteenth-Century Britain*. Woodbridge: Boydell Press, 2015.

Maclean, Ian. *Scholarship, Commerce, Religion: The Learned Book in the Age of Confessions, 1560–1630*. Cambridge: Harvard University Press, 2012.

Maitzen, Rohan Amanda. *Gender, Genre, and Victorian Historical Writing*. New York: Garland Publishing, 1998.

Mason, Nicholas. *Literary Advertising and the Shaping of British Romanticism*. Baltimore: The Johns Hopkins University Press, 2013.

McDowell, R. B. *Alice Stopford Green: A Passionate Historian*. Dublin: Allen Figgis, 1967.

McLachlan, Jean O. "The Origin and Early Development of the Cambridge Historical Tripos." *The Cambridge Historical Journal* 9, no. 1 (1947): 78–105.

Melman, Billie. "Gender, History and Memory: The Invention of Women's Past in the Nineteenth and Early Twentieth Centuries." *History and Memory* 5, no. 1 (1993): 5–41.

Mitchell, Rosemary. "Women in the *English Historical Review*." *English Historical Review* (2018): 2–3. https://doi.org/10.1093/her/cey395.

Mitchell, Rosemary Ann. "'The Busy Daughters of Clio': Women Writers of History from 1820 to 1880." *Women's History Review* 7, no. 1 (1998): 107–134.

Opitz, Donald L. "'Not Merely Wifely Devotion': Collaborating in the Construction of Science at Terling Place." In *For Better or For Worse? Collaborative Couples in the Sciences*, edited by Annette Lykknes, Donald L. Opitz, and Brigitte Van Tiggelen, 33–56. Basel: Birkhäuser, 2012.

Perkin, Harold. *The Rise of Professional Society*. New York: Routledge, 1989.

Perronne, Fernanda. "Women Academics in England, 1870–1930." *History of Universities* 12, no. 1 (1993): 339–367.

Pettigree, Andrew. *The Book in the Renaissance*. New Haven: Yale University Press, 2011.

Porter, Roy. "Gentlemen and Geology: The Emergence of a Scientific Career, 1660–1920." *The Historical Journal* 21, no. 4 (1978): 809–836.

Powell, W. Raymond. *John Horace Round: Historian and Gentleman of Essex*. Chelmsford: Essex Record Office, 2001.

Reader, W. J. *Professional Men: The Rise of the Professional Classes in Nineteenth-Century England*. London: Weidenfeld and Nicolson, 1966.

Scott, David. "Signs in the Text: The Role of Epigraphs, Footnotes and Typography in Clarifying the Narrator-Character Relationship in Stehdahl's *Le Rouge et le noir*." In *Ma(r)king the Text: The Presentation of Meaning on the Literary Page*, edited by Joe Bray, Miriam Handley, and Anne C. Henry, 26–34. Aldershot: Ashgate, 2000.

Secord, James A. *Victorian Sensation: The Extraordinary Publication, Reception, and Secret Authorship of* Vestiges of the Natural History of Creation. Chicago: University of Chicago Press, 2000.

Shevlin, Eleanor F. "'To Reconcile *Book* and *Title*, And Make 'em Kin to One Another': The Evolution of the Title's Contractual Functions." *Book History* 2 (1999): 42–77.

Shteir, Ann B. "Elegant Recreations? Configuring Science Writing for Women." *Victorian Science in Context*, edited by Bernard Lightman, 236–255. Chicago: University of Chicago Press, 1997.

Slee, Peter H. *Learning and a Liberal Education: The Study of Modern History in the Universities of Oxford, Cambridge and Manchester, 1800–1914*. Manchester: Manchester University Press, 1986.

Smith, Bonnie G. *The Gender of History: Men, Women and Historical Practice*. Cambridge, MA: Harvard University Press, 1998.

Soffer, Reba N. *Discipline and Power: The University, History, and the Making of an English Elite, 1870–1930*. Stanford: Stanford University Press, 1994.

Spongberg, Mary. *Women and the Nation's Past: Empathetic Histories*. London: Bloomsbury, 2018.

Stapleton, Julia. *Political Intellectuals and Public Identities in Britain since 1850*. Manchester: Manchester University Press, 2001.
Stone, Lawrence. "The Size and Composition of the Oxford Student Body 1580–1910." In *The University in Society*, edited by Lawrence Stone, 3–110. Vol. 1. Princeton: Princeton University Press, 1974.
Tanselle, G. Thomas. *Book-Jackets: Their History, Forms, and Use*. Charlottesville: Bibliographical Society of the University of Virginia, 2011.
Thirsk, Joan. "The History Women." In *Chattel, Servant or Citizen: Women's Status in Church, State and Society*, edited by Mary O'Dowd and Sabine Wichert, 1–11. Belfast: Institute of Irish Studies, 1995.
Trettien, Whitney. "Title Pages." In *Book Parts*, edited by Dennis Duncan and Adam Smyth, 41–49. Oxford: Oxford University Press, 2019.
Turner, Frank M. *Contesting Cultural Authority: Essays in Victorian Intellectual Life*. Cambridge: Cambridge University Press, 1993.

CHAPTER 3

Champions of a Virtuous Historian

When the *English Historical Review* was launched, Acton remarked with satisfaction that "At least half the great *names* are there, and I discern the makings of a sacred band of university workers."[1] He had recognized as early as 1859 how distinguished names radiated certain "éclat," and how such names helped the *Review* to gain a reputation as an organ of the scientific historians.[2] Acton was right in claiming the immense importance names can have in scholarship. They mark textual and intellectual ownership, encapsulate individual and disciplinary properties, and are invested with meanings that authorize scientific texts. Names thus have significance for their bearers and, as Acton envisioned, for the scholarly community at large. This chapter uses as a starting point this intricate relationship between names, reputations, and scholarly communities and argues that historians turned proper names into models of virtues and vices to promote their scholarly persona. As they had only limited opportunities to perform the persona to a live audience, their names assumed the role of a *factual paratext* that exemplified and sanctioned the persona.

The names of the leading historians evoked associations of skills and qualities, virtues and vices: Freeman symbolized profuseness, accuracy, and pedantry; Acton embodied cosmopolitan intellectualism; and Froude, the nemesis of scientific history, stood either for dishonesty or literary elegance depending on whom one happened to ask. But it was the name of William Stubbs that came to mark an ideal scientific historian and to

E. Garritzen, *Reimagining the Historian in Victorian England*, https://doi.org/10.1007/978-3-031-28461-8_3

invite the kind of heroization that undermined the warnings historians sounded about the epistemic perils of hero worship. He was to George Prothero the "greatest living" historian, to John Horace Round "our supreme authority," and to one anonymous writer "one of those men to honour whom reflects honour on oneself ... a great man." Freeman divided the historiographical time between a "præ-Stubbian æra" and the present.[3] These testimonials are only a modest sample of the outpouring of panegyrics Stubbs inspired, yet they already give a reason to ask how and for what purposes historians erected a heroic monument out of Stubbs, what made Stubbs an ideal model, and what consequences the heroization had for the discipline.

An author's name and the information readers associate with it is, according to Genette, a factual paratext. The knowledge readers gain from earlier publications, reviews, advertisements, and other media outlets creates expectations about authors and texts.[4] The paratextuality of a name is capitalized in sciences, where names are like brands that either grant or detract value. As Mario Biagioli and James Secord have emphasized, a name had exceptional power in authorizing scientific texts in the nineteenth century.[5] A recognized name signified peer approval and validation from a scholarly community that fixed to it normative qualities such as truthfulness, novelty, or relevance. The immensely popular and anonymously published *Vestiges of the Natural History of Creation* showed how names shaped textual reception and how readers struggled to ascribe meanings to a text without guidance from an author's name.[6] Indeed, a scientist's name was an interpretative compass, and the critics of the culture of anonymous reviewing used this to their benefit. Anthony Trollope found anonymity particularly harmful in "the treatises of science" where the value derived from the fact that complicated subjects were studied by "capable minds comprehending them."[7] A writer in the *Glasgow Herald* confirmed this in a review of Freeman's historical essays by complaining how the great evil of anonymity forced readers to wade through "a mass of uncredited matter, in the hope of finding one article out of a dozen that is really good" rather than being able to rely on a name like Freeman as an assurance of the "sterling worth" of the content.[8] The abundant attention that was given to historians' names and to the meanings ascribed to them highlights the paradox of history being detached and embodied at the same time. The methodological strictures demanded the utmost disinterestedness, but the constant use of names to sanction and judge reliability created a strong tie between histories and the properties of their writers.

The paratextuality of a historian's name entailed an investment of a repertoire of qualities that the public associated with the name and its bearer. Reputation was critical for the paratextuality of a name, and therefore historians energetically constructed, managed, and circulated stories about Stubbs. They were rather successful in their project, and Stubbs's reputation came to function like a paratext that vouched for the quality of his work, those who were associated with him, and the entire discipline which he exemplified. Taking the meaning of factual paratext in its most encompassing sense, I show how Stubbs was turned into a towering figure that historians needed for grounding their new scientific status and how the reputation they assigned to him was an interpretation and an expression of the sum and the parts of their scholarly persona. Whatever contradicted the persona in Stubbs's work or character was silently omitted.

Historians constructed the Stubbs myth in a favorable climate. The Victorian "cult of the personal" and the nascent modern celebrity culture provided fertile ground for portrayals of exemplary lives. Victorians were fascinated with exceptional lives that they could imitate to instigate personal growth. Biographies and self-help manuals with inspirational and dramatized life stories recounting internal battles, extraordinary talent, and the cultivation of character and moral virtues were steady sellers. Historians contributed to this cultural practice on two temporal levels as they celebrated virtuous historical characters and sculpted a heroic figure out of Stubbs for others to emulate. The Victorians were, moreover, absorbed by the modern celebrity culture and had a voracious appetite for news about the private lives of public figures. Even historians' personalities, abilities, dispositions, and countenances came under increasing scrutiny. Their names spilled out from the pages of their books and surfaced in newspapers, book reviews, publishers' advertisements, celebrity interviews, and gossip columns, and popped up in dinner conversations and private correspondence. The recurrent reports about their personal traits reduced their multidimensional personalities into easily digestible stereotypes that roused curiosity, amazement, admiration, or rejection. It was often through encounters with historians' names in the press, not through their books, that they became meaningful to the public. Whether historians agreed with their public portrayals or not, the publicity nonetheless had consequences for how they viewed themselves, their readers, and the inescapable public function the writing of history entailed. When historians set their minds to carving a heroic historian out of Stubbs, they tapped into the same mechanism of repetition and selection of personal qualities which the journalists used.

Stubbs may at first appear an unlikely candidate for stardom, as he seemed to lack the qualities that a celebrity usually needed, but it was this unlikeliness that made him an ideal role model for historians, as they were able to develop his reputation around the "right stuff": the virtues, skills, and dispositions they considered crucial for the scholarly persona. Stubbs rose into the public consciousness when he was unexpectedly appointed Regius Professor of Modern History in Oxford in 1866. Unlike the better known contestants Freeman and Froude, he lacked a colorful public persona, but after the extravagant Goldwin Smith, the Cabinet preferred a sober and prudent candidate.[9] His Conservatism and High Church Anglicanism were kept at bay during the nomination process. It was this early impression of him as the first true scholar holding a professorship in history together with his assumed neutrality as a contrast to his Oxbridge predecessors that gave an initial boost to his heroization. Yet Stubbs was not catapulted into notoriety out of nowhere: he had been a Lambeth librarian, an editor for the Rolls Series, and complimented for his manly and erudite work.[10] These, however, were not positions that conferred popular fame, and his relative unfamiliarity to the general public is encrypted in the news about his appointment: hardly anyone was able to say anything else about him than to list his former positions and publications.[11] Three years after Stubbs's appointment, the *Bath Gazette* quoted *London Society* in an article with a title "The English Historical School" to identify Lord Stanhope and Froude as the two leading historians of the time. The article mentioned Freeman in passing as someone with high repute "in the world of letters" and left out Stubbs entirely.[12]

The contradiction between Stubbs's limited popular and considerable scholarly fame was a reoccurring theme in the press from his inaugural lecture to the obituaries. The *Daily Telegraph* professed that after the publication of *Select Charters* (1870) "Dr. Stubbs's fame as a historian may be said to have passed beyond the walls of the University and to have established itself firmly among the larger world of educated men."[13] Others were less sanguine. While they agreed that appealing to scholars was more valuable than appealing to the readers of historical novels, there was nonetheless an undertone that regretted that Stubbs's "genius" was not "showy."[14] Yet Stubbs's limited fame was more a construction than reality, and the simultaneous visibility and invisibility was an important backdrop of the Stubbs cult. It placed the heroization in the world of scholarship and targeted the message first and foremost at historians from antiquaries to experts, as well as at the educated segment of the public. In these circles, the curiosity was directed at Stubbs's scholarly accomplishments, not at his private habits, wardrobe, or the color of the wallpaper in the Stubbs household. From learned circles Stubbs's reputation spread gradually to the newspaper-reading public, and even here historians gained desirable visibility for their model persona as journalists reported Stubbs's scholarly accomplishments instead of his personal quirks. The heroization project was so successful that Stubbs continued to radiate scholarly virtuosity even after he withdrew from active research in 1884 when he was nominated Bishop of Chester and then Bishop of Oxford in 1889.

Acceptable Hero Worship

> The mightily grasp with which he [Stubbs] holds his vast and multifarious learning, the largeness of view, the loftiness and majesty with which he passes on his way, the matchless precision of language, the vigorous manliness, the graceful playfulness, the profoundly critical insight—and all this wonderful combination of the noblest characteristics of genius controlled by absolute surrender of himself to the fearless quest of truth, must have made themselves felt hardly less in his younger than they have done in his mature years.[15]

Jessopp was one of the many who majestically eulogized Stubbs as an incarnation of scholarly virtuosity and intellectual masculinity. Strikingly, this panegyric was a by-product of his Jubilee tribute-cum-review of Burrows's *The Family of Brocas*.

Stubbs was, for his fans, like an Aristotelian hero who possessed a remarkable mind and above-ordinary powers and his greatest admirers appeared rather naïve in their enthusiasm. But the heroization was certainly not accidental or in the hands of a few fanatics. His public image was a result of a collaborative endeavor from different reputational entrepreneurs who constructed, validated, and circulated the myth of an infallible and flawlessly heroic historian. It was not a coincidence that Freeman used the word "Froudish" when he wanted to signal dishonesty and "Stubbist" when he wished to indicate honesty and truthfulness. Froude was, Freeman proclaimed, just as incapable of copying his authorities right as Stubbs was unable to copy them wrongly.[16]

The public image of Stubbs was not an authentic depiction. His reputation was a construction of real and imagined qualities and characteristics that historians ascribed to him. They highlighted certain aspects of his character and played down others to craft the collective persona they needed to ground history as a branch of science. According to Geoffrey Cubitt, heroic reputations allow communities to define and articulate their values and assumptions and to share social and cultural identities. Heroic figures facilitate identification as they enable communities to convey what they consider important and how they think that the meanings they assign to them differ from those given by other groups. This strengthens the sense of community and the construction of boundaries with "others" who interpret things differently.[17] The name "Stubbs" became the glue that tied historians together, and even the most obstinate paid attention when "Stubbs" was evoked to encourage the use of the scientific method or dignified scholarly conduct.[18]

The construction of heroic reputations engages multiple communities of reputational entrepreneurs. The worship "may become an integral part of their own subjective identities" according to Simon Morgan.[19] At least four groups staked a claim on Stubbs's sanctification. First, there were the members of the scientific community who were also the guardians of his—and their discipline's—reputation. Second, the younger historians promoted the myth with great enthusiasm. They opportunistically calculated that an association with Stubbs would help them to brand themselves as supporters of scientific history. Third, the myth was developed and circulated outside of the professional set. Reviews in periodicals and newspapers were instrumental in endorsing and grounding it in the public imagination. The reviewers did not spare the superlatives when they praised his books. As Jessopp's panegyric implied, the myth also gained force from the popular practice of invoking Stubbs's name to evaluate other historians. The economist J. Shields Nicholson's critical review of Thorold Rogers's *A History of Agricultural Prices in England* (1889) gives us some taste of how reviewers used Stubbs's name as a qualitative category, for Nicholson accentuated Rogers's weaknesses by calling him "the greatest contrast possible to Bishop Stubbs."[20] Nicholson, trusting on the audience's familiarity with Stubbs's virtuosity, did not need more than a few words to effectively undermine Rogers's competence and buttress the saintly image of Stubbs.

Lastly, we have Stubbs himself. He was one of the most vigorous promoters of the Stubbs myth. He employed paratexts and lectures to fashion a public image that contributed to his heroization. Since vanity was considered a false motive for conducting historical research, he carefully disguised his self-praise in feigned modesty, ironic self-deprecation, and self-parody, although his carefully upheld image gave away here and there when he could not resist the temptation to blow his own horn. In his inaugural lecture, he admitted that he was still searching for "the tone of authority which befits the professorial character" and apologized for the conflict between modesty and self-regard that was characteristic of inaugurals as a genre. He switched in the lecture between the first and third person singular to distance himself from direct self-praise: "he," meaning Stubbs himself, "feels that he must introduce himself, in a way that is of necessity painful both to his self-respect and to his modesty, to an audience from whom he has at least as much to learn as he can ever hope to teach them."[21] Later on, apologies for self-promotion were standard elements in his lectures and he enforced the message with an ingenious use of paratexts. One of the running heads in his last statutory lecture, for instance, stated "Excuse for Egotism" while the corresponding narrative begged readers not to think that he voluntarily talked about himself or wished to set "an especially high value" on his services.[22] He wanted readers to understand that he talked of himself not out of vanity, but because the occasion demanded self-reflection. The self-eulogies fixed him so successfully as a sober scholar and a fair colleague that many who learned to know him personally were astonished about the discrepancy between the public and the private historian. He was not in fact "like a nice *tame cat* ... Always amiable, never on the way" as Mrs. Hook described him, but fond of "gossip & scandal," full of irony and witty sarcasm, and ready to give "a bad drubbing" to careless historians.[23]

Historians condemned hero worship as a scholarly vice that led to exaggeration and partiality. They associated it with Romantic historians such as Carlyle, who had famously framed the admiration of great men as a transcendental experience. There was not, Carlyle professed, any "nobler feeling than this of admiration for one higher than himself." Such worship was "the vivifying influence in man's life."[24] When it came to applauding Stubbs, many however assumed a Carlylean pose. William Holden Hutton, who later edited *Letters of William Stubbs*, was a devoted fan and expressed his unconditional admiration in a review of *Willelmi Malmesbiriensis* which Stubbs had edited for the Rolls Series. According to Hutton, readers had learned to anticipate from his editions "a most complete study and analysis of the manuscript sources" as well as "a laborious collection of everything that bears in any way upon the life of the author, or the work in hand; and … a luminous and philosophic criticism of the book, and a judicial, but by no means dogmatic statement of its position among other similar works." Hutton was convinced that Stubbs, the "indefatigable" editor of medieval manuscripts, was endowed with supra-human capacities, although he was compelled to report that there were a few minor errors in the book despite its editor's "fondness for correcting proofs." Yet he abstained from the common habit of listing the blunders and instead closed with a more hyperbolic eulogy: "There is, indeed, no occupation for the critic in reading two volumes like these. It is the office of the *English Historical Review* rather to congratulate the editor on the completion of a great work which will never need to be done again."[25] This, if anything, was genuine hero worship.

Freeman censured hero worship and even cautioned against the sanctification of Stubbs. Historians, he demanded, had to object to any attempts to make "an idol of himself" or "of anything else," because idolization discouraged the critical evaluation of a historian's work.[26] Moreover, he feared that the heroization might mislead readers to believe that *all* historians were unerring authorities. He had encountered too many readers who perceived historians as oracles who knew everything about their subject. Such a lack of critical edge was alarming, and hence "It is important to insist—as a rather abstract thing—on the fallibility of Stubbs himself."[27] There was, however, dissonance between Freeman's words and deeds, for he was known as one of the most prominent contributors to the heroization of Stubbs. Presumably the collective need for a model that would fortify history's scientific status justified this for him.

Freeman's posture as a Carlylean hero worshiper inspired gentle parody of him as the most "disinterested admirer" of Stubbs and the most unceasing popularizer of "his brother historian's fame."[28] The *Derby Mercury* explained how Freeman and Stubbs's "mutual self-laudations afforded much fool for merriment among the profane jesters" and the *Daily Telegraph* even repeated in Stubbs's obituary the satirical couplet that Thorold Rogers, the historian of wages and prices, had invented:

> Here ladling flatteries from alternate tubs
> Stubbs butters Freeman, Freeman butters Stubbs.[29]

Freeman's Stubbs worship generated compliments as well. Those who regretted that Stubbs was not as well known as he should have been were grateful to Freeman for amplifying his visibility. *The Times* thanked him for sparing "no opportunity of telling the world that he [Stubbs] was a great historian."[30] The *Daily Telegraph* maintained that it had been a great favor to the entire discipline that Freeman had used his authority to assure the general audience about Stubbs's erudition and virtuosity. There had not been any other scholar except Freeman since the Renaissance who had held on to his opinions with more "cocksureness" or "who [had] defended them with a more arrogant dogmatism," and when such a man praised Stubbs, then there was no doubt that Stubbs was worth all the praise.[31] Whether the writers were amused by or appreciative of Freeman's raptures about his "brother-historian," they did not contest or question Stubbs's heroic status. Instead, they helped to confirm the image of a humble and devoted scholar by observing how modesty stopped Stubbs making a noise about himself.

The Founder of the Oxford School

Stubbs's heroic reputation was dramatized on two different stages where he was interpreted as playing the leading act: in Oxford and in the field of constitutional history. Both contexts enabled the reputational entrepreneurs to showcase how the virtues, skills, and conduct which Stubbs cultivated were pivotal for the new scholarly persona. Historians perceived Stubbs's professorship as a great turning point in the development of history into a research-based discipline and a corrective to the devastating appointments of Burrows and Kingsley. The Oxford School which Stubbs embodied should not be mistaken for the Oxford Modern School of History. It was more comprehensive than the curriculum or syllabus: it embraced teaching and learning and formed a network of Oxford graduates either with loose ties to the history community or with more formal connections through fellowships and examinations. Despite its hybridity and geographical dispersity, the school was conceived as an established entity and Gladstone claimed in 1875 with confidence how it was the only "truly historical school in England" and how it owed its excellence and existence to Stubbs.[32]

The Oxford School was associated with Stubbs's ideas about scientific history. For him, history meant at its plainest the study of original authorities, their critical evaluation, and the search for novel historical facts. Historians' duty was to produce fresh knowledge, not a "*rachauffé*" of what was already known.[33] These two aims—examination of original authorities and production of knowledge—became the principles that characterized the School, as Stubbs was seen to steer history away from its age-old allegiances with literature. Another strand in the narrative was that Stubbs rescued English historians from their backwardness and connected them with developments abroad.[34] Germany was for Stubbs the most obvious point of reference and, with Ranke, he shared disdain for the positivist search for historical laws and an appreciation of uniqueness and individuality in history.[35] In the insular fashion of English historians, Stubbs's actual contacts with German historians were limited. This did not stop the reputational entrepreneurs from comparing him with German historians. Freeman contrasted him to Ranke's famous disciple Georg Waitz, calling him "the Waitz of England" and Waitz "the Stubbs of Germany."[36]

The status of Stubbs and the Oxford School as the harbingers of scientific history was enhanced by the relative invisibility of Cambridge and its historians in the narrative of scientification. The insignificance of the Cambridge historians was an unfair assessment, but it is not the reality but the images and impressions which count in the mythmaking. The idea of the ascendancy of the Oxford School was recognized and endorsed in Oxford and Cambridge alike. In Cambridge, history was personified in Seeley. His nomination as Regius Professor of Modern History in 1869 was tagged as political and he was welcomed at Cambridge rather as an outsider. The *Saturday Review* introduced him, if not "a full guild-brother of the order of historical scholars," then "an outsider" to that guild.[37] Seeley had fostered doubts about his suitability for the position; he had been trained as a classicist, had taught Latin at University College London, and had published in that field, not in history. He was not familiar with any modern period of history or with any type of historical source. To compensate for this, he set to work on the recent history of Germany and France after his appointment and metamorphosed into an adamant defender of scientific history, opposing any attempts to make history literary, popular, or entertaining.[38]

Seeley had cheerfully written to Reinhold Pauli in 1876 about "the little company of historical students that is forming itself in Cambridge" and welcomed Browning to Cambridge with a prospect of "a great opportunity in History in Cambridge just now."[39] Yet he knew that he and the other Cambridge historians were being sidetracked as the discipline was dominated by what he called a strong Oxford "clique."[40] He identified two major reputational obstacles to the Cambridge historians: first, their invisibility in the press, and second, the lack of any truly groundbreaking publications from them. He claimed bitterly how the press was mainly "in the hands of Oxford men" who "try to keep me down as much as they write each other up." The Cambridge men who wrote to papers were of no help because they cultivated "scientific impartiality to such an extent that I scarcely remember ever to have been helped by any one of them."[41] In other words, Seeley implied that the Cambridge historians did not unify behind him to promote him in the press as a heroic figure and a model for those who would have preferred contemporary history and historians' alignment with political science.

When Seeley received a slightly larger budget for the tripos in the early 1880s, he saw in it an opportunity to increase the visibility of Cambridge historians by recruiting an experienced historian outside of the university to elevate their reputation. For Seeley, the question was about both the stature and the future of history in Cambridge. Hence, his choice fell on H. C. Pearson, who had been lecturing in Cambridge but had since then returned to the political life of his native Australia. This decision infuriated Browning, who saw himself as a spokesperson of the tutors. He demanded that Seeley should hire one of their own young men who had proven himself as a teacher. Seeley refused. He was on the lookout for a senior scholar with a strong publishing record:

> I think it very important that our school should have at least one historian of established reputation, who will send out from Cambridge books that will be attended to. We produce too little and we can hardly produce much until our teachers are more comfortably settled & have leisure.[42]

Seeley had clearly grown tired waiting for the young men to gain recognition. Until this happened, Pearson's "name … will be useful to our reputation." He knew that this meant that he had to discontinue the contract of one of the younger tutors. Browning was irritable about this, but Seeley showed no understanding of the tutors' challenges when he replied that it would be possible to rehire that person a couple of years later. Surely, they could somehow manage for that period, he added. He allowed that he was not enthusiastic about Pearson, but it was the interest of the department that mattered most:

> I still think the principle of calling in distinguished men from outside a good one, and would not exchange it for the principle you seem to lay down of appointing only our own men. Anyhow, I think most of us are sufficiently enthusiastic in the cause of History to be ready to endure some temporary loss if we are really considering that we can produce at that expense some gain to the study itself.[43]

Seeley could not have made the importance of reputation any plainer. He was so concerned about the competition between Oxford and Cambridge that he was willing to sacrifice one of his young tutors to gain publicity for the Cambridge historians. Pearson, however, did not return to Cambridge and the scholarly reputation of Cambridge historians continued to vacillate. Ward regretted in 1885 how Seeley had failed to turn out good enough men to produce "any historical work of high merit."[44] When Acton succeeded Seeley in 1895, he admitted that Cambridge had "no great school of history."[45]

Meanwhile, the Oxford School of History became a recognizable trademark and symbol of scientific history rapidly after Stubbs's nomination to the professorship. But historians kept drumming the message tirelessly even decades later.[46] Jessopp announced triumphantly how the school under Stubbs had won for history "a throne on which she may take her seat without fear of supercilious slight or contemptuous comparison."[47] There was no doubt where this kingdom of history lay and who was its sovereign. Maitland used a similar metaphor in an obituary of Stubbs to mark how the English historians had "had a king and now are kingless."[48]

Stubbs was not ignorant of the connection between his name and the Oxford School. He contributed to the narrative by offering several interpretations about the part he played in the founding of the Oxford School and shaping the course of historical research in England. He presented in his inaugural lecture a dawn of a new, scientific era in historical research in England and highlighted the collaborative nature of history by calling himself an able assistant in the founding of "an historical school in England, which shall join with the other workers of Europe in common task [of unearthing historical truths]." The flood of hitherto unknown sources had rendered history too vast for any individual to master it alone.[49] The tone of his last statutory lecture in 1884 was different. He pushed collaboration aside and instead took the full credit for establishing "for History its proper place among the studies of Oxford" and representing "our Oxford School of History before the outside world." Forgetting the scholar's humility, he added how his accomplishments were so evident that it would have been "mock humility" not to mention his part in events.[50] He reassessed his role once more in 1900 when he wrote a new preface for the third edition of his *Seventeen Lectures*. The preface was his legacy to the historical community, and he reversed his position, allowing that his accomplishments had depended on a "*felix opportunitate operis*." When he had arrived in Oxford in 1866, the change had already been taking place, and he had been able to learn from those who before him had "fought their way through the times of discouragement." The conciliatory mode was accentuated by an acknowledgment of the achievements of the Cambridge historians. Oxford was no longer portrayed as the solitary paragon of scientific history, but "Hand in hand with Cambridge under Sir John Seeley and Lord Acton, our [Oxford] workers have undertaken and have gone no small way toward the accomplishment of the ideal which charmed me, and laid deep and broad foundations for a great future."[51] This was a significant adjustment to the narrative that had projected Stubbs and the Oxford School alone as the saviors of the integrity of historical research in England.

Commendable Constitutional Historian

If the Oxford School allowed historians to formulate original research with high acclaim as their noble goal, then constitutional history gave them a chance to define which epistemic virtues and practical skills they needed to acquire truthful knowledge. Stubbs was not depicted as the inventor of constitutional history—as he had powerful predecessors such as Hallam—but as its foremost champion in his own age.[52] Constitutional history in the Stubbsian fashion entailed extraordinary perseverance, industry, and patience. It required exceptional intellectual abilities and methodological competence as well as a specific narrative aptitude that distinguished it from literary history. Historians mythologized constitutional history as a topic that only the most gifted historians comprehended and were able to study, granting it a role in regulating the boundaries of the scholarly community. The links between constitutional history and historians' persona were evoked wherever Stubbs's histories were discussed. The essence of his historical *oeuvre* consisted of nineteen volumes of medieval chronicles which he edited for the Rolls Series and the two key works that came to define the field of constitutional history: *Select Charters* (1870) and *The Constitutional History of England* (1874–1878). Freeman called the latter "the greatest monument of English historical scholarship" and Sir James Ramsay promoted it as "a truly monumental work, not likely to be ever displaced."[53]

Constitutional history suited the persona project exceptionally well because it had high contemporary relevance that elevated it into a noble topic for historical investigation. As Stubbs phrased it, constitutional history explained the becoming of modern-day Britain and gave readers "a personal hold on the past and a right judgment of the present."[54] It had attracted interest already in the eighteenth century when the constitution and constitutional institutions had been identified as key components of Englishness. Unlike in the late-Victorian era, it had not been treated as an exceptionally challenging topic and it had captured the attention of readers of different ages, sexes, and occupations.[55] In the nineteenth century, constitutional history continued to provide the keys for understanding the English national character and the development and uniqueness of English institutions. As the institutions shaped the habits of people and people shaped the institutions, they illustrated what was unique in the English national character.[56] The Stubbsian form of constitutional history traced the origins of modern institutions such as Parliament to the Anglo-Saxon communal roots and explained how and why the English had enjoyed steady progress and an unbroken continuity from early English history to the present. This optimistic idea of progress brought unity to history as historians drew on the common notion of continuity even when they explored other periods than Stubbs. Even those who did not specialize in the history of institutions considered the identification of continuity and progress as a historian's main duty.

Moreover, constitutional history's popularity derived from its appeal to historians with different political backgrounds: conservatives and liberals alike justified their politics with the steady and unrevolutionary development of the constitutional institutions and found consolation in the institutions' ability to adjust to the changing conditions as their own political environment altered in the aftermath of the Reform Acts of 1867 and 1884.[57] Although Stubbs's work was framed as too demanding for an ordinary reader, the imagined continuity of the English nation was a powerful historical pretext for the English throughout the century. The narrative of peaceful continuity comforted the English amidst the social, economic, political, and religious transformations, as it attested that "England remained England" no matter what. The elevation of continuity and the examination of institutions, laws, and customs gave a unique flavor to the nineteenth-century Whig history that Stubbs embodied.[58]

Since constitutional history had high public utility, Stubbs demanded its integration into the English education system. It was inconceivable that students were "acquainted with the state machinery of Athens and Rome" but "ignorant of the corresponding institutions of our own forefathers."[59] Yet Stubbs did not see himself supplying such education or producing educational histories for schools. Quite the contrary, he promoted constitutional history as a branch of science and an intellectual endeavor that suited the most competent historians alone. As he listed the qualities which a constitutional historian needed, he came to define a historians' persona as well and to demarcate it from competing models such as antiquarianism. According to him, a constitutional historian required "a strong instinct for the investigation of continuities and coincidences which lead men to the study of chronology and genealogy for the pleasure of exercise." In the preface to *Select Charters*, he elaborated how the aim was to establish the "causes and consequences" instead of just collecting "a multitude of facts and views," as antiquaries were blamed for doing. A proper historian pieced "the links of a perfect chain." Later on in life, Stubbs did not hesitate to admit that he possessed all these qualities.[60]

Constitutional history allowed historians to cast Stubbs as an incarnation of the manly virtues of perseverance, industry, and courage which were constitutive both for their persona and for the character of a Victorian middle-class man. Stubbs epitomized a nineteenth-century man of strong will and a sense of duty when he professed how constitutional institutions could not be studied without an effort, extraordinary courage, or "an unlimited capacity for taking pains," for the subject lacked the usual charms of history—the "play of personal character" and picturesqueness—and the sources were "clothed in the repulsive form of Medieval Latin." To further applaud his work ethic, Stubbs drew on the classic motif of scholarly asceticism as he confessed how he was unable "to reconcile myself with smoking, late hours, dinner-parties, Sunday breakfasts, or University sermons."[61] Such comments found fertile ground among his admirers, and newspapers reported how he devoted his leisure for improving activities such as "making out pedigrees and correcting proof-sheets."[62]

The gospel of hard work which Stubbs sang was important also because it allowed him to undermine the role of an individual genius in historical research. Genius was a popular explanation for scientists' gifts throughout the Victorian period, but the idea of genius evolved from the Romantic idiosyncratic spirit into a more sober possession of exceptional intellectual endowments during the latter half of the century. The new definition owed much to Francis Galton, a biostatistician and eugenicist, who advocated a genius that materialized when hereditary abilities were combined with zeal, dedication, and hard work.[63] Stubbs's work ethic concurred with this late-Victorian notion of genius, but his energetic attempts to downplay genius suggest that he viewed it in its Romantic garb. He was backed in this by Seeley and Acton, who similarly questioned the value of genius in historical research.

A Romantic genius was infected with qualities that were inconsistent with scientific history. They were driven by an inner creative force and governed by their passions, while historians emphasized detached rationality. Instead of emotional rashness, exceptionalism, and eccentricity, historical research was about arduous work, meticulousness, patience, and the cultivation of research skills.[64] Stubbs allowed that genius might help in the reconstruction of men and their character, but historians did not only paint pictures of men's virtues and vices. They reconstructed events or forces that had influenced them, and in this genius had no role. It demanded an intervention from a historian who molded the accumulated facts into a historical account.[65] Industry, perseverance, and dedication distanced history from the ideals of the Romantic sages and grounded it into the modern world of science which the assiduous middle-class men dominated. Stubbs saw no reason to build a bridge between the scientific historian and the hereditary genius in its Galtonian dress. The heroization of his exceptionalism suggests that his followers, on the contrary, might view him as an embodiment of a genius.

Another thread in the story about the heroic historian tied together Stubbs, constitutional history, and the new narrative style that was believed to reflect history as a branch of science. Stubbs campaigned against the dramatization of history and warned the Oxford audience in 1877 about the "sensational and picturesque writing" which destroyed and corrupted "the more valuable features of painstaking and conscientious truthfulness." His words had a moral undertone, as he feared that picturesque histories would encourage intellectual laziness among historians and readers: it was too easy to produce "the romances of history … when the principal object is to attract the half-educated to read."[66] Thus, he strove to set an example by adopting a style befitting history's novel scientific character.

Stubbs was described by his colleagues and the media as an excessively dry writer and, according to Freeman, *The Constitutional History* was therefore "more like a German than an English book."[67] However, Stubbs's dryness was an invention that reflected historians' disciplinary anxieties and the persona project more than Stubbs's actual literary powers. Robert Brentano has argued that Stubbs was a more engaging writer than his contemporaries had us believe. He played with phrases, words, and letters in a way that gave his message unique gravity and persuasiveness.[68]

The literary qualities of history were reassessed in the 1890s, and Ian Hesketh has shown how the new generation of historians disassociated themselves from the narrative preferences of Stubbs, Freeman, and the other precursors of scientific history.[69] They dismissed dryness and accommodated an appreciation of literary talent in their persona. Accordingly, both historians and journalists now sought to establish Stubbs in the obituaries as a first-rate narrator who adorned his texts with sober eloquence. The *Daily Telegraph* complained first how history was too often regarded as a mere application of the scientific method and not of artistry and hastened to add that Stubbs should not be blamed for this misapprehension. He had been "the very reverse of a Dryasdust."[70] Tout, too, lamented the misconception of Stubbs's deficient eloquence and how it had eclipsed his popularity. His historical introductions to the Rolls Series testified to "an historical narrator of the first order." According to Tout, the popular belief in Stubbs's dryness originated from Constitutional History and even he had to assent that it was indeed a heavy book to digest and repelling to "the beginner." Yet he clarified that the heaviness was not caused by Stubbs's literary shortcomings, but by the grim subject matter that "gave few opportunities for the remarkable narrative and pictorial gifts he displayed."[71]

Nevertheless, for Stubbs' contemporaries, the image of Stubbs's books as unattractive to the general reader was extremely useful, and they exploited it to separate the talented historians from the mass of untalented ones. Creighton, who wanted to improve history teaching for women, assigned Stubbs to Ella Paese, one of his students, and explained how it would help her to build her character as a historian. He prefaced the assignment with a warning and an encouragement: "any reading worth mentioning must be hard work, and the subject must be stiff. I have chosen for you the hardest subject and the stiffest book that I know of ... Do not object to its being dull at first, but go at it steadily."[72]

The Constitutional History and *Select Charters* formed the backbone of history teaching in Oxford, and they were presented as daunting tasks for the undergraduates. Constitutional history was expected to instill in the students an appreciation of national history and to bestow upon the School of Modern History the prestige of liberal education because, as a technical topic, the study of constitutional institutions and their development equaled the effort and fortitude which a command of Greek grammar or mathematical formulae demanded. In Cambridge, on the contrary, students were not expected to master the details of constitutional history and this was considered one of the program's great flaws. Creighton was appalled that *Select Charters* seemed "almost unknown" among the students there. Instead, they had to read political economy and jurisprudence, as Seeley was convinced that a pure "love for history" would not suffice to make history a necessity for the university. But the large share of political sciences in the history tripos puzzled historians with strong commitment to the Stubbsian style of history, and Creighton wrote Freeman, "I cannot make neither head nor tail of Seeley. I don't know what he means of History."[73]

In Oxford, constitutional history strengthened the students' acculturation, as struggling with Stubbs's work created a sense of shared experience and a rite of passage for them. Oman recollected in his autobiography his undergraduate encounters with Stubbs's prose. Upon his entry to Oxford, Hereford B. George, the history tutor at New College, had advised him to devote his time to constitutional history because "this sort of technical knowledge was precisely the matter to which general reader of history would probably have paid only a passing attention." Oman had followed the advice and quickly noticed how the *Charters* was among the students "a sort of bible, from which a candidate was expected to identify any paragraph without its context being given." This was so formidable a task that Oman, just like many others, decided to attend a series of lectures called "Steps to Stubbs" that A. L. Smith from Balliol gave.[74]

A branch of educational publishing spawned in the 1870s to capitalize on the reputation of Stubbs's work and to further consolidate it. According to Tout, the challenges which students experienced with

Stubbs's austere style were "mitigated by a whole literature of easy introductions to its doctrines, some good, more indifferent, none original, nearly all useful."[75] One of these was Forrest Fulton's *A Manual of Constitutional History*, which the *Sunday Times* applauded as "invaluable" to the student.[76] Fulton marketed the book in his preface as an easy entrance to constitutional history to those who were frightened of Stubbs:

> A very general feeling exists that Hallam is [a] very hard book, and it must be confessed that it is not the lightest possible reading; but if Hallam is hard Professor Stubbs is ten thousand times harder.[77]

Other attempts to make Stubbs comprehensible for ordinary readers were equally warmly welcomed. Green, who used Stubbs as a reference in *A Short History of the English People*, was congratulated for quickening "with a magic or Macaulayan wand" the constitutional history's "dry bones."[78] Even Stubbs's one-of-a-kind attempt to popularize history, *The Early Plantagenets* (1876), was perceived as a convenient aid for approaching *Select Charters* and *The Constitutional History*.[79]

The flood of manuals that were sold as remedies for Stubbs was incessant. In 1894, the *Athenaeum* wrote critically that "There are, in fact, almost too many 'easy introductions' to Stubbs" and that their number should no longer be "lightly increased."[80] The manuals, popular histories, and their reviews strengthened the impression of Stubbs, his histories, and constitutional history as the embodiments of scientific history that only the scholarly elite were able to understand, practice, and appreciate. Newspapers and magazines mythologized Stubbs as "an historical anatomist" as they claimed constitutional history was located beyond the reach "of popular approbation and applause" or "hasty and superficial readers."[81] Those who persistently worked their way through *The Constitutional History* and *Select Charters* were a step closer to becoming genuine historians and assuming the persona of a scientific historian.

Borrowing Virtuosity in Paratexts

If the heroization of Stubbs produced a model historian for the discipline, his name also had paratextual value for individual historians who used it to sanction their persona and the reliability of their studies. While they used Stubbs's name as a scholarly celebrity endorsement in their paratexts, they generated publicity for both the Stubbs myth and for the scholarly persona. Some showed admirable creativity in how they applied his name. George Prothero, lecturer at King's College in Cambridge, ingeniously wrote in the preface to *The Life of Simon de Montfort* (1877) how "it was with hearty satisfaction that on reading his [Stubbs's] pages I found I was in the main in agreement with the greatest living authorities."[82] If Prothero boldly implied equal footing with Stubbs, it was more common to highlight the closeness of the acquaintance to ensure that readers grasped the authenticity and value of the association. Freeman thanked Stubbs "for his personal readiness to correct and to suggest on all points" and William Bright had "enjoyed the special advantage of repeatedly consulting the Professor himself." For Burrows, Stubbs had been "a true friend and judicious critic," and for Creighton "an unfailing refuge in the case of difficulties."[83]

Historians drew on the classic mechanism of "gifting," or what Gladys Engel Lang and Kurt Lang call "the satellite effect," which assumes that an association with a renowned name confers authority. According to Emily Apter, the "poetics of names operates as a currency of transference, enabling embodied self-properties to travel from one gifted subject to another."[84] The transferable qualities could be both positive and negative. When Acton composed the list of authors for *The Cambridge Modern History*, he was careful not to include in the volume with an anticipated essay from Stubbs names that could have put "Stubbs" into disrepute.[85]

Acknowledgment created a stage for a multidirectional transmission of actual and imagined qualities between the sender and receiver as well as between all the other acknowledged people in the same textual sphere. Acknowledgments were normatively regulated exclusionary and inclusionary acts and read as evidence of networks and collaborations. They were invested with messages about attachment, collective values, hierarchies, and codes of conduct. Historians were well versed in the strategic and symbolic part acknowledgments played; they scrutinized prefaces and footnotes to search for their names, composed thank you notes to those who had mentioned them, and dissected the acknowledgments in their colleagues' books privately in letters and publicly in book reviews. The popular practice of addressing acknowledgments in reviews increased their visibility exponentially.

The unwritten rule was that an acknowledgment required personal affinity between the sender and the receiver. This increased their symbolic value as confirmations of scholarly connections and peer approval. The etiquette should have protected historians such as Stubbs from fictitious acknowledgments, but often those who could not boast his acquaintance bent the rules by mentioning him and his work as great sources of inspiration. Such misappropriation of names was rather common and was sometimes condemned both as an exploitation of reputable names and as an insult to the historians whose names were mentioned in subpar histories without their consent.[86] As acknowledgments were written not just to recognize collegial help but to boost status and credibility too, it was crucial to avoid an impression of selfish opportunism or misuse of a colleague's name. Lewis T. Dibdin wrote to Stubbs with this purpose in mind because otherwise his acknowledgment "might be misconstrued into an attempt to obtain the advantage of your *imprimatur* unjustifiably."[87]

The acknowledgments flattered Stubbs and he interpreted the prefatorial limelight as a consolidation of his station. He was "overwhelmed with the honour" when Freeman elevated him "to the literary Peerage" in the first volume of *Norman Conquest* and he told with pride in his last statutory lecture how his name had appeared "as the name of a helper in many prefaces" and how these acknowledgments had proven his influence.[88] Strikingly, he rarely returned the favor. In *The Constitutional History* (1874) he acknowledged "friends in England, Germany, and America," creating an illusion of an internationally well-connected historian. He expressed "A more special debt" to "the two Scholars" who had helped him with "counsel and criticism," and to whom he had "drawn by their association with my early Oxford ambitions, and whose patient kindness and acquaintance now nearly thirty years has not exhausted."[89] Although many guessed the identities of the unnamed Oxford scholars, the anonymization reduced the symbolic value of the recognition for those acknowledged.

Historians' acknowledgments established intentional and unintentional webs or relations which readers interpreted as reflections of the disciplinary community. Stubbs shared the prefatorial space with a mix of other names and his name radiated authority to the other acknowledged ones. Freeman was exultant when the American scholar and lawyer Hannis Taylor had regarded him and Stubbs as equals in his preface to *The Origin and Growth of the English Constitution*. Taylor praised the achievements of Stubbs, Freeman, and Seeley in elevating "the scientific study of history … to as high a point in England as it has ever reached upon the Continent," and then complemented specifically Freeman and Stubbs for their valuable work for historical science. Freeman was so honored by the preface that he wrote to Taylor, "it is pleasant to be so comfortably bracketed with Bishop Stubbs."[90] Footnotes created similar associations and responses. When Round found out that he and Stubbs were mentioned together in several footnotes in Maitland's and Frederick Pollock's *The History of English Law Before the Time of Edward I* (1895) he composed a self-congratulatory verse which he sent to Maitland as a token of his gratitude: "Their case they ground at times with Round, who shoulders rubs (in notes) with Stubbs."[91]

But prefaces (and footnotes) also misrepresented collegial loyalties. Round's preface in *Feudal England* (1895) illustrates how historians could mislead readers with cleverly crafted acknowledgments. Round was a divisive figure: his talents as a historian were not questioned, but his fondness for controversy was unanimously condemned. He prided himself on his utmost accuracy and minute knowledge of historical records. It was his mission to correct errors and dispose misconceptions. He had begun to fulfill his goal when still an undergraduate in Oxford in 1877 by compiling "a little souvenir" for his tutor James Franck Bright—a long list of errors in the latest volume of Bright's *A History of England*.[92] Such malignant tricks came to define his scholarly ethos. His constant waving of a battleax, however, wore out his supporters. Maitland complained in 1899 how Round's *The Commune of London* once again "gives his readers too much controversy and too little history." Readers' interest, Maitland went on to state, "is always being distracted by the castigation of some unfortunate being who lived in the nineteenth."[93] Round's exclusion from the scholarly community only fueled his vindictiveness and bitterness toward those who failed to understand that all he wanted was to improve the standards of scientific history by exposing everyone who undermined its principles with blunders, carelessness, and methodological incompetence.

One of Round's main targets was Freeman, and in the preface to *Feudal England* he continued his crusade against the late renowned historian. With a clever acknowledgment, Round made Stubbs look like his accomplice in the campaign to undermine Freeman's authority. Round was outraged that Freeman enjoyed the reputation of an accurate historian while Freeman was, according to him, someone who could "contrive within the compass of six words to make three mistakes."[94] He feared that Freeman's endless blunders would jeopardize the credibility of the scientific method. Thus, he thought that he was on a noble mission to salvage the discipline from the threat that Freeman's careless and biased application of the method posed for its credibility as a branch of science. Arch-conservative Round's resentment was further incited by Freeman's liberalism.[95] As he continued his assault on the late historian, Freeman's friends became outraged. Even Gardiner, who did his best to avoid confrontations, grew tired of the "excessively acrimonious" attack. He went as far as suspecting that Round's ill temper derived from the fact that "His father was out of his mind" and that this "heritage shows itself in his method of dealing with men superior to himself, as Freeman was."[96]

Round wrote the preface for *Feudal England* to discredit Freeman's scholarly authority, and if the name of Stubbs helped him to achieve his goal, all the better. The preface indeed created an illusion that Stubbs supported Round's allegations about Freeman's methodological defectiveness. Round first cataloged Freeman's many shortcomings and then announced that he had derived "encouragement" and "approval" from "our supreme authority—I mean the Bishop of Oxford." This gave the readers a reason to conclude that Stubbs in fact agreed with Round's claims that Freeman's *name* was surrounded by "an obstinate and mischievous superstition" about his "pre-eminent accuracy and authority on matters of fact." Or that Freeman and his nemesis Froude were equally guilty of partiality: "Just as his bias against the Roman Church led Mr. Froude to vindicate Henry in order to justify the breach with Rome, so Mr. Freeman's passion for democracy made him an advocate on behalf of Harold."[97] The name of Stubbs, known for undisputed virtuosity, was an ideal tool for validating methodological claims and Round used it cleverly to create an impression that he had Stubbs's backing for his crusade against Freeman.

Hero Worship and Its Epistemic Consequences

The heroization of Stubbs as an infallible historian discouraged critical scrutiny of his research procedures. Instead of acknowledging his flaws, historians portrayed him as a man who triumphed in everything from solving complex historical problems to composing a perfect index. Half-jokingly, Freeman claimed that the only thing Stubbs was not able to do was to make "a silk purse out of a cow's ear."[98] The flip side was that the heroization allowed Stubbs to cultivate scholarly habits that would have been classified as scholarly vices in other instances. His gravest sin was his revulsion at revising his published histories. This derived from his dislike for scholarly confrontations and from his conviction that there was nothing to correct in them.

Stubbs refrained from publicly criticizing other historians' work, as he believed that it would only strengthen the perception of academia as a nesting ground for petty disputes and overcharged egos. He also referred to the unflattering image of a "professorial character" that embodied unfairness and bullying and stressed that he had never given in to the temptation of intrigues.[99] His methodical cultivation of a persona that was above petty confrontations had consequences for his research practices and for the field of constitutional history. In practice, he refused to review books from allies or opponents—or, as a matter of fact, from anyone. Apparently, he wrote only one review during his entire career.[100] He also applied self-censorship. When he assisted Prothero in the preparation of *Select Statutes and other Constitutional Documents Illustrative of Elizabeth and James I*, he recommended alterations to the historical introduction to "eliminate words of controversy—perhaps more than you would care?—but I think safer."[101] Furthermore, he suppressed critical debate with other historians. Nowhere in *The Constitutional History*, an unusually critical reviewer complained, Stubbs "disputes a point with Hallam; yet there is a substantial difference between his theory and Hallam's with respect to the progress of the Constitution in the fourteenth century."[102]

Stubbs' aversion to controversies and his confidence in his flawlessness manifested in his reluctance to revise his published histories. This was not a minor error. Although historiographers have treated nineteenth-century histories as static texts, editing was an integral part of Victorian scientific, literary, and publishing cultures. As publishers kept their print runs low, historians received steady invitations to prepare new editions.[103] Indeed, they worked constantly side by side with their past, present, and future projects, and revising had both practical and scholarly significance for them. New editions extended copyright and promised some profit without the burden of composing something entirely new. The new editions of scholarly titles were rarely great business opportunities though, and P. L. Gell from Oxford University Press wrote to Stubbs that the greatest reward was the sense of fulfilling the scholarly duty of updating and improving one's own texts.[104] As historians embraced accuracy, perfection, and completeness as crucial epistemic virtues and components of their persona, they used feedback from readers and reviewers, the latest archival discoveries, and new studies to improve and keep their publications up to date. The mindset of pursuing perfection one edition at a time was captured by a writer who observed mercifully how nobody "will expect a first edition to be perfect; happy [are] those who can boast of perfection in [a] second or even a third edition."[105] Apart from Stubbs, historians found great solace in the possibility of fixing their errors and improving their texts, and William Edward Hartpole Lecky expressed this common mindset when he confessed to John Tyndall amidst revisions that "there is a comfort in getting one's books as perfect as one can."[106]

Editing could produce profoundly altered knowledge, narratives, and paratextual portraitures of the persona. If Hallam had maintained in 1849 that historical knowledge had a more enduring quality than scientific knowledge and that readers were less inclined to understand the continuous editing of a medieval historian than of chemists or geographers "whose science is continually progressive," the surge of archival discoveries and new studies effectively surpassed such a notion during the latter half of the century.[107] Lorraine Daston has pointed out that, in the nineteenth century, scientists realized that knowledge was not only cumulative but could contest and alter previous discoveries. Just like modern life itself, knowledge was perpetually moving at an accelerated tempo.[108] The mutability of knowledge rendered definitiveness an unattainable goal and history books were infinitely expandable and open to additions and alterations. Admitting and correcting one's own errors was not considered humiliation or a sign of insufficient research abilities: on the contrary, it was manly and courageous to admit and rectify flaws. Erring was a human quality, for no man was an encyclopedia "neatly bound in cloth."[109] According to Freeman, it was a "mark of true greatness" in a historian if he "could allow himself to have been wrong."[110]

Stubbs deviated from the common pattern of revising and gave only modest touches to his later editions. The new editions of *The Constitutional History* which were issued in the 1890s attest to his reluctance to alter the narrative. He acknowledged Vinogradoff, Maitland, and Pollock in the footnotes, but did not modify his text or address their discoveries or criticism. Phrases such as "But see" or "c.f." implied disagreement, but Stubbs left it entirely to the reader to determine what the differences might be. He similarly shifted the burden of judgment to readers when he admitted that later historians differed with him in some points without elaborating the contradictions. Yet he eagerly expanded existing footnotes when fresh studies and document editions provided additional proof for his views.[111] It is evident that he knew that he was not following the rules of the discipline, and he continuously defended himself and his historical understanding against mounting criticism in the 1890s. He affirmed that the views he had held in the 1870s did not need revision. He boldly contended in the 1895 edition of *Select Charters* that his theories about constitutional history rested "on arguments so sound … as those on which Copernicus and Kepler worked out their astronomical conclusions." He repeated this in 1900 in *Seventeen Lectures*.[112] His need to comment on the topic shows how he was aware of the new trends in constitutional history and how he needed to defend himself as he was, against good disciplinary conduct, unwilling to integrate them into his histories.

Since editing was at the heart of historical research and the scholarly persona, Stubbs's dislike of revisions contradicted his exemplary status. As Minna Bromberg and Gary Alan Fine suggest, once a positive reputation is firmly established, the audience may ignore or not even see inappropriate or conflicting actions.[113] This helps to explain why Stubbs's insufficient revisions came under critical scrutiny only posthumously. It was Maitland who was among the first to draw attention to the issue in the obituary he wrote about Stubbs for the *English Historical Review*. The fact that it was the only critical comment in the obituary gave additional weight to the protest. Maitland deplored how Stubbs's "desire to avoid controversy" had impeded him from updating his books. Consequently, readers who yearned for his guidance were left "doubting whether he is deliberately maintaining in the nineties a position he held in the seventies" and wondering what he thought about the present state of constitutional history. Maitland was inclined to think that Stubbs had intentionally remained silent about some of the most recent histories which had challenged his conclusions.[114] Although Maitland formulated his criticism discretely, it nonetheless revealed disappointment at Stubbs, who had breached the disciplinary guidelines.

The Declining Aura of a Heroic Historian

Jackson's Oxford Journal reported in 1884 how Freeman had announced during his inaugural lecture that "he must be a very bold man indeed who would call in question his [Stubbs's] historical decisions."[115] Freeman was right; Stubbs's monumental reputation was so petrifying that it discouraged critical appraisal of his work. Those who exposed slips in his books did it with the risk of being scolded by reviewers even when they cautiously balanced the criticism with apologies and praise. It was not until the 1890s that the new generation of historians submitted his interpretations to critical scrutiny.

Freeman dared once to admit that he and Stubbs differed in one historical detail. In a footnote in *The Growth of the English Constitution* (1872), he revealed that they disagreed "altogether as to the constitution of the Witenagemót" and added in the same note how "It is dangerous to set oneself up against the greatest master of English constitutional history."[116] When a revised edition appeared four years later, Freeman expanded the footnote and announced delightfully how in fact their interpretations diverged only in gradation.[117] Other historians were equally prudent when they found themselves opposing some of Stubbs's theories. Even Round was courteous when he questioned Stubbs's accuracy. In the preface to *Geoffrey de Mandeville* (1892) he wrote

> if, in the light of new evidence, I have found myself compelled to differ from the conclusions even of Dr. Stubbs, it in no way impeaches the accuracy of that unrivalled scholar, the profundity of whose learning and the soundness of whose judgment can only be appreciated by those who have followed him in the same field.[118]

Round's guardedness is understandable because questioning Stubbs could irritate reviewers. George Child's *Church and State under the Tudors* (1890) contained a footnote that disputed Stubbs's interpretation of the "judicial murder" of Archbishop Scrope and Henry IV's role in it.[119] Mary Bateson, who reviewed the book for the *English Historical Review*, treated the criticism as evidence of Child's prejudices. She accused him of unfairly charging Stubbs with misplaced clericalism. Whenever Child had relied on "his pilot"—Stubbs's *Appendices* of 1883—he had been on a firm ground, but as soon as he had dropped the pilot, he appeared "to lose his compass."[120] Bateson's spirited defense shows how a disagreement with Stubbs could antagonize reviewers. Hubert Hall from the Public Record Office faced similar reproof after calling Stubbs's categorization of taxable commodities "unsatisfactory" and claiming that his technical errors "deface an important section of the ablest historical text-book of this generation." The result was that *The Times* found Hall guilty of unreasonable and unjust criticism of Stubbs.[121]

In this climate, reproof of Stubbs's historical views or research procedures was far and apart. Yet the narrative of an infallible historian began to falter during the 1890s when younger legal historians started to read Stubbs's work with a new critical eye. The first serious attempt to re-evaluate Stubbs's historical views came from Maitland, who was then the Downing Professor of the Laws of England at Cambridge and an expert in early English law. He was named the torchbearer of Stubbs and the *Edinburgh Review* declared in 1896 his research marked "a new era in our legal literature."[122] Maitland formulated his criticism of Stubbs gingerly. His prefatorial message in *Domesday Book and Beyond* (1897) was that while he had done his utmost to reach historical truth, he had abandoned "as little as may be of what we learnt from Stubbs".[123] Even this, Maitland was told, had made the old man "sour."[124] He was, hence, surprised when Reginald Lane Poole asked him to write an obituary of Stubbs for the *English Historical Review*. Knowing the eulogizing tendencies and demands of the genre, he doubted his suitability for the task. He reminded Poole that he had had "the boldness to dissent from Stubbs about the Canon Law and that by so doing I made a little noise and even (which is much harder) a little money."[125] Poole was nonetheless convinced that Maitland was the right person for the task. Maitland accepted and politely abstained from any criticism of Stubbs's historical views and only disapproved of his unwillingness to engage in revisions.

Perhaps emboldened by Maitland's example, more discordant voices appeared. One of them belonged to Dudley Medley, a tutor at Keble College in Oxford and Professor of Modern History at the University of Glasgow from 1899. He published in 1894 *A Student's Manual of English Constitutional History* and politely maintained in the preface how "The great work of Dr. Stubbs … will probably always be our starting point," but added that in light of recent research his conclusions "require modification if not restatement."[126] When he worked on a revised edition four years later, his new preface showed how the center of gravity in constitutional history had shifted. According to Medley, everyone had been unprepared for Maitland's "lavish suggestiveness" and therefore entire sections in the new edition "have been re-written and the views expressed on many points have been largely modified." The "great work" of Stubbs had been overshadowed by the "masterly studies" of Maitland.[127]

The posthumous reputation of Stubbs was forged in the spirit of the intensifying criticism of his research practices and historical thinking. In

1908, Manchester University Press published *Studies and Notes Supplementary to Stubbs' Constitutional History*. It was a translation of the critical remarks the French Professor Charles Petit-Dutaillis had attached to *Histoire Constitutionnelle de l'Angleterre*, a French translation of the first volume of *The Constitutional History*. Petit-Dutaillis admitted that the book had its great merits but needed revising and updating. His extensive commentaries reflected the novel approach to constitutional history which the French historians in particular had adopted. According to Petit-Dutaillis, the minor revisions Stubbs had carried out did not give "an accurate idea of the progress made by research," nor were they "executed with all the attention to details which is desirable." He could only wonder how a historian like Stubbs had been so repelled by the duty of revising and, despite all his other virtues, had failed in a crucial responsibility in historical research.[128] James Tait, Professor of Ancient and Medieval History at Manchester, prefaced *Studies and Notes* and agreed with Petit-Dutailles's criticism adding how the outdated shape of *The Constitutional History* was a sheer embarrassment to teachers and students of English constitutional history.[129] Stubbs's obstinacy about revisions shaped his immediate posthumous reputation, as it was widely agreed that he had failed to fulfill the historian's duty to keep publications abreast with the latest development in the field.

Stubbs continued to have supporters, but they too admitted that his work was not as flawless as many of his contemporaries had claimed. Tout wrote about Stubbs in the 1912 supplement to the *Dictionary of National Biography* and balanced between admiration and disapproval. He certainly did not spare the superlatives when describing Stubbs's exceptional virtuosity. The biography reads at parts like an obituary and its praise is worth quoting here as an example of the stature Stubbs still enjoyed among historians. Tout presented Stubbs as the "foremost scholar of his generation," a historian who "worked with extraordinary rapidity, accuracy, and sureness," and whose range of "historical vision was enormous." His work showed "extraordinary mastery of the mass of material which had to be dealt with" and he moved "easily under all this mass of learning" using it "with accuracy, precision, and insight." He had been a "most distinguished editor" of the Rolls Series. His volumes were "in every respect models of what the 'editio princeps' of an original authority should be" and "much more than ideal examples of editorial workmanship." Furthermore, Tout praised Stubbs for "Self-suppression, impartiality, accuracy, sympathy, sobriety of judgment, and sense of proportion" and repeated some of Stubbs' pet topics: his dislike of controversy and preference for studies and libraries over dinner parties.[130]

Yet Tout could not escape admitting that Stubbs's Anglo-Saxon and Norman chapters needed careful rewriting because, as he explained, these were the passages where Stubbs had "looked into facts with the eyes of his German guides."[131] Tout conveniently cast at least part of the blame of Stubbs's outdatedness on the German school of history, which had become unpopular in England since Stubbs and his contemporaries had written their histories. Tout and his generation were inspired by the French historians and refuted the German nationalist idea of primitive German institutions as the source of human dignity and political independence. Stubbs had interpreted with patriotic enthusiasm the English constitution to be a unique expansion of these first seeds of self-government.[132] Tout, moreover, marked Stubbs's aversion for revisions and explained it with his conservatism, which had prevented him from "having little sympathy now for historical novelties" and from engaging with Maitland's "damaging criticism." Rather, he had "contented himself with affirming" that what he had written constituted "true history."[133]

As the standards of historical research were constantly under debate, the persona and the models that embodied it shifted accordingly. The posthumous criticism reduced the value of Stubbs as a model historian and his *Constitutional History* as a shining example of scholarly virtuousness. Nonetheless, his name had a long afterglow. The following generations inherited and embraced his grand historical vision of continuities and historical evolution. The ideas of progress and continuity guided historical thinking in England up to the Second World War.[134]

At the same time, Stubbs became a link in the historiography's great chain rather than an uncontested model for younger historians to emulate. In 1939, Oman noted how history was just as progressive a science as chemistry and therefore it was about time to move Stubbs's *Constitutional History* from the reference library to the "more inaccessible shelves" which housed histories that exemplified the style and mentality of their own age.[135] Well before that, new names had already surpassed Stubbs as bywords for a model persona as historians continued to draw the boundaries of proper history. Tout was to be one of them, as his research on medieval administrative history developed an entirely new direction for institutional history in England. Stubbs joined Newton, Faraday, Darwin, and many others who lent their names to the scientific communities that interpreted and reinterpreted them time and again in the context of shifting scientific procedures, virtues, and values.[136] Disciplines need their heroes to denote the prevailing persona, and as personae are continuously contested, so the meanings ascribed to these proper names have been—and continue to be—in constant movement.

Names as paratexts are indeed indispensable in modern sciences. They grant authority and reliability, but also serve the discipline collectively as models that capture desirable qualities and moral values. The paratextuality of names thrives on the reputations their bearers enjoy, making the reputational entrepreneurs the key interpreters of what matters in sciences and what it takes to be a good scientist. Although historians were instrumental in the heroization of Stubbs, reviewers and journalists were crucial in disseminating beyond scholarly circles the message about Stubbs as a model of the new scientific history. Various communities tend to construct divergent reputations, but Stubbs's public image was astonishingly cohesive and uncontested. In this sense, historians' attempt to forge an exemplary model was successful. As the next chapter shows, when attention was shifted to the particularities of historians' scientific method, a gap began to emerge between historians and non-expert commentators and their notions about proper historical research and historians.

Notes

1. Creighton, *Life and Letters*, 339.
2. Acton to Richard Simpson, October 15, 1859, in Altholtz, McElrath, and Holland, *Correspondence*, 2:25.
3. Prothero, *Life of Simon de Montfort*, viii; Round, *Feudal England*, xiv; "Notes of the Week," *North Devon Journal*, February 6, 1879, 6; Freeman to Macmillan, November 23, 1883, Add MSS 55052, BL.
4. Genette, *Paratexts*, 7–8.
5. Biagioli, "Rights or Rewards?" 254–257, 274; Secord, "Progress in Print," 375.
6. Secord *Victorian Sensation*, 24–25.
7. Trollope, "On Anonymous Publishing," 495–496.
8. "Literature," *Glasgow Herald*, October 26, 1871.
9. Williams, "Stubbs's Appointment," 121–124.
10. "Chronicles and Memorials of the Reign of Richard First," *Athenaeum*, June 4, 1864, 774.
11. "Regius Professor of History at Oxford," *Morning Post*, August 7, 1866, 4; "Ecclesiastical and Religious," *Newcastle Courant*, August 10, 1866; "Church and Universities," *Illustrated London News*, August 11, 1866, 131; "University and Collegiate," *John Bull*, August 11, 1866, 537.

12. "English Historical School," *BCWG*, November 5, 1868, 6.
13. "Death of the Bishop of Oxford," *Daily Telegraph*, April 23, 1901, 7.
14. "Oxford," *BCWG*, March 14, 1867, 6; "Professor Stubbs's Inaugural Lecture," *PMG*, March 29, 1867, 10–11; "Modern School of Historians," *Times*, February 5, 1879, 4; "Death of the Bishop of Oxford," *Times*, April 23, 1901, 3.
15. [Jessopp], "Brocas Book," 231.
16. Freeman to Macmillan, June 1, 1879, Add MS 55052, BL; Freeman to Thompson, January 29, 1879, U DX9/75, HHC.
17. Cubitt, "Introduction: Heroic Reputations," 2–3; Goldsmith, "Celebrity and the Spectacle of Nation," 22.
18. Macmillan to Stubbs, December 22, 1883, Add MSS 55415, BL.
19. Morgan, "Heroes," 178.
20. Nicholson, "Review of J. E. Thorold Rogers," 168.
21. Stubbs, *Seventeen Lectures*, 1.
22. Stubbs, *Seventeen Lectures*, 374.
23. Dean Hook to Freeman, July 24, 1866, FA 1/7/297–320, JRL; Oman, *Memories*, 106; Freeman to Thompson, January 29, 1879, U DX9/75, HHC; Freeman to Thompson, August 15, 1882, U DX9/98, HHC.
24. Carlyle, *On Heroes,* 11.
25. Hutton, "Review of William Stubbs," 560–562.
26. Freeman, *Methods*, 274, 278–279.
27. Freeman to Green, March 26, 1876, FA1/8/59–59a, JRL; Freeman, *Methods*, 275–277.
28. [Escott], "A Joke or a Job?" 104; "Death of the Bishop of Oxford," *Daily Telegraph*, April 23, 1901, 7; "Dr. Stubbs and the Large 'B'," *Manchester Courier and Lancashire General Advertiser*, April 23, 1901, 6.
29. "What the *World* says," *Derby Mercury*, March 23, 1892, 7; "Death of the Bishop of Oxford." *Daily Telegraph*, April 23, 1901, 7.
30. "Death of the Bishop of Oxford," *Times*, April 23, 1901, 3.
31. "Death of the Bishop of Oxford," *Daily Telegraph*, April 23, 1901, 7.
32. Gladstone to Stubbs December 27, 1875, in Hutton, *Letters*, 147–148.
33. Stubbs to Freeman, April 13, 1858, in Hutton, *Letters*, 42–43.
34. "London," *Times*, March 21, 1892, 9.
35. Warren, "Rankean Tradition," 32–33.
36. Freeman to Green, January 21, 1875, FA 1/8/31–60, JRL; Freeman, *Methods*, 289.
37. Quoted in Hesketh, *Science of History*, 77–78.
38. Hesketh, *Victorian Jesus*, 157–158.
39. Seeley to Reinhold Pauli, January 5, 1876, and to Browning, April 28, [1877], GBR/0272/OB/1/1455/A, King's Cam.

40. Seeley to Browning, March 17, s.a., GBR/0272/OB/1/1455/, King's Cam.
41. Seeley to Browning, December 26, c. 1885–1886, GBR/0272/OB/1/1455/, King's Cam.
42. Seeley to Browning, September 26, c. 1882, GBR/0272/OB/1/1455/, King's Cam.
43. Seeley to Browning, September 26, c. 1882, GBR/0272/OB/1/1455/, King's Cam.
44. Ward to Freeman, March 4, 1885, FA 1/7/778–801, JRL.
45. Acton to Gladstone, Easter Sunday 1893, in Figgis and Laurence, *Selections*, 172.
46. "Modern School of Historians," *Times*, February 5, 1879, 4; "Mr. Freeman, the Historian," *Gloucester Citizen*, January 18, 1884, 3; "New Regius Professor of History," *Bristol Mercury*, March 18, 1884, 8; "Evening News," *Edinburgh Evening News*, March 18, 1884, 2; "Notes of the Week," *BCWG*, March 20, 1884, 5.
47. Jessopp, "Brocas Book," 225.
48. Maitland, "William Stubbs," 417.
49. Stubbs, *Seventeen Lectures*, 11–12.
50. Stubbs, *Seventeen Lectures*, 374, 387.
51. Stubbs, *Seventeen Lectures*, 3rd ed., ix–xi.
52. Bentley, "Henry Hallam Revisited," 453–473.
53. Freeman, *NC*, 5:v–vi; Ramsay, *Lancaster and York*, vii.
54. Stubbs, *Constitutional History*, 1:iii–iv.
55. Towsey, *Reading History*, 104–110.
56. Mandler, *English National Character*, 36–38, 45–47.
57. Soffer, *Discipline and Power*, 63–64.
58. Readman, "Place of the Past," 189, 197–199; Bentley, *Modernizing England's Past*, 19, 23–25.
59. Stubbs, *Select Charters*, v.
60. Stubbs, *Select Charters*, v; Stubbs, *Registrum*, 2nd ed., ix.
61. Stubbs, *Select Charters*, v–vi; Stubbs, *Constitutional* History, 1:iii; Stubbs, *Seventeen Lectures*, 81, 386.
62. "Talk of the Day," *Academy*, July 27, 1901, 64; Cumberbirch, "Late Bishop of Oxford," 4.
63. Browne, "Inspiration to Perspiration," 78, 87; White, *Thomas Huxley*, 6–7.
64. Stubbs, Seventeen Lectures, 91–92; Seeley, "History and Politics [IV]," 32; Hesketh, *Science of History*, 44; Fara, *Newton*, 167–171.
65. Stubbs, *Seventeen Lectures*, 91–92.
66. Stubbs, *Seventeen Lectures*, 99.
67. Freeman to Hook, December 20, 1874, in Stephens, *Life and Letters*, 2:88; "Simon the Montfort," *Examiner*, March 17, 1877, 336.

68. Brentano, "Sound of Stubbs," 7–10, 14.
69. Hesketh, *Science of History*, 133–152.
70. "Death of the Bishop of Oxford," *Daily Telegraph*, April 23, 1901, 7.
71. Tout, "Stubbs," 446–447.
72. Creighton to Paese, September 30, 1878, in Creighton, *Life and Letters*, 192.
73. Creighton to Freeman, March 18, 1885, FA1/7/122, JRL.
74. Oman, *Memories*, 104–105.
75. Tout, "Stubbs," 447.
76. "New Books," *Sunday Times*, January 31, 1875, 7.
77. Fulton, *Manual*, v.
78. "New Edition," *Glasgow Herald*, January 24, 1878, 2; "Current Literature," *Daily News*, May 3, 1875, 3; "New Books," *PMG*, December 1, 1875, 11.
79. "Epochs of History," *Nottinghamshire Guardian*, October 6, 1876, 6.
80. "Elements of English Constitutional History," *Athenaeum*, June 9, 1894, 740.
81. "Talk of the Day," *Academy*, April 27, 1901, 356; "Modern School of Historians," *Times*, February 5, 1879, 4.
82. Prothero, *Life of Simon de Montfort*, viii.
83. Freeman, *NC*, 2:vii; Bright, *Chapters of Early English*, v; Burrows, *Worthies of All* Souls, viii; Creighton, *History of the Papacy*, 1:viii.
84. Apter, "Celebrity Gifting," 87, 97; Engel Lang and Lang, "Recognition and Renown," 95–96.
85. Acton to Creighton, November 26, 1896, Add MSS 6871/110, CUL.
86. "Review of Alfred J. Church," *EHR*, 607–608.
87. Dibdin to Stubbs, April 10, 1885, in Hutton, *Letters*, 268.
88. Stubbs to Freeman, March 5, 1867, MS. Eng. Misc. e. 148, Bodleian; Stubbs, *Seventeen Lectures*, 386.
89. Stubbs, *Constitutional History*, 1:iv.
90. Taylor, *Origin and Growth*, viii, xii–xiii; Freeman to Taylor, November 8, 1889, in Stephens, *Life and Letters*, 2:410.
91. Round to Maitland, March 28, 1895, Add MSS 7006 (52), CUL.
92. Powell, *John Horace Round*, 38–39; Procter, "*Red Book of the Exchequer*," 512.
93. Maitland, "Round's 'Commune of London'," in Cam, *Selected Historical Essays*, 259.
94. Round, "Scientific History," 560.
95. Powell, *John Horace Round*, 94–102, 116–122.
96. Gardiner to Acton, July 1892, Add MSS 8119 (2)/I/923, CUL.
97. Round, *Feudal England*, x–xi.
98. Freeman to Macmillan, August 13, 1877 and July 25, 1878, Add MSS 55051, BL.

99. Stubbs, *Seventeen Lectures*, 11–12, 385–386.
100. Tout, "Stubbs," 450.
101. Stubbs to Prothero, January 10, 1894, in Hutton. *Letters*, 362.
102. "Constitutional History of England," *Athenaeum*, January 8, 1876, 47.
103. Secord, *Victorian Sensation*, 152; Dooley, *Author and Printer*, 87; Garritzen, "Revise, Edit, and Improve," 289–294; Weedon, "Analysis of the Cost of Book Production," 61.
104. Macmillan to Price, August 14, 1875, Add MSS 55397, BL; Gell to Stubbs, June 10, 1897, Letter Books 69, OUP.
105. "A. N.," "Review of M. de Maulde," 162.
106. Lecky to Tyndall, February 3, 1891, in *Memoir*, 227–228.
107. Hallam, *Supplemental Notes*, v–vi.
108. Daston, "History of Science," 527–528.
109. "English Historical School," *BCWG*, November 5, 1868, 6.
110. Freeman, *Methods*, 281.
111. Stubbs, *Constitutional History*, 4th ed., 2:187, 2:265, 2:279, 2:281.
112. Stubbs, *Select Charters*, 8th ed., viii; Stubbs, *Seventeen Lectures*, 3rd ed., vii.
113. Bromberg and Fine, "Resurrecting the Red," 1139.
114. Maitland, "William Stubbs," 424.
115. "Modern History," *Jackson's Oxford Journal*, October 18, 1884, 5.
116. Freeman, *Growth*, 167.
117. Freeman, *Growth*, 3rd ed., 176.
118. Round, *Geoffrey de Mandeville*, vii.
119. Child, *Church and State*, 32.
120. Bateson, "Review of George W. Child," 35.
121. Hall, *History of the Custom-Revenue*, 4–7; "Custom Revenue in England," *Times*, September 21, 1885, 3.
122. [Roscoe], "Early History of English Law," 429.
123. Maitland, *Domesday Book*, vi.
124. Maitland to Poole, September 17, 1898, GBR/0012/MS Add 7474/14, CUL.
125. Maitland to Poole, April 21 and 29, 1901, in Fifoot, *Letters*, 225.
126. Medley, *Student's Manual*, vi.
127. Medley, *Student's Manual*, 2nd ed., vi, ix.
128. Petit-Dutaillis, *Studies and Notes*, xii–xiii.
129. Tait, "Preface to the English translation," in Petit-Dutaillis, *Studies and Notes*, vii–viii.
130. Tout, "Stubbs," 444–447.
131. Tout, "Stubbs," 447.
132. Petit-Dutailles, *Studies and Notes*, xiii.
133. Tout, "Stubbs," 447–448.
134. Soffer, *Discipline and Power*, 62, 106–107.

135. Oman, *On the Writing of History*, 219–220.
136. Fara, *Newton*, 12–13; Miller, "'Puffing Jamie'," 2–5; Cantor, "Scientist as Hero," 172.

References

Unpublished Primary Sources

Bodleian Library, Oxford: Stubbs Papers.
British Library: The Macmillan Papers.
Cambridge University Library: Acton Papers.
Cambridge University Library: Frederic William Maitland Papers.
Cambridge University Library: Reginald Lane Poole Papers.
Hull History Centre: Letters from Edward Augustus Freeman to Edith Thompson.
John Rylands Library, Manchester: E. A. Freeman Archive.
King's College Archive Centre, Cambridge: Oscar Browning Papers.
Oxford University Press Archive: Letter Books.

Printed Primary Sources

Altholz, Josef L., Damian McElrath, and James C. Holland. *The Correspondence of Lord Acton and Richard Simpson*. Vol. 2. Cambridge: Cambridge University Press, 1973.
[Anon.]. "Chronicles and Memorials of the Reign of Richard First." *Athenaeum*, June 4, 1864, 774.
[Anon.]. "The Regius Professor of History at Oxford." *Morning Post*, August 7, 1866, 4.
[Anon.]. "Ecclesiastical and Religious." *Newcastle Courant*, August 10, 1866.
[Anon.]. "Church and Universities." *The Illustrated London News*, August 11, 1866, 131.
[Anon.]. "University and Collegiate." *John Bull*, August 11, 1866, 537.
[Anon.]. "Oxford." *Bath Chronicle and Weekly Gazette*, March 14, 1867, 6.
[Anon.]. "Professor Stubbs's Inaugural Lecture." *Pall Mall Gazette*, March 29, 1867, 10–11.
[Anon.]. "The English Historical School." *Bath Chronicle and Weekly Gazette*, November 5, 1868, 6.
[Anon.]. "Literature." *Glasgow Herald*, October 26, 1871.
[Anon.]. "New Books and Editions." *The Sunday Times*, January 31, 1875, 7.
[Anon.]. "Current Literature." *Daily News*, May 3, 1875, 3.

[Anon.]. "New Books and New Editions." *Pall Mall Gazette*, December 1, 1875, 11.
[Anon.]. "The Constitutional History of England." *Athenaeum*, January 8, 1876, 47–48.
[Anon.]. "Epochs of History." *Nottinghamshire Guardian*, October 6, 1876, 6.
[Anon.]. "Simon de Montfort." *Examiner*, March 17, 1877, 336–337.
[Anon]. "The New Edition of Mr. Green's English History." *Glasgow Herald*, January 24, 1878, 2.
[Anon.]. "The Modern School of Historians." *The Times*, February 5, 1879, 4.
[Anon.]. "Notes of the Week." *North Devon Journal*, February 6, 1879, 6.
[Anon.]. "Mr. Freeman, the Historian." *Gloucester Citizen*, January 18, 1884, 3.
[Anon.]. "The Evening News." *Edinburgh Evening News*, March 18, 1884, 2.
[Anon.]. "New Regius Professor of History." *Bristol Mercury*, March 18, 1884, 8.
[Anon.]. "Notes of the Week." *Bath Chronicler and Weekly Gazette*, March 20, 1884, 5.
[Anon.]. "The Modern History Professor's Inaugural lecture." *Jackson's Oxford Journal*, October 18, 1884, 5.
[Anon.]. "The Customs Revenue in England." *The Times*, September 21, 1885, 3.
[Anon]. "Review of Alfred J. Church's *Early Britain*." *English Historical Review* 5, no. 19 (1890): 607–608.
[Anon.]. "London." *The Times*, March 21, 1892, 9.
[Anon.]. "What the *World* Says." *Derby Mercury*, March 23, 1892. 7.
[Anon.]. "The Elements of English Constitutional History." *Athenaeum*, June 9, 1894, 740.
[Anon.]. "Death of Bishop of Oxford." *The Times*, April 23, 1901, 3.
[Anon.]. "Death of the Bishop of Oxford. A Great Historian." *Daily Telegraph*, April 23, 1901, 7.
[Anon.]. "Dr. Stubbs and the Large 'B'." *Manchester Courier and Lancashire General Advertiser*, April 23, 1901, 6.
[Anon.]. "The Talk of the Day." *Academy*, April 27, 1901, 356.
[Anon.]. "The Talk of the Day." *Academy*, July 27, 1901, 64.
Bateson, Mary. "Review of Gilbert W. Child's *Church and State under the Tudors*." *English Historical Review* 6, no. 22 (1891): 381–383.
Bright, William. *Chapters of Early English Church History*. London: Macmillan, 1878.
Burrows, Montagu. *Worthies of All Souls. Four Centuries of English History Illustrated from the College Archives*. London: Macmillan, 1874.
Cam, Helen M. *Selected Historical Essays of F. W. Maitland*. Cambridge: Selden Society & Cambridge University Press, 1957.

Carlyle, Thomas. *On Heroes, Hero-Worship and the Heroic in History*. Lincoln: University of Nebraska Press, 1966.

Child, Gilbert W. *Church and State under the Tudors*. London: Longman, 1890.

Creighton, Louise. *Life and Letters of Mandell Creighton, Sometime Bishop of London*. Vol. 1. London: Longman, 1904.

Creighton, Mandell. *A History of the Papacy during the Period of the Reformation*. Vol. 1. London: Longman, 1882.

Cumberbirch, John. "The Late Bishop of Oxford." *Yorkshire Evening Post*, April 30, 1901, 4.

[Escott, T. H. S.]. "A Joke or a Job?" *The Fortnightly Review*, July 1885, 102–107.

Fifoot, C. H. S. *The Letters of Frederic William Maitland*. London: Selden Society, 1965.

Figgis, John Neville and Reginald Vere Laurence. *Selections from the Correspondence of the First Lord Acton*. Vol. 1. London: Longman, 1917.

Freeman, Edward A. *The History of the Norman Conquest of England, Its Causes and Its Results*. Vol. 2. Oxford: Clarendon Press, 1868.

Freeman, Edward A. *The Growth of the English Constitution from the Earliest Times*. London: Macmillan, 1872.

Freeman, Edward A. *The Growth of the English Constitution from the Earliest Times*. 3rd ed. London: Macmillan, 1876.

Freeman, Edward A. *The History of the Norman Conquest of England*. Vol. 5. Oxford: Clarendon Press, 1876.

Freeman, Edward A. *The Methods of Historical Study*. London: Macmillan, 1886.

Fulton, Forrest. *A Manual of Constitutional History*. London: Butterworths, 1875.

Hall, Hubert. *A History of the Custom-Revenue in England from the Earliest Times to the Year 1827*. London: Elliot Stock, 1885.

Hallam, Henry. *Supplemental Notes to the View of the State of Europe during the Middle Ages*. London: John Murray, 1848.

Hutton, William H. "Review of William Stubbs's *Willelmi Malmesbiriensis de gestis regum Anglorum: Historiæ novella libri tres*." *English Historical Review* 6, no. 23 (1891): 560–562.

Hutton, William Holden. *Letters of William Stubbs Bishop of Oxford 1825–1901*. London: Archibald Constable, 1904.

[Jessopp, Augustus]. "The Brocas Book." *Edinburgh Review*, July 1887, 225–253.

Maitland, Frederic William. *Domesday Book and Beyond. Three Essays in the Early History of England*. Cambridge: Cambridge University Press, 1897.

Maitland, F. W. "William Stubbs, Bishop of Oxford." *English Historical Review* 16, no. 63 (1901): 417–426.

Medley, Dudley Julius. *A Student's Manual of English Constitutional History*. Oxford: B. H. Blackwell & Simpkin, Marshall, Hamilton, Kent, & Co., 1894.
Medley, Dudley Julius. *A Student's Manual of English Constitutional History*. 2nd ed. Oxford: B. H. Blackwell & Simpkin, Marshall, Hamilton, Kent & Co., 1898.
A Memoir of The Right Hon. William Edward Hartpole Lecky, by his wife. London: Longman, 1909.
Nicholson, J. Shield. "Review of J. E. Thorold Rogers's *A History of Agriculture and Prices in England*." *English Historical Review* 4, no. 13 (1889): 167–170.
Oman, Charles. *On the Writing of History*. London: Methuen, 1939.
Oman, Charles. *Memories of Victorian Oxford and Some Early Years*. London: Methuen, 1941.
Petit-Dutaillis, Charles. *Studies and Notes Supplementary to Stubbs' Constitutional History*. Manchester: University Press, 1908.
Prothero, George Walter. *The Life of Simon de Montfort Earl of Leicester with Special Reference to the Parliamentary History of His Time*. London: Longman,1877.
Ramsay, Sir James H. *Lancaster and York. A Century of English History (A.D. 1399–1485)*. Vol. 1. Oxford: Clarendon Press, 1892.
[Roscoe, E. S.]. "The Early History of English Law." *Edinburgh Review*, April 1896, 428–448.
Round, J. H. "Scientific History." *Athenaeum*, October 26, 1889, 560.
Round, J. H. *Geoffrey de Mandeville. A Study of the Anarchy*. London: Longman, 1892.
Round J. H. *Feudal England: Historical Studies on the XIIth and XIIIth Centuries*. London: Swan Sonnenschein, 1895.
Seeley, J. R. "History and Politics [IV]." *Macmillan's Magazine*, November 1879, 23–32.
Stephens, W. R. W. *The Life and Letters of Edward A. Freeman D.C.L., LL.D.* Vol. 2. London: Macmillan, 1895.
Stubbs, William. *Select Charters and Other Illustrations of English Constitutional History from the Earliest Times to the Reign of Edward the First*. Oxford: Clarendon Press, 1870.
Stubbs, William. *The Constitutional History of England and Its Origin and Development*. Vol. 1. Oxford: Clarendon Press, 1874.
Stubbs, William. *Seventeen Lectures on the Study of Medieval and Modern History and Kindred Subjects*. Oxford: Clarendon Press, 1886.
Stubbs, William. *Select Charters and Other Illustrations of English Constitutional History from the Earliest Times to the Reign of Edward the First*. 8th ed. Oxford: Clarendon Press, 1895.
Stubbs, William. *The Constitutional History of England in Its Origin and Development*. Vol. 2. 4th ed. Oxford: Clarendon Press, 1896.

Stubbs, William. *Registrum Sacrum Anglicanum. An Attempt to Exhibit the Course of Episcopal Succession in England from the Records and Chronicles of the Church*. 2nd ed. Oxford: Clarendon Press, 1897.

Stubbs, William. *Seventeen Lectures on the Study of Medieval and Modern History and Kindred Subjects*. 3rd edition. Oxford: Clarendon Press, 1900.

Tait, James. "Preface to the English Translation." In *Studies and Notes Supplementary to Stubbs' Constitutional History*, by Charles Petit-Dutaillis, vii–viii. Manchester: University Press, 1908.

Taylor, Hannis. *The Origin and Growth of the English Constitution. An Historical Treatise*. Boston and New York: Houghton, Mifflin, 1889.

Tout, T. F. "Stubbs, William." *National Biography, 1912 Supplement*. https://en.wikisource.org/wiki/Stubbs,_William_(DNB12). Accessed July 2, 2020.

Trollope, Anthony. "On Anonymous Publishing." *Fortnightly Review*, July 1, 1865, 491–498.

Secondary Sources

Apter, Emily. "Celebrity Gifting: Mallarmé and the Poetics of Fame." In *Constructing Charisma: Celebrity, Fame, and Power in Nineteenth-Century Europe*, edited by Edward Berenson and Eva Giloi, 86–102. New York: Bergham Books, 2010.

Bentley, Michael. *Modernizing England's Past: English Historiography in the Age of Modernism 1870–1970*. Cambridge: Cambridge University Press, 2005.

Bentley, Michael. "Henry Hallam Revisited." *The Historical Journal* 55, no. 2 (2012): 453–473.

Biagioli, Mario. "Rights or Rewards?" In *Scientific Authorship: Credit and Intellectual Property in Science*, edited by Mario Biagioli and Peter Galison, 254–279. New York: Routledge, 2003.

Brentano, Robert. "The Sound of Stubbs." *Journal of British Studies* 6, no. 2 (1967): 1–14.

Bromberg, Minna and Gary Alan Fine. "Resurrecting the Red: Pete Seeger and the Purification of Difficult Reputations." *Social Forces* 80, no. 4 (2002): 1135–1155.

Browne, Janet. "Inspiration to Perspiration: Francis Galton's *Hereditary Genius* in Victorian Context." In *Genealogies of Genius*, edited by Joyce E. Chaplin and Darrin M. McMahon, 77–95. Basingstoke: Palgrave Macmillan, 2016.

Cantor, Geoffrey. "The Scientist as Hero: Public Images of Michael Faraday." In *Telling Lives in Science: Essays on Scientific Biography*, edited by Michael Shortland and Richard Yeo, 171–193. Cambridge: Cambridge University Press, 2008.

Cubitt, Geoffrey. “Introduction: Heroic Reputations and Exemplary Lives.” In *Heroic Reputations and Exemplary Lives*, edited by Geoffrey Cubitt and Allen Warren, 1–26. Manchester: Manchester University Press, 2000.

Daston, Lorraine. “The History of Science as European Self-Portraiture.” *European Review* 14, no. 4 (2006): 523–536.

Dooley, Allan C. *Author and Printer in Victorian England*. Charlottesville: University Press of Virginia, 1992.

Engel Lang, Gladys and Kurt Lang. “Recognition and Renown: The Survival of Artistic Reputation.” *American Journal of Sociology* 94, no. 1 (1988): 79–109.

Fara, Patricia. *Newton: The Making of Genius*. London: Macmillan, 2002.

Garritzen, Elise. “Revise, Edit, and Improve: Writing and Publishing History as an Unending Process in Victorian Britain.” *Clio: A Journal of Literature, History, and the Philosophy of History* 45, no. 3 (2016): 289–314.

Genette, Gerard. *Paratexts: Thresholds of Interpretation*. Cambridge: Cambridge University Press, 1997.

Goldsmith, Jason. “Celebrity and the Spectacle of Nation.” In *Romanticism and Celebrity Culture, 1750–1850*, edited by Tom Mole, 21–40. Cambridge: Cambridge University Press, 2009.

Hesketh, Ian. *The Science of History in Victorian Britain: Making the Past Speak*. London: Pickering & Chatto, 2011.

Hesketh, Ian. *Victorian Jesus: J.R. Seeley, Religion, and the Cultural Significance of Anonymity*. Toronto: University of Toronto Press, 2017.

Mandler, Peter. *The English National Character: The History of an Idea from Edmund Burke to Tony Blair*. New Haven: Yale University Press, 2006.

Miller, David Philip. “‘Puffing Jamie’: The Commercial and Ideological Importance of Being a ‘Philosopher’ in the Case of the Reputation of James Watt (1736–1819).” *History of Science* 38 (2000): 1–24.

Morgan, Simon. “Heroes in the Age of Celebrity: Lafayette, Kossuth, and John Bright in 19th-Century America.” *Historical Social Research* 32 (2019): 165–185.

Powell, W. Raymond. *John Horace Round: Historian and Gentleman of Essex*. Chelmsford: Essex Record Office, 2001.

Procter, Margaret. “The *Red Book of the Exchequer*: A Curious Affair Revisited.” *Historical Research* 87, no. 237 (2014): 510–532.

Readman, Paul. “The Place of the Past in English Culture c. 1890–1914.” *Past and Present* 186 (2005): 147–199.

Secord, James. “Progress in Print.” In *Books and the Sciences in History*, edited by Marina Frasca-Spada and Nick Jardine, 369–389. Cambridge: Cambridge University Press, 2000.

Secord, James A. *Victorian Sensation: The Extraordinary Publication, Reception, and Secret Authorship of* Vestiges of the Natural History of Creation. Chicago: University of Chicago Press, 2000.

Soffer, Reba N. *Discipline and Power: The University, History, and the Making of an English Elite, 1870–1930*. Stanford: Stanford University Press, 1994.
Towsey, Mark. *Reading History in Britain and America, c.1750–c.1840*. Cambridge: Cambridge University Press, 2019.
Warren, John. "The Rankean Tradition in British Historiography, 1840 to 1950." In *Writing History: Theory & Practice*, edited by Stefan Berger, Heiko Feldner, and Kevin Passmore, 23–41. London: Hodder Arnold, 2003.
Weedon, Alexis. "An Analysis of the Cost of Book Production in Nineteenth-Century Britain." In *Book Publishing*, edited by John Feather, 216–237. Vol. 2. New York: Routledge, 2011.
White, Paul. *Thomas Huxley: Making the "Man of Science."* Cambridge: Cambridge University Press, 2003.
Williams, N. J. "Stubbs's Appointment as Regius Professor, 1866," *Historical Research* 33, no. 87 (1960): 121–125.

CHAPTER 4

Almost Antiquaries

Readers judged books by their size. A large format was a conventional sign of scholarly value.[1] John Richard Green was on a crusade against the common practice of measuring a historian's worth more "by size of one's book than by its intrinsic value." According to him, this foolish idea had misled historians to believe that to be a great scholar one had to publish in "three volumes octavo."[2] Green's remonstrance was inspired by his personal fear that *A Short History of the English People* which he was writing was going to be too popular and too small to qualify him as a person of serious scholarly consequence. Even if Green's protest was motivated by selfish considerations, he correctly observed that size as a paratext signaled genre, value, authorial qualities, and anticipated audiences. The writing of multi-volume histories was ingrained in the scientific historians' persona, and they regarded the monumental volumes as testimonies of their scientific competence. This forged a tight bond between historians and their comprehensive scholarly undertakings. Readers, by contrast, had reservations about big histories whose minute recording of facts bore resemblance to chronicles or almanacs. The *Quarterly Review* condemned historians' obsession with detail and maintained that "The mania of saying everything, of picking up every glove, of guarding every point, is one to which a certain school of historians at the present day is peculiarly subject, and which has resulted in the bulging of their histories

E. Garritzen, *Reimagining the Historian in Victorian England*, https://doi.org/10.1007/978-3-031-28461-8_4

into most amorphous amplitude."[3] The "certain school" was of course that of scientific history.

These contradicting opinions made a book's size a convenient pretext for negotiating the method, style, and aim of history, as this chapter shows. Three strands with consequences to historians' persona emerged from these dialogs. The first originated from the interpretation of the cardinal virtue of completeness: did every historical detail serve a truth-bearing function? The second addressed the practical and methodological side of the same issue: were historians obliged to recount everything they discovered from the sources, what was the meaning of selection and generalization—particular and general—in historical research, and did selection generate biased histories? The third strand directed attention to the origins of the fascination for detail: was the passion for detail inspired by the flourishing antiquarian culture or was it spurred on by the inductive method? The first part of this chapter explores these discordant voices and illustrates how reconciling the particular and general in historians' persona was anything but straightforward. The purpose of historical research, whether it was the minute chronicling of facts or their synthesizing and generalizing, entailed a unique combination of skills and qualities and gave rise to a model for a historian's persona that drew on the antiquarian tradition.

These debates highlighted the research method as a marker of intellectual and disciplinary boundaries, as historians maintained that they pursued different goods to the antiquaries. Method had multiple definitions in historians' lexicons. Herman Paul has traced at least three meanings which were used in late nineteenth-century England: a posture of the inquirer, a mental process for acquiring truthful knowledge, and the course of the research process.[4] The method, moreover, held symbolic force in establishing history as a distinct discipline. Just like mid-Victorian scientists who had used method-talk to remedy the image that scientific knowledge was attainable with common sense, historians now referred to the inductive method to correct the impression that history was something that anyone was capable of doing.[5] Toward the end of the century, scientists reoriented their focus. The "cult of scientific method," as Beatrice Webb called the incessant talk about methods, foregrounded once again issues that overlapped with historians' intentions: the demarcation of disciplinary boundaries and the demonstration that competence in a scientific method benefitted society as it advanced informed citizenship. Minds attuned to a scientific method could resist passionate responses to social and political challenges.[6] This latter was one of the key arguments of those history tutors who wished to integrate the method into the curricula. From these multiple meanings ascribed to the method, this chapter draws attention to its use for explaining why history and antiquarianism should not be confused.

The critical commentary of bulky histories brought attention to the unstable boundary between scientific history and antiquarianism, and the second part of this chapter revisits the alliance between history and antiquarianism which historians' disciplinary ambitions made a burning issue. Philippa Levine has argued in her seminal study on history, antiquarianism, and archeology that the professionalization of history entailed a split between history and antiquarianism, and that historians had achieved a separate status by the 1880s.[7] However, the thoroughness of this separation—or historians' professional ethos—should not be exaggerated. If one condition for a scientific discipline was a specific enough object of research, historians struggled to claim monopoly over the past.[8] Another challenge was that the precursors of scientific history tolerated a persona that combined history and antiquarianism. The following generations advocated a more radical break between the two approaches to the past, but the long careers of Freeman and his contemporaries who confounded the distinctions between history and antiquarianism ensured that the boundary remained porous. Browning informed James Bryce with alarm as late as 1899 how in King's College the antagonism was growing between historical and antiquarian "types."[9] It is significant that Browning evaluated his colleagues according to their affinity either to antiquarianism or history, as it implies that the two continued to overlap. Indeed, the persona of a scientific historian was inspired both by history and antiquarianism, and historians trod a fine line between the two approaches to the past. More crucial than denying the debt to antiquarianism was ensuring that antiquarian traits did not dominate the scholarly persona. It was, thus, a matter of gradations. Scores of reviews, reports, and public lectures confirmed this vexing kinship.

The public, meanwhile, accepted that history and antiquarianism were different activities but struggled to recognize a clear gulf between the two, as the scientific historians endorsed practices that resembled the antiquarian passion for detail; they appeared as much hunters of undiscovered documents and chroniclers of insignificant details as the antiquaries were.[10] The last part of this chapter brings together historians and the readers who struggled to appreciate the bulky histories. Historians were agonized by readers who satisfied their curiosity with more popular and entertaining forms of history because they realized that the fascination for light reading was problematic for the persona project. First, historians were committed to sharing knowledge and second, history could become useful only if it had readers who found their histories relevant. Without a large readership, they resembled antiquaries, whose pursuits were perceived as meaningless and only fulfilling personal curiosity and selfish intellectual ambitions. Third, a lack of public acclaim also complicated the promotion of the persona, as historians were confronted with an audience who were unwilling to follow their enactment of expertise through multiple volumes of dense narration. The production of bulky histories put historians' relationship with readers under a magnifying glass, and the readers' dwindling interest in large histories unmasks important developments in the culture of reading during the later Victorian era. The responses of readers, publishers, and historians to big histories highlight how the forging of the scholarly persona was affected by the literary marketplace.

While historians, readers, and publishers evaluated books according to their physical proportions, there were also those who reminded that size was an unreliable and imprecise indicator of quality or style. Thomas Russell directed attention to this when he reviewed Thomas Craig-Brown's *The History of Selkirkshire* (1886). Craig-Brown, an amateur historian and wool merchant, had apparently produced a book that bordered on "grotesque," but impressed the audience with its monumental size. Hence, Russell begged readers to be cautious, as "a big book is apt to be taken by many for a big authority, and those who follow it without sufficient knowledge to apprehend and rectify its mistakes may thus unintentionally propagate serious historical error."[11] But Russell was in the minority—despite the constant difficulties of using bulkiness to divorce historical and antiquarian publications.

Completeness: Generalizations or Particularities

For scientific historians, completeness was a thorny kind of epistemic virtue since it embodied parallels with the antiquarian approach. Creighton praised Francisco Ehrle from the Vatican Secret Archives for his "monumental completeness" and encouraged his fellow historians to "aim at completeness." Readers of histories, he believed, "must know the whole truth as far as it can be known."[12] Readers failed to agree and instead complained how such complete histories appeared like annals and reports of historians' research processes rather than clearly organized narratives. For historians, completeness, a comprehensive assemblage of particulars, was a vision that had a direct bearing on their methods and narrative techniques. Although they knew that uniformity was essential for sustaining their persona, there were subtle differences in how they interpreted completeness. The meaning and value they ascribed to detail captured these differences. Those who prioritized completeness as a research practice that was intimately related to an assiduous gathering of facts and fascination for details came closest to the antiquarian tradition.

Completeness had long roots in Western science, where it had inspired scientists to share their full knowledge with their audiences. In the nineteenth century, the obsession with completeness continued to epitomize the scientific ethos. As one writer critically observed, "the modern method, 'now exacted in every branch of science' ... delights in the analysis of particulars ... and would surely be infallible if genius were indeed the faculty of taking pains ad infinitum."[13] Completeness belonged to a cluster of virtues that were considered fundamental to attaining truthful knowledge: perfection, thoroughness, industriousness, and impartiality were all its close allies. Striving for completeness ensured impartiality by eliminating the risk of subjectiveness which a selection of facts would have entailed. Historians, as Gardiner remarked, were never able to agree "to the relative importance of any set of facts."[14]

The virtue of completeness set the target for research, as historians aspired for the elusive "definite" history. Ideally, they produced histories that were so complete and perfect that no one ever needed to return to the same topic again.[15] However, the quest for completeness was rendered unrealistic by the flood of recently discovered primary sources and the ceaseless manufacturing of special studies. Andrew Lang, a critic of scientific history, parodied the situation by remarking how histories became outdated by the date they came out of the press.[16] Historians were aware of the impact which the opening of archives had on their craft and on the permanence of historical knowledge. Froude observed, when reviewing Gardiner's *The Fall of Monarchy of Charles I*, how English history "was rewritten for us at intervals of ten years." Gardiner, on his behalf, explained that since the opening of the Vatican Secret Archives in 1880, transcripts had been sent monthly, even weekly, from Rome to the Public Record Office, allowing historians to "clear up" many unsettled points.[17] These new conditions of historical research should have scattered any illusions about the perfection or permanence of historical knowledge, but historians nonetheless continued to chase the holy grail of completeness as they composed new studies and enlarged their earlier ones.

If completeness drew on the industrious gathering of facts, what was then the status and use of generalization of these facts in historical research? The appeals for generalization alarmed historians as they associated the notion with positivism and Buckle. Comtean positivism found support in England among scientific and philosophical writers, novelists, social reformers, and politicians, but historians resisted positivist thinking. Edward Beesly was one of the rare exceptions to this rule.[18] For Buckle, the epistemic goal was the formulation of general laws that governed historical facts. Only in this way could history claim to be science. As the majority of historians were steeped in Victorian individualism, they rejected causality, universal laws, and determinism. They embraced instead empiricism and stressed coincidence, free will, and individual actors as historical forces. While they hailed progress that linked the past to the present, their understanding about progress did not mean pre-ordained development. As historians fostered hostility toward the theories and philosophies of history, they were attracted to empiricism, which appeared to them as a theory-free orientation.[19] Of course this was an illusion and empiricism was a philosophy of history in its own right.

Historians concurred that generalization was a necessary heuristic operation of moving beyond singular facts and drawing broader pictures of historical events and actors. How far this should be pushed at the expense of the particulars was then a different matter. The value and meaning ascribed to details opened up one avenue for dealing with the issue: was every detail equally valuable and how did one define which facts were significant and which were insignificant? How did one select what to include and what to exclude? How did one restrain the fascination for detail and the urge to indiscriminately share the discoveries with readers, as uncontrolled passion for detail was a vice that historians attributed to antiquaries?

At the bottom of these methodological conundrums lay the question of the meaning of details and the manner of treating them. Stubbs set a model for those who placed historical value on every detail as well as for those who favored generalization. He had written to Freeman as a young man that "All chronological minutiae are the pebbles of the concrete in which the foundation of histories must be laid" and revisited the topic in his professorial lectures.[20] He talked about going to the sources, "fountains of historical refreshment," and accorded importance to every historical detail, proposing that even pebbles which first appear insignificant might gain significance when historical knowledge expands. Yet he also stressed that the details had to be critically weighed, judged, selected, arranged, and then generalized.[21] He knew that the process was not this straightforward, as he struggled to hit the right balance between the particular and general when writing *The Constitutional History*. He confessed privately to Freeman that he was not altogether "sanguine" about the result. George Kitchin had found "the Early English part too much compressed," while Stubbs himself had feared that it was "both long and tough & and yet not long enough." It was difficult "to know what to leave out" and he added that he had "just killed Henry II. for the n.th time."[22] Despite his methodological pronouncements and efforts toward generalization, the *Athenaeum* diagnosed him as suffering from "a wholesome dread of historical generalizations" caused by his aversion to positivism.[23]

The book reviews in the *English Historical Review* provided a venue for a discussion about the significance of detail. Mary Robinson belonged to the camp that defended the value of every fact as she argued for "passion of fact for fact's sake."[24] John Andrew Doyle, a fellow of All Souls College, similarly embraced the value of every piece of information.[25] This resonated with the contention that seemingly insignificant details may gain significance as historical knowledge expands. This line of thinking gained force from the understanding that the torrent of new sources had propelled history into a transitional phase between the age of old myths and half-truths and the future of science and truth. As historians were engulfed by the continuous flow of new sources, they were unable to evaluate the value of the details or establish lasting truths. Therefore, as *The Times* wrote in 1879 in a piece titled "The Modern School of Historians," historical research was "at present in the phase rather of the laborious collection of materials than of their skillful display."[26] The thought held on, and Bury and Oman proposed in their inaugural lectures in 1903 and 1906, respectively, that historians should not waste time on writing major studies but instead concentrate on collecting and publishing sources to the benefit of future historians. The following generations could then use the material as a scaffolding for erecting the permanent structures of historical knowledge when enough groundwork had been done by their predecessors.[27] This offered an alternative definition for a transitional period to those who considered the late-Victorian era as a period of wrestling between history as art and history as science.

Many of the writers in the *Review* however advocated a more rigorous selection and summarizing of facts. All details were not equally valuable, and Stanley Lane-Poole argued that historians needed "the power of generalisation" to place "these crowded details in their proper relations to each other."[28] The ecclesiastical historian George Perry was more outspoken when he lectured about the fundaments of historical research to Canon Dixon, an amateur historian and poet. A proper historian, Perry explained, did not "seize upon every scrap of gossip which he can pick up, swaying hither and thither according to his last discovery." Instead, a qualified historian will get to the meaning of the events and then "group and arrange his narrative in subordination to the main issues."[29] Arthur Ropes, fellow of King's College, made it even plainer that it was the amateurs who drowned their books in trivialities, and he established an explicit connection between generalization and scientific history:

> This work of generalisation within proper limits is both the right and the duty of the scientific historian—his right if he has made himself acquainted with all the facts of the period he is studying; his duty if he is to rise above the rank of chronicler and compiler.[30]

Research meant for these historians a critical digestion, generalization, and harmonization of facts into a coherent whole. The antithetical pursuits would have been, according to Ropes, chronicling and compiling—both activities commonly associated with antiquarianism. The scientific method which demanded generalization prevented historians from sliding into antiquarian pedantry. Yet, as will be shown below, non-expert critics in periodicals and newspapers pointed out how Freeman, Gardiner, and many others with scientific credentials struggled to rise above the details.

The opening of archives and the expansion of available source material compelled historians to reassess their research practices and the nature and purpose of historical pursuits more broadly. English historians shared concerns with their continental brethren, but the adaptation to these altering conditions and research infrastructures was shaped by national scholarly cultures. The universal concern was whether a historian by profession was either a meticulous archival scholar who unearthed minute historical truths from the records or someone who rescued himself from drowning in a bottomless sea of original sources by exercising self-discipline and generalization. In other words, should a historian cultivate a persona of what Jacob Burckhardt slightingly called an *Urkundionen*, or nurture skills and qualities that gave him courage to select, generalize, and write monographs with clear arguments. In central Europe, two rival camps emerged: one around source-editing projects such as *Monumenta Germaniae Historica* and the other around historians who published monographs.[31] Such a dichotomy would have been unfathomable in England, where many of the leading historians were involved in editing records either for the Public Record Office or for the historical societies that undertook such projects. Stubbs contributed to the Rolls Series, Maitland was active in the Selden Society, Creighton in the Church Historical Society, and Laughton in the Navy Records Society. Rather, it was the unstable border between history and antiquarianism that was evoked in the debates about completeness and historical detail, as the persona and pursuits of a historian and an antiquary appeared strikingly similar to the public.

Antiquarianism, History, and the Invisible Boundary

Antiquarianism was a popular form of leisure among Victorian genteel society. The conventional understanding is that the heyday of antiquarianism was earlier in the nineteenth century, but Paul Readman has convincingly argued for a strong antiquarian sentiment still existing in the 1880s and 1890s.[32] History and antiquarianism had traditionally shared common features, and the Victorian antiquaries' growing interest in local history made this entanglement even more pronounced. The overlapping qualities and the persistence of antiquarian pursuits gave a sense of urgency to historians' attempts to consolidate their persona and win popular acceptance for their markedly distinct disciplinary status. Indeed, the taxonomical or terminological demarcations between history and antiquarianism during the last quarter of the nineteenth century were indefinite. The tendency was to admit that history and antiquarianism bore similarities in terms of their aims, methods, and epistemologies, but not to treat them as one and the same. The historians' persona, ideally, embodied qualities of an antiquary. *The Times*, using Freeman as a model, explained, "A historian is not always an antiquary, and even less frequently is an antiquary a historian." The two could, and should, however, be fruitfully combined, as antiquarianism redeemed "historical writings from the danger of shallowness and inaccuracy" and history rescued "antiquarianism from pedantry and dryness."[33] The Victorians were fond of categorizations, but the attempts to characterize historians and antiquaries show how ambiguous their taxonomies could be and how that ambiguity allowed historians to reconcile certain aspects of the antiquarian tradition with scientific history.

Antiquarianism was traditionally associated with material artifacts; collecting and cataloging and history with textual sources, critical analysis, and narrativity. The *Athenaeum* simplified this by claiming that antiquaries "take but little interest in the grander questions of history."[34] E. Blanch Hamilton reinforced historians' intellectual superiority in the *English Historical Review* by maintaining that antiquarianism rested on omnivorous collecting while history required discriminating judgment.[35] However, Victorian antiquaries were not just passionate collectors. Their activities ranged from barrow-digging and curiosity-collecting to authoring local and family histories. The archive turn excited antiquaries as well, and their zeal for textual sources gained expression in the many printing societies which they established in the 1830s and 1840s to boost the publication of historical records. Despite the diversity of antiquarian diversions, the public image of antiquaries retained the classical motives of absurd pedantry, dilettantism, and unchecked enthusiasm for collecting meaningless curiosities. If Scott's Dr. Dryasdust was a Romantic archetype of antiquarian satire, then the recognizable characters of forlorn antiquaries which George Eliot, Thomas Hardy, Thomas Hughes, and many others created kept the tradition of antiquarian parody alive throughout the century. The popularity of such caricatures speaks for the widespread acceptance of antiquarianism as a comically meaningless pursuit.[36] Yet antiquarianism was a popular form of intellectual leisure and the frequent reports of antiquarian discoveries in newspapers and periodicals only added to its appeal as an appropriate gentlemanly diversion. The new local societies that sprang up during the early-Victorian period altered the aristocratic outlook. For instance, the university-educated clergy found in antiquarianism a meaningful channel for their intellectual curiosity. Their immersion in local history set an example for fusing antiquarianism and amateur history as they edited local records and compiled histories with a regional scope.[37]

A major challenge for the attempts to categorize antiquarianism and history was that they pursued similar goods and had a significant methodological affinity, as both cherished empirical research and chasing unexplored historical records. The split into antiquaries who hunted down material traces and historians who explored textual sources was too simplistic. Just like antiquaries, historians mixed material, visual, and textual sources. Oman developed numismatic methods, and the topographical method which Freeman, Green, Gardiner, and Lecky embraced owed much to the antiquarian interest in physical and geographical evidence. The multimodality of historians' sources challenged the conventional notion that the use of either textual or material traces established the dividing line between history and antiquarianism.

Freeman is a good example of how futile it was to try to draw a clear distinction between historians and antiquaries. His historical ethos was shaped by the ecclesiological antiquarianism that had been popular in Oxford of his undergraduate years. As Ian Hesketh argues, the antiquarian background, not so much Rankean treatises, spurred his inductive tendencies. Although recognizing the reputation of antiquarianism as an amateur obsession, he framed his ideas within the Rankean vocabulary, which was more in tune with the claims of scientific history.[38] Freeman was, furthermore, influenced by the fascination for Gothic architecture that was fostered both at Trinity College and in the antiquarian community. Architectural antiquarianism was instrumental in furthering the development of nomenclature and precise terminology of historical architecture and this, too, left a lasting mark on Freeman, who became renowned for his fixation on correct dates and archaic spelling.[39] His persona of an antiquary surfaced in his texts, topographical and architectural field trips, as well as in the meetings of his local archeological society. Although he and the other historians erased such memberships from their title pages, the newspaper reports about their engagement in societies made their antiquarian interests public. The *Western Daily Press* reported how Freeman participated in the "scientific picnic parties" of the local archeological society, and how, upon Somersetshire ruins he discoursed with zeal about the history of the surroundings, letting his passion for acquiring and spreading knowledge shine to the great amusement of the participants.[40]

Freeman's fourth volume of *The History of the Norman Conquest of England* inspired a reviewer in the *Pall Mall Gazette* to draw a map of historical pursuits and to try to place Freeman on it. The writer used as coordinates three types of historian: an antiquarian-chronicler, a picturesque historian, and a historian proper. Without the antiquarian-chronicler, scientific history would have become "moonshine," but a good historian was more than that: he was an "annalist and the interpreter, the accurate and painstaking narrator and the generalizing thinker, who groups, co-ordinates, deduces, and theoretically applies." Freeman leaned too much toward the chronicler, although the writer did not demand from him the posture of "that monster, picturesque historian." Just enough artistry "to give the fair value to perspective and proportion" would have sufficed. But contrary to this, Freeman pushed too far the "democratic formula" of every fact being equally valuable, and blurred the clear view of historical events with the profuseness of detail.[41] The portrait was a caricature, although not entirely unfounded. It appalled Freeman, who protested vehemently against the idea that he would have been "a mere 'laborious' annalist."[42] But many agreed with the reviewer's assessment and *Norman Conquest* was for them a physical testimony of Freeman's compulsion for minutiae.

It is not hard to see why readers thought so, as *Norman Conquest* grew from an anticipated single volume to five volumes and the first edition alone contained almost 3800 pages of historical details. One critic summed up Freeman's ethos fittingly: his "historical conscience" kept him miserable unless he told every historical occurrence.[43] Freeman was certainly pedantic and minute, but it would be unfair to brand him a mere antiquary. He considered the antiquarian attention to detail a good handmaiden for history but not an end itself. He was critical about the antiquaries' collecting mania and narrow perspective on the past. Rather, he was a historian with a strong antiquarian bent. The cultivation of such personae bridged the gap between history and antiquarianism and impeded assertions of a clearly defined disciplinary identity of history.

As many of the historians' and antiquaries' endeavors overlapped, it was impossible to pinpoint the exact boundary between them. Perhaps because of this, reviewers were intrigued about the similarities and contributed to the blurring of the borderline by either contrasting or paralleling history and antiquarianism or by applying similar standards and vocabulary when evaluating historical and antiquarian publications. Exhaustiveness, lack of a clear narrative, and an obsession for recording even the most trifling details were regularly used to describe both historical and antiquarian publications. Bulkiness was the paratextual expression of these qualities. Thomas Kebbel's review of Georg Birkbeck Hill's *Letters of Samuel Johnson* (1892) in the *Quarterly Review* illustrates these reviewing habits. Hill had established his reputation as an editor of Boswell's *Life*, which had appeared in 1887 in six volumes. He had continued since then to collect *Johnsoniana*. Kebbel questioned the historical value of the documents in *Letters* and began his review by characterizing how a historian constructs a narrative that "sets before us the public life of the period," while an antiquary, following in the wake of the historian, fills in gaps with trifling details without adding anything to the general picture. In other words, if a historian built an edifice, an antiquary described its gateways, staircases, and buttresses. Hill represented to Kebbel the latter type. He had indiscriminately amassed together superfluous details, presuming that readers cared about "every scrap of additional information" he had discovered.[44] Reviews like this, rich with popular stereotypes about antiquaries, instituted a hierarchy of intellectual capacities between history and antiquarianism.

Yet that border between history and antiquarianism was erased by reviewers' habit of similarly censoring scientific historians, too, for substituting complex intellectual commitments for antiquarian tendencies, such as profuseness and a mania for detail. Gardiner, who was held as a model of the inductive method, was nevertheless sanctioned for a superabundant use of details. The *Athenaeum* played with the symbolism of book sizes when it introduced the first installment of Gardiner's *History of England* with a warning: it is a "big book" which will tire readers. Such bulkiness implicated drowsiness and the reviewer expected the public to discard the book altogether.[45] Francis Warre-Cornish reviewed several of Gardiner's later books for the *Quarterly Review* and found them symptomatic of the scientific history that seemed to institutionalize antiquarian vices. He accused the "university schools of history" for creating this "rage for accuracy." Historians had replaced the cultivation of judgment, instruction of readers, and philosophizing with a chasing of hidden details without realizing that it was not the industrious collector, in other words the antiquary, who enlarged readers' knowledge about the past. Gardiner was an incarnation of the type of historian who immersed himself in the details and failed to realize that readers expected from a historian instruction; enlightenment and, above all, judgment upon facts, not a mere recording of them. Warre-Cornish was astonished that historians insisted on writing books that lacked "a sense of proportion" and which posterity would judge as mere pedantry.[46] Gardiner understood that he was expected to check his enthusiasm, but the prefatorial inventories of the English, Scottish, Spanish, French, Italian, and other archives which he had visited betray an eagerness to share his findings with the readers.[47] The spell of documentary discoveries was hard to resist even when converting the carefully transcribed notes into narratives.

The reviewers did not oppose thoroughness as such, but their main objection was historians' habit of leaving readers to their own devices to construe the meaning of historical details. Historians settled for amassing particularities and transformed the burden of interpretation to the reader as if this resignation from interpretative responsibility guaranteed the construction of impartial knowledge. Hugh Eliot identified wider implications in this as he connected the matter to the benefits which historical knowledge was expected to have for English society. It should have been this usefulness that legitimized historical research, but the scientific historians failed to fulfill their duty according to any standards. Modern statesmen, Eliot wrote, needed histories that acquainted them with "the currents which direct human action" so that they could abstain from repeating past mistakes. But instead of instructiveness, historians offered them isolated incidents. Such a lack of historical instruction posed a great danger to British society, Eliot warned.[48] Historians' bulky books were just as unhelpful for ensuring the future prosperity of England as were those of the antiquaries. Such a mania for useless antiquarian minutiae about "chipped flints and old pots and pans" or the "rubbish" from "the old law-books" was considered fine for Germans, but not for the English, who were destined to live lives with political inclinations.[49]

Taking a Distance from Antiquarianism

The growing disciplinary territorialism and specialization made the distinction between history and antiquarianism urgent for historians. Their attempt to forge separate personae for historians and antiquaries was a textbook example of impression management that painted blue as green. Accordingly, they strove to frame history as a branch of science and antiquarianism as an amateur diversion that was an adjunct to history. As part of this disciplinary self-fashioning, they stressed their moral superiority and methodological finesse, which antiquaries apparently lacked. They also projected their endeavors as more ambitious and valuable than antiquarianism entailed. Moreover, they claimed that the two pursued different goods: while antiquaries listed, cataloged, and described discoveries, historians evaluated and selected facts to form chronological narratives. They repeatedly criticized antiquarians for stopping their investigation at the objects themselves. Yet when antiquaries pushed their analysis further, they charged them with a lack of critical edge that rendered them susceptible to guessing and quixotism. Bury chided antiquarians in his inaugural lecture for surrendering to "fanciful speculation."[50] Finally, historians considered themselves morally superior to antiquaries. They wrote histories that contained reliable and valuable knowledge that benefitted British society. Antiquaries, on the contrary, collected historical traces only out of the selfish pleasure of discovery. When Kebbel reviewed Birkbeck Hill's *Letters of Dr. Johnson*, he appealed to the popular image of a morally dubious antiquary who was driven by an egotistic urge to show off the width of his research at the expense of readers' patience and comprehension.[51] To draw a distance from antiquarianism, historians then erected a moral boundary between scholarly altruism and antiquaries' selfishness and stressed how they were doing a service for society with their histories. They effectively ignored the constant complaints about the little practical value of their bulky histories.

Historians substantiated these claims with concrete examples to confirm the deficiencies of antiquarian research. First, they addressed the antiquaries' fascination for regional and local history. This was an important nineteenth-century departure from classical antiquarianism. The antiquaries' appreciation for national heritage encouraged the use of English history for generating patriotic sentiments. The ownership of local historical knowledge was a source of pride for antiquaries, who gained moral tenor for their pursuits from the engagement with the past of their own communities.[52] Historians were aware of this. Freeman and Stubbs complimented antiquaries for reviving interest in local history and heritage. Yet Freeman found fault in their approach, which stimulated the mania for curiosities without benefitting local and national history. Local history gained significance only when it was placed in dialog with national history and helped to reveal general patterns and continuities, Freeman explained. He tested the method of merging local and general history in *History of the Cathedral Church of Wells* (1870) to raise awareness of the need to enlarge the scope of local history and pay attention to "the history of particular places" in general history.[53]

Antiquarianism was also portrayed as too narrow in the sense that antiquaries tended to be thoroughly familiar with a single historical period while a historian possessed a wider understanding of historical development. Historians had abandoned the Romantic ideal of a generalist who knew something about everything. They were experts who knew everything about something, though they resented the tendency toward excessive specialization which they detected as a German vice. Familiarity with a time-period of one's own simply was not enough.[54] Freeman was famous nationwide for scolding Froude for a lack of a broad historical grasp and Gardiner instructed in *Introduction to the Study of English History* (1881) how a wider understanding was necessary because historians' subjects were always "a portion of a living whole." They could not be chronologically "isolated and examined like a piece of inorganic matter."[55] History was not antiquarianism, nor was it like the sciences with their microscopic perspectives. It was uniquely a discipline of its own.

Historians' sense of superiority and their hierarchical vision of cognitive ventures into the past could inspire arrogance. Stubbs assigned to historians the highest duty, the building of "a treasure-house of knowledge"—the unraveling of historical progress, generalization, and production of cohesive narratives. He placed antiquaries on a lower ladder, to gather material for historians' building projects. This, he generously agreed, was an important task and therefore antiquaries should not be treated as "the parasites of historical study, as they are too often regarded."[56] Stubbs extended this ranking to his private dealings with antiquaries and amateur historians as his correspondence with Jessopp shows. Although the eager amateur historian published studies on medieval parish life and composed well-received editions of records for the Camden Society, Stubbs made it plain that they were not equals. He showed interest in Jessopp's archival finds and shared his interest in parochial history, but not for entirely unselfish reasons. He tended to approach Jessopp when he needed his friend's help in looking up something for him from the parochial records to which Jessopp had access. When Jessopp asked his correspondent to borrow his name to advance some publishing scheme of his, Stubbs refused.[57] To protect his name, he was unwilling to lend it for causes which he considered either unpromising or unwise for his reputation.

Jessopp, who fostered antiquarian tendencies, was nonetheless critical of antiquarianism. In his Jubilee eulogy, he placed antiquarianism into the childhood of historical studies and granted as much value to antiquarian curiosities as to the "specimens in a child's museum."[58] Victorians interpreted "childlike" as opposite to manliness, and by comparing antiquaries' collections to a museum of toys he evoked a conventional image of antiquarianism as an unmanly diversion. Such an association with effeminacy posed yet another reputational threat to historians, who took pride in manly virtues such as intellectual courage and independence.

However, historians were careful not to conjoin femininity and antiquarian vices, because some features that made antiquarianism appear feminine were not foreign to history either: mainly the passion for particulars. As Naomi Schor argues, particulars have been associated with femininity and domestic triviality since antiquity.[59] W. L. Courtney pronounced in 1904 in *The Feminine Note in Fiction* that "The feminine intellect has a passion for detail" and that this was "the distinguishing mark" of women writers.[60] Early- and mid-Victorian women historians were censored for their uncontrolled fascination with trivialities and inability to distinguish the significant from the insignificant.[61] Women historians continued to be evaluated according to this gendered matrix throughout the century, but when similar sins were attributed to male historians, the reviewers were careful not to imply unmanliness. Instead, the dizziness of details was presented as a sign of industriousness and thoroughness. A stout volume symbolized a high work ethic; there was something noble in a historian who spared "neither time nor pains in research of documents and materials from all quarters."[62] This was a neat way to foreground gender. After all, the infinite capacity for taking pains was at the heart of post-Romantic manly genius.

Moreover, the constant talk about passion for detail indicated an emotional investment that should have been alien to scientific endeavors. Historians promoted restraint of emotions as a condition for impartial knowledge and accordingly Creighton assured in the preface to his *A History of the Papacy* (1887) that he had exercised "self-restraint at every turn."[63] The antiquaries' perceived morbid cognitive pleasure in detail was conceived as a sign of missing self-restraint which encouraged excessiveness and subjective responses to the past.[64] The control of passions had even wider implications, because the stern discipline of emotions was a universal Victorian moral value. Middle-class men were not expected to suppress their emotions, but to govern them. Uncontrolled emotions undermined manliness, citizenship, and scientific reliability because they suggested incapability for taking charge in the way Victorian men were expected to do.[65] Mixing emotions and historical research did not then only erode the boundary between history and antiquarianism or threaten historians' culturally prescribed manliness; it could have also undermined the reliability of historical knowledge, diminished its utility for society, and endangered the validity of historical research.

Paradoxically, historians could be blind to what they labeled as antiquarian vices among their own ranks. They condemned antiquaries' compulsion for collecting, but when a historian committed a similar sin in the name of completeness and perfection, it was turned into a virtue. Edwin Guest pledged the utmost perfection and exhaustion of *all* relevant source material to such an extent that he never completed his grand history of early English settlement. Guest was a historian, philologist, and master of Caius College who devoted his life to scholarship. In Freeman's words, Guest was a mix of "the in-door scholar and of the out-door antiquary." He was highly esteemed by Freeman and Stubbs, but unknown to the broader audience because his "morbid" love of perfection and completeness prevented him from publishing anything but scattered papers in the proceedings of the Archaeological Institute. His "history of Britain and its inhabitants until the completion of the conquest by the Angles and Saxons" was far from completion at his death in 1880.[66] Stubbs and C. Deeds prepared the unfinished manuscript for publication with the title *Origines Celticae*.

The editors enforced an image of a dilettante antiquary in their biographical preface. They highlighted Guest's "untiring zeal and determination" and his compulsive perfectionism, and perhaps accidentally reproduced the popular satirical image of a forlorn antiquary conducting field work:

> careless of weather, or hunger, or any physical inconveniences, he would walk mile after mile with a keenness of perception almost amounting to an instinct, tracing the lines of ditch and bank which marked the boundaries of the settlements of the early tribes, and measuring and mapping the earthworks which formed their places of shelter and defence; and how under his potent touch these mysterious remains became vocal with the history of the past.[67]

It is likely that the editors only sought to accentuate Guest's intellectual zeal and manly physical strength, but the preface mediated a portrait of an eccentric antiquary immersed in his thoughts and oblivious to his physical surroundings. It also attached Guest to the long antiquarian tradition of hearing the voices of the past through physical remains. Stubbs and Deeds certainly wished to celebrate Guest as an accomplished and devoted scholar, but the copious references to his passion for detail, perfection, industry, and completeness together with the vivid descriptions of his excursions created an unavoidable impression of an antiquarian dabbler. If Stubbs and Deeds showed admiration for Guest's commitment to perfection, some twenty years later Acton's inability to write his magnum opus only puzzled his obituarists, as Ian Hesketh has written.[68] Acton's scholarly consciousness was tuned in to such extreme perfection and thoroughness that it prevented him from realizing the grand work he had envisioned. Oman found nothing amusing in this kind of antiquarian obsessiveness and, in an unusually vitriolic inaugural lecture, painted a grim picture of Acton deluded by infallibility and fixation on "an epoch-making book."[69]

Conceptual flexibility surrounded history and antiquarianism throughout the nineteenth century. They were seen as both different and similar at the same time. As the historians themselves collated the two branches of inquiry, it is not surprising that reviewers and readers failed to see them as categorically distinct. For historians, nevertheless, the demarcation was not an insignificant issue and by the end of the century they showed less tolerance of a persona that blended components of the two. Bury, who famously insisted in his inaugural lecture in 1903 that "history is a science, no less and no more," deplored how some historians still had not forsaken the old "irresponsible" ways of literature, antiquarianism, and superfluous erudition which prevented history from evolving into a fully fledged branch of science. He did not deny the value of collecting details, but scientific history rested on application of an exact method and generalizations were part and parcel of that. Antiquarian fact-gathering was not history, for the truth "can be attained only through the discovery, collection, classification, and interpretation of facts," Bury professed.[70]

Big Books and the New Geographies of Reading

The rise of scientific history coincided with the emergence of new cultures of reading, and this prompted historians to wonder who their readers were and how to approach them. Without a sufficiently substantial audience, their performance of the persona would have lost its drift. If historians struggled to understand the new cultures of reading, the non-scholarly world failed to comprehend why the scientific historians were so off-pitch on the pulse of modern readers—and of modern times. The *Athenaeum* elaborated this presumed gap between historians and the realities of the literary marketplace by asserting, "In these brisk and rattling times, when a man can lunch in London and sup in Paris, when a message may be sent to San Francisco and answered between soups and dessert," it was inconceivable that historians wrote books that belonged to an age when readers had had "six or seven hundred years leisure." There was no longer need for histories that were a "hotch-potch" of everything in a "thousand pages of closely-packed type."[71]

The link between history books and cultures of reading came to the fore in reviews where the writers mixed criticism of unwieldy histories with nostalgia for the "good old days" of ample leisure for reading or with contempt for what was interpreted as the modern reader's desultory reading habits. These reviewers tended to be firmly pitched between the past and the present. On the one hand, they harked back to an age when readers had had time and patience for traversing through thousands of pages of slowly progressing histories and had been unanimously seduced by a few luminous writers such as Voltaire, Hume, and Gibbon.[72] They glorified the eighteenth-century masters and the uniformity of the Enlightenment audience, ignoring the fact that many of their contemporaries had found Hume and Gibbon tedious or that there had not been just one set of readers then.[73] On the other hand, these same critics demanded concise histories that were tailored to this "irregular and impetuous age."[74]

Yet the critics were not unreserved about the blessings of the modern era either, as they reproved the tempestuous reading patterns among the middle classes. Reading on trains became a symbol for these new bad habits, and John Forster remarked in his preface to *Walter Savage Landor* (1869) how railways had encouraged "the desire to read without the trouble of thinking."[75] These commentaries were underpinned by fears that the multiplication of reading communities were shaking existing social hierarchies. The literary and scholarly elite attempted to retain their cultural authority by patronizing the middle-class readers whose taste and patience for digesting serious books apparently had been ruined by circulating libraries, periodicals, and potted learning, as Agnes Clerke complained when reviewing a ponderous German biography of Copernicus.[76] The theoretical backing for these discussions came from literary theorists who argued that literature shaped readers and interpreted the fondness for shallow stories and impulsive reading as indicators of cultural degeneration. A serious student, one writer cried, was becoming "as rare as a dodo."[77] These ideas were not foreign to historians. If Freeman hovered between condemning "the vulgar public" and admitting that a general reader was not entirely a fool, Adolphus Ward warned against lowering scholarly standards just to appeal to the "inoffensive animal the general reader" whose tastes derived from his ignorance, and Seeley bemoaned "the want of good readers" in England. Had there only been enough such readers, he would have brought "twice as much ardour" to his work.[78]

The question then was that of for whom historians wrote big histories, the testimonies of their learning and embodiments of the scholarly persona. The struggle to win approval from the fragmented readership frustrated many. Alexander Macmillan identified three kinds of readers of non-fiction: the ordinary reader who longed for broad and generalized works, "A thoughtful class" that appreciated originality, and a learned class with the digestion for specialized books.[79] Seeley was confident that historians should give precedence to the last two types. He cautioned historians against seeking popular appeal. Accordingly, he envisioned two "distinct publics" and two classes of books—literature and science—which lived "under a wholly different constitution." Science was, essentially, difficult and it did not matter what "the general public" thought about it. However, as Seeley discovered, the public did not recognize these differences and judged scientists against entirely false standards of literary worth, condemning them merely because they "write books which cannot be read in the easy-chair or in the railway-carriage!" Yes, scientific books were often bulky, but he asserted that "in other departments we admire the thoroughness and laboriousness indicated by such bulky works." He saw as a solution an "aristocracy of students" who would stand between historians and readers, translating historians' words into a popular format. In this way, the public would have access to accurate historical knowledge and historians would be relieved of tedious popularizing.[80]

Seeley did not find much support for his scheme among historians. Freeman feared that if historians stopped writing for the public, the public would seek instruction from more unreliable sources such as Froude. Instead of choosing to be dry and unattractive, historians should educate the general reader to appreciate scientific history.[81] Of course, Freeman knew how utopian this was. He had envisioned *Norman Conquest* pleasing girls, curates, and scholars alike, but when the dry pedantry of his narration was widely ridiculed in reviews, he announced that he would focus on writing histories that delighted no one else than "myself and Stubbs."[82] Beneath this sardonic proclamation lay genuine pain, as Freeman realized that he had not become famous for a talent to tell a story but for an exhausting pedantry which apparently rendered his books impenetrable for a common reader. His name came to signal uncontrolled profuseness in similar fashion as "Stubbs" denoted virtuosity. If historians pursued immortality, the qualities that survived Freeman were his missing sense of proportion, and a style that drove "truth into men's souls with the force of [a] sledgehammer."[83]

There was, though, an important exception to this image of Freeman as a literary failure or of the persona eclipsed by the fixation for completeness. The *Western Daily Press* contested the idea that thorough historical studies had to appeal to the general reader and gave voice to a small group of writers who thought, just like Seeley, that historians had a right and responsibility to write for an educated audience. The writer in the *Western Daily Press* reacted to an obituary of Freeman in *The Times* where the writer had complained about Freeman's missing sense of proportion and gift of selection and had concluded that because of this, *Norman Conquest* would not pass into posterity as a great history that general readers enjoyed. This, according to the *Western Daily Press*, betrayed a flawed premise: historians such as Freeman did not—and should not—care about the tastes of the general reader. Each generation produced a few exceptional talents, and their duty was to write studies which serious students "can accept with confidence." It was crucial that in an age of cheap eloquence historians prioritized learning. Such "men neither can nor should compress their work," because their object is.

> to be exhaustive, to treat their subject completely, and to enrich the nation's literature with work that will always be a standard, that will always be referred to by the historians of the future, that will always have readers fit if few.[84]

If anything, the two obituaries show how histories were written, published, and read at a junction of diverging expectations. Those who believed in an educated audience with an appetite for exhaustive histories did not consider historians' completeness problematic at all. Such an educated audience, though, remained small until the turn of the century, when a larger readership for extensive scientific histories began to emerge as the number of students at universities grew and new universities with history departments were founded. This created a public that was familiar with the scientific forms of history.[85] Until then, as Macmillan knew, the market was so small that it could almost be exhausted with presentation copies alone.[86] The reality was that most readers abhorred big histories. Reading was a mental, emotional, and physical activity, and unwieldy books caused discomfort in each of these sectors. As William Longman once explained to Freeman, "When a book is very long in hand" it is "apt to get stiff and cold like a man lying in the snow."[87] The truth was that it was nearly impossible to lure general readers to indulge in big histories and, to gain support for the persona project, historians then began to search for alternative bookish stages on which to perform the persona to a broader audience, as the next chapter will demonstrate.

*

The physically considerable size of history books was connected to several overarching features that were significant for crafting a historians' persona and establishing history as a scientific discipline. The size of a book as a material object invited historians, reviewers, publishers, and readers to discuss the methods and research practices, the truth-bearing function of historical detail, the unresolved tension between history and antiquarianism, and the aims and audiences of histories. The virtues of completeness, perfection, thoroughness, and industriousness guided historians' responses to these issues. The deluge of published and unpublished source material made the question about completeness acute, as historians, armed with a mass of primary sources, were tempted to pour all their discoveries into their books at the cost of generalization and concise narrative. Historians held tight to both the idea of completeness and the desire to write big books which were material testimonies of the virtuosity of their persona. It is not surprising that Creighton announced, "life is not complete unless one has a 'great work' in the stocks" and went on to state, "one's happiness is increased if the work is so great that one has no hopes of ever finishing it."[88] While historians held on to that almost sentimental bond to big histories, they risked veering toward antiquarianism. This necessitated a constant negotiation between the two approaches to the past and their respective personae.

Notes

1. Genette, *Paratexts*, 17–18; McKitterick, *History of Cambridge University Press*, 3, 110.
2. Green to Olga Von Glehn, August 2, 1873, in Stephen, *Letters*, 357.
3. [Saintsbury], "Municipal London," 4.
4. Paul, "Habits of Thought," 277.
5. Yeo, *Science*, 72–73.
6. Yeo, *Science*, 259–263.
7. Levine, *Amateur and the Professional*, 1–6, 39.
8. Karstens, "Bopp the Builder," 104.
9. Bryce to Browning, May 11, 1899, GBR/0272/OB/1/268/C, King's Cam.
10. [Morris], "M. de Beaucourt's *Charles VII. Of France*," 170–172.
11. [Russell], "Ettrick Forrest and the Yarrow," 3–4, 14.
12. Creighton to Hodgkin, October 25, 1882, and to Alice Gardner, January 2, 1889, in Creighton, *Life and Letters*, 231, 410; Creighton, "Review of Francisco Ehrle," 379.
13. [Barry], "Discovery of America," 2; Blair, *Too Much to Know*, 177.
14. Gardiner, "Introduction to English History," xxii.
15. Doble to York Powell, August 13, 1894, Letter Books, 60, OUP; Doble to Osmund Airy, December 6, 1894, Letter Books 61, OUP; Poole, "Review of C. W. Baird," 390; Doyle, "Review of William B. Weedon," 597.
16. Lang, "History as She Ought to Be," 269.
17. [Froude], "Gardiner's *Fall of the Monarchy*," 295; Gardiner, *History of England*, 1, vi.
18. Dixon, *Invention of Altruism*, 64–65.
19. Hesketh, *Science of History*, 14–19, 36–45; Heyck, *Transformation of Intellectual*, 134–138; Fuchs, "English Positivism," 232–235; Goldstein, "J. B. Bury's Philosophy," 896.
20. Stubbs to Freeman, [April 13, 1858], MS. Eng. Misc. e. 148, Bodleian.
21. Stubbs, *Seventeen Lectures*, 11–12, 81–82.
22. Stubbs to Freeman, September 13, [1873], MS. Eng. Misc. e. 148, Bodleian.
23. "Lectures on the Study of Mediæval and Modern History," *Athenaeum*, March 26, 1887, 413.
24. Robinson, "Review of J. Dufresne de Beaucourt," 167.
25. Doyle, "Review of Hubert H. Bancroft," 595.
26. "Modern School of Historians," *Times*, February 5, 1879, 4.
27. Bury, "Science of History," 16–17; Oman, *Inaugural*, 29.
28. Lane-Poole, "Review of Jurien de la Graviére,", 785.
29. Perry, "Review of Richard Watson Dixon," 166.

30. Ropes, "Review of Herbert Tuttle," 584.
31. Paul, "Heroic Study of Records," 76–77.
32. Readman, "Place of the Past," 186.
33. "Italian Architecture," *Times*, December 19, 1876, 4.
34. "History of Agriculture and Prices in England," *Athenaeum*, June 23, 1866, 825–827.
35. Hamilton, "Review of L. Dussieux," 179.
36. Levine, *Amateur and the Professional*, 17–18, 40–45; Buchanan, "Science and Sensibility," 169–171; Peltz and Myrone, "Introduction," 3–5, 8–9; Speight, "Gentlemanly Pastime," 144; Saunders, *Tin Trumpet*, 37–38.
37. Levine, *Amateur and the Professional*, 19–20, 55–56; Speight, "Gentlemanly Pastime," 145–148; Hingley, "Society," 174–175; Briggs, "Prehistory," 243–244.
38. Hesketh, *Science of History*, 52.
39. Hesketh, *Science of History*, 52; Kirby, "From Tractarian to Democrat," 33; Burrow, *Liberal Descent*, 160.
40. "Professor Freeman," *Western Daily Press*, March 18, 1892, 5.
41. "Mr. Freeman's Norman Conquest," *PMG*, March 2, 1872, 851–852.
42. Freeman to Macmillan, March 3, 1872, Add MSS 55050, BL.
43. "Mr. Freeman's Norman Conquest," *PMG*, March 2, 1872, 851.
44. [Kebbel], "Dr. Johnson's *Letters*," 394–395.
45. "History of England," *The Athenaeum*, March 21, 1863, 392–393.
46. [Cornish], "Presbyterians and Independents," 492–493; [Cornish], "Gardiner's *Protectorate*," 446.
47. Gardiner, "Introduction to English History," xxii; Gardiner, *Prince Charles*, 1, vii–x; Gardiner, *History of England*, 10, vii; Gardiner, *History of the Great Civil War*, 2, vi–vii.
48. [Elliot], "Lecky's *England in the Eighteenth Century*," 203–204.
49. "History of Agriculture and Prices," *Athenaeum*, May 20, 1882, 630.
50. Bury, "Science of History," 20–21.
51. [Kebbel], "Dr. Johnson's *Letters*," 395.
52. Levine, *Amateur and the Professional*, 61–63.
53. Freeman, *History of the Cathedral Church of Wells*, ix–x; Freeman, *Exeter*, v–vi; "Royal Archæological Meeting at Cardiff," *Western Daily Press*, July 27, 1871, 3; "The Archæological Institute," *Times*, July 27, 1871, 12; Stubbs, *Seventeen Lectures*, 12.
54. Ihne to Freeman, February 12, 1879, FA1/7/401a–401b, JRL.
55. "How History Is Written," *Chelmsford Chronicle*, May 3, 1878, 4–5; Gardiner, "Introduction to English History," xxii.
56. Stubbs, *Seventeen Lectures*, 75–81, 101–102.
57. Stubbs to Jessopp, April 2, 1882, September 1 and 3, [1882], January 18, 1884, April 17, 1885, October 3, 1895, October 18, 1895, and March 6, 1896, Augustus Jessopp Papers, 1–10 c.1, RUB Bay 0039:07, RUB.

58. [Jessopp], "Brocas Book," 233–234.
59. Schor, *Reading in Detail*, 4, 17.
60. Courtney, *Feminine Note*, x–xi, xxxii.
61. Maitzen, *Gender, Genre*, 13–17.
62. [Wilson], "Forster's *Life of Swift*," 42; Brown, "Review of Henry Simonsfeld," 563; [Wace], "Early Christian Biography," 201; [Gairdner], "Sir James Ramsay's *Lancaster and York*," 92.
63. Creighton, *History of the Papacy*, 3, vi.
64. [Simpkinson], "Tyerman's *Life of John Wesley*," 56–58; [Palgrave], "Masson's *Life of Milton*," 396–399; [Laughton], "Low's *History of the Indian Navy*," 344; [Kebbel], "*Dr. Johnson's Letters*," 395.
65. Langford, *Englishness Identified*, 250–251; Forth, *Masculinity*, 22–23; White, "Darwin's Emotions," 826; Yeo, *Science*, 259.
66. Freeman, "Late Dr. Guest," 1551–1552; Guest, *Origines Celticae*, xii, xix, xxiv.
67. Guest, *Origines Celticae*, xvii.
68. Hesketh, *Science of History*, 149–151.
69. Oman, *Inaugural*, 26–30.
70. Bury, "Science of History," 3–4, 13.
71. "Prince Charles and the Spanish Marriage," *Athenaeum*, May 8, 1869, 629–630.
72. [Morris], "M. de Beaucourt's *Charles VII. Of France*," 170–171.
73. Towsey, *Reading History*, 5–9.
74. [Oliphant], "Modern Light Literature," 437–438.
75. Forster, *Walter Savage Landor*, 1.
76. [Clerke], "Leopold Prowe's *Life of Copernicus*," 297–298; Hammond, *Reading*, 7; Jones, "Victorian Literary Theory," 242–243; Dettmar, "Bookcases," 10–13; Mays, "Disease of Reading," 165–166.
77. "Some Recent Books" *Contemporary Review*, February 1881, 312.
78. Freeman to Macmillan, May 21, 1876, Add MSS 55051, BL; Freeman to George Macmillan, May 17, 1891, Add MSS 55053, BL; Ward to Freeman, April 10, 1877, FA 1/7/778–801, JRL; Seeley to Furnivall, October 12, [1883], FU 810, Furnivall Papers, Huntington Library.
79. Macmillan to David Watt, March 6, 1860, in Macmillan, *Letters*, 39.
80. Seeley, "Historical Society," 43–47, 50–52.
81. Freeman. "On the Study," 326–327.
82. Freeman to Macmillan, February 24, 1867, Add MSS 55049, BL; Freeman to Macmillan, March 3, 1872, Add MSS 55050, BL.
83. "Liberal Historian," *Daily News*, March 18, 1892, 4–5.
84. "Professor Freeman," *Western Daily Press*, March 18, 1892, 5; "Death of Professor E. A. Freeman," *Times*, March 17, 1892, 5.
85. Howsam, *Past into Print*, 49; Bentley, "Shape and Pattern in British Historical Writing," 204–205.

86. Macmillan to Price, November 23, 1875, Add MS 55,398, BL.
87. Longman to Freeman, January 8, 1874, FA 1/7/476–492, JRL.
88. Creighton to Browning, February 10, 1879, OB/1/423/A, King's Cam.

References

Unpublished Primary Sources

Bodleian Library, Oxford: Stubbs Papers.
British Library: The Macmillan Papers.
Huntington Library, San Marino, California: Furnivall (F.J.) Papers.
John Rylands Library, Manchester: E. A. Freeman Archive.
King's College Archive Centre, Cambridge: Oscar Browning Papers.
Oxford University Press Archive: Letter Books.
Rubinstein Library, Duke University: Augustus Jessopp Papers.

Printed Primary Sources

[Anon.]. "History of England." *Athenaeum*, March 21, 1863, 392–393.
[Anon.]. "A History of Agriculture and Prices in England." *Athenaeum*, June 23, 1866, 825–827.
[Anon.]. "Prince Charles and the Spanish Marriage." *Athenaeum*, May 8, 1869, 629–630.
[Anon.]. "Royal Archæological Meeting at Cardiff." *Western Daily Press*, July 27, 1871, 3.
[Anon.]. "The Archæological Institute." *The Times*, July 27, 1871, 12.
[Anon.]. "Mr. Freeman's Norman Conquest—Vol. IV." *Pall Mall Gazette*, March 2, 1872, 851–852.
[Anon.]. "Italian Architecture." *The Times*, December 19, 1876, 4.
[Anon.]. "How History is Written." *Chelmsford Chronicle*, May 3, 1878, 4–5.
[Anon.]. "The Modern School of Historians." *The Times*, February 5, 1879, 4.
[Anon.]. "Some Recent Books." *Contemporary Review*, February 1881, 312–332.
[Anon.]. "A History of Agriculture and Prices in England." *Athenaeum*, May 20, 1882, 630–631.
[Anon.]. "Lectures on the Study of Mediæval and Modern History." *Athenaeum*, March 26, 1887, 412–414.
[Anon.]. "Death of Professor Freeman." *The Times*, March 17, 1892, 5.
[Anon.]. "A Liberal Historian." *Daily News*, March 18, 1892, 4–5.
[Anon.]. "Professor Freeman." *Western Daily Press*, March 18, 1892, 5.

[Barry, William]. "The Discovery of America." *Quarterly Review*, July 1893, 1–41.

Brown, Horatio W. "Review of Henry Simonsfeld's *Der Fondaco dei Tedeschi in Venedig*." *English Historical Review* 3, no. 11 (1888): 563–564.

Bury, J. B. "The Science of History." In *Selected Essays of J. B. Bury*, edited by Harold Temperley, 3–22. Cambridge: Cambridge University Press, 1930.

[Clerke, Agnes M.]. "Leopold Prowe's *Life of Copernicus*." *Edinburgh Review*, October 1883, 295–332.

[Cornish, F. W.]. "Presbyterians and Independents." *Quarterly Review*, October 1889, 492–527.

[Cornish, F. W.]. "Gardiner's *Protectorate*." *Quarterly Review*, April 1898, 446–470.

Courtney, W. L. *The Feminine Note in Fiction*. London: Chapman & Hall, 1904.

Creighton, Louise. *Life and Letters of Mandell Creighton, Sometime Bishop of London*. Vol. 1. London: Longman, 1904.

Creighton, Mandell. *A History of the Papacy during the Period of the Reformation*. Vol. 3. London: Longman, 1887.

Creighton, Mandell. "Review of Francisco Ehrle's *Historia Bibliotheca Romanorum Pontificum I*." *English Historical Review* 6, no. 22 (April 1891): 378–379.

Doyle, J. A. "Review of Hubert H. Bancroft's *The History of the Pacific States*." *English Historical Review* 3, no. 11 (1888): 595.

Doyle, J. A. "Review of William B. Weedon's *Economic and Social History of New England 1620–1789*." *English Historical Review* 6, no. 23 (1891): 592–597.

[Elliot, Hugh F.]. "Lecky's *England in the Eighteenth Century* (vols. III and IV)." *Edinburgh Review*, July 1882, 203–240.

Forster, John. *Walter Savage Landor. A Biography*. Vol. 1. London: Chapman and Hall, 1869.

Freeman, Edward A. *History of the Cathedral Church of Wells as Illustrating the History of the Cathedral Churches of the Old Foundation*. London: Macmillan, 1870.

Freeman, Edward A. "The Late Dr. Guest." *Spectator*, December 4, 1880, 1551–1552.

Freeman, Edward A. "On the Study of History." *Fortnightly Review*, Spring 1881, 320–339.

Freeman, Edward A. *Exeter*. London: Longman, 1887.

Froude, J. A. "Gardiner's *Fall of the Monarchy of Charles I*." *Edinburgh Review*, October 1882, 295–346.

[Gairdner, James]. "Sir James Ramsay's *Lancaster and York*." *Edinburgh Review*, January 1893, 92–128.

Gardiner, Samuel Rawson. *Prince Charles and the Spanish Marriage. 1617–1623*. Vol. 1. London: Hurst and Blackett, 1869.

Gardiner, Samuel R. "Introduction to English History." In *Introduction to the Study of English History*, edited by Samuel R. Gardiner and J. Bass Mullinger, xxi–199. London: Kegan Paul, 1881.

Gardiner, Samuel Rawson. *History of England from the Accession of James I. to the Outbreak of the Civil War.1603–1642.* Vol. 1. London: Longman, 1883.

Gardiner, Samuel Rawson. *History of England from the Accession of James I. to the Outbreak of the Civil War.1603–1642.* Vol. 10. London: Longman, 1884.

Gardiner, Samuel Rawson. *History of the Great Civil War. 1642–1649.* Vol. 2. London: Longman, 1889.

Guest, Edwin. *Origines Celticae (a Framgment) and other Contributions to the History of Britain*, edited by William Stubbs and C. Deedes. Vol. 1. London: Macmillan, 1883.

Hamilton, E. Blanche. "Review of L. Dussieux's *Le Château de Versailles.*" *English Historical Review* 2, no. 5 (1887): 177–180.

[Jessopp, Augustus]. "The Brocas Book." *Edinburgh Review*, July 1887, 225–253.

[Kebbel, T. E.] "Dr. Johnson's *Letters.*" *Quarterly Review*, October 1892, 394–422.

Lane-Poole, S. "Review of Vice-Admiral Jurien de la Graviére's *Doria et Barberousse.*" *English Historical Review* 2, no. 8 (1887): 784–786.

Lang, A. "History as She Ought to Be Wrote." *Blackwood's Edinburgh Magazine*, August 1899, 266–274.

[Laughton, John Knox]. "Low's *History of the Indian Navy.*" *Edinburgh Review*, October 1878, 343–379.

Macmillan, George A. *Letters of Alexander Macmillan.* Glasgow: Printed for Private Circulation, 1908.

[Morris, W. O'. C.]. "M. de Beaucourt's *Charles VII. Of France.*" *Edinburgh Review*, July 1883, 170–198.

[Oliphant, Margaret]. "Modern Light Literature—History." *Blackwood's Edinburgh Magazine*, October 1855, 437–451.

Oman, Charles. *Inaugural Lecture on the Study of History.* Oxford: Clarendon Press, 1906.

[Palgrave, Francis Turner]. "Masson's *Life of Milton.*" *Quarterly Review*, April 1872, 393–423.

Perry, G. G. "Review of Richard Watson Dixon's *History of the Church of England.*" *English Historical Review* 2, no. 5 (1887): 165–169.

Poole, R. L. "Review of C. W. Baird's *History of the Huguenot Emigration to America.*" *English Historical Review* 1, no. 2 (1886): 389–391.

Robinson, Mary F. "Review of J. Dufresne de Beaucourt's *Histoire de Charles VII.*" *English Historical Review* 4, no. 13 (1889): 161–167.

Ropes, Arthur R. "Review of Herbert Tuttle's *History of Prussia under Frederick the Great.*" *English Historical Review* 4, vol. 15 (1889): 582–588.

[Russell, John]. "Ettrick Forrest and the Yarrow." *Edinburgh Review*, July 1887, 1–34.

[Saintsbury, George]. "Municipal London." *Quarterly Review*, July 1884, 1–39.

Saunders, Jefferson (ed.). *The Tin Trumpet; or, Heads and Tales, for the Wise and Waggish*. Vol. 1. London: Whittaker, 1836.

Seeley, J. R. "A Historical Society." *Macmillan's Magazine*, November 1881, 43–55.

Simpkinson, J. N. "Tyerman's *Life of John Wesley*." *Edinburgh Review*, January 1872, 56–88.

Stephen, Leslie. *Letters of John Richard Green*. London: Macmillan, 1902.

Stubbs, William. *Seventeen Lectures on the Study of Medieval and Modern History and Kindred Subjects*. Oxford: Clarendon Press, 1886.

[Wace, Henry]. "Early Christian Biography." *Quarterly Review*, January 1890, 201–225.

[Wilson, John]. "Forster's *Life of Swift*." *Quarterly Review*, January 1876, 42–80.

Secondary Sources

Bentley, Michael. "Shape and Pattern in British Historical Writing, 1815–1945." In *The Oxford History of Historical Writing*, edited by Stuart Macintyre, Juan Maiguashca, and Attila Pók, 205–224. Vol. 4. Oxford: Oxford University Press, 2011.

Blair, Ann M. *Too Much to Know: Managing Scholarly Information before the Modern Age*. New Haven: Yale University Press, 2010.

Briggs, C. Stephen. "Prehistory in the Nineteenth Century." In *Visions of Antiquity: The Society of Antiquaries of London 1707–2007*, edited by Susan Pearce, 227–265. London: The Society of Antiquaries of London, 2007.

Buchanan, Alexandrina. "Science and Sensibility: Architectural Antiquarianism in the Early Nineteenth Century." In *Producing the Past: Aspects of Antiquarian Culture and Practice 1700–1850*, edited by Martin Myrone and Lucy Peltz, 169–186. Aldershot: Ashgate, 1999.

Burrow, J. W. *A Liberal Descent: Victorian Historians and the English Past*. Cambridge: Cambridge University Press, 1981.

Dettmar, Kevin J. H. "Bookcases, Slipcases, Uncut Leaves: The Anxiety of the Gentleman's Library." *Novel* 39, no. 1 (2005): 5–24.

Dixon, Thomas. *The Invention of Altruism: Making Moral Meanings in Victorian Britain*. Oxford: Oxford University Press, 2011.

Forth, Christopher E. *Masculinity in the Modern West: Gender, Civilization and the Body*. Basingstoke: Palgrave Macmillan, 2008.

Fuchs, Eckhardt. "English Positivism and German Historicism: The Reception of 'Scientific History' in Germany." In *British and German Historiography 1750–1950: Traditions, Perceptions, and Transfers*, edited by Benedikt Stuchtey and Peter Wende, 229–250. Oxford: Oxford University Press, 2000.

Genette, Gérard. *Paratexts: Thresholds of Interpretation*. Cambridge: Cambridge University Press, 1997.

Goldstein, Doris S. "J. B. Bury's Philosophy of History: A Reappraisal." *The American Historical Review* 82, no. 4 (1977): 896–919.

Hammond, Mary. *Reading, Publishing, and the Formation of Literary Taste in England, 1880–1914*. London: Routledge, 2019.

Hesketh, Ian. *The Science of History in Victorian Britain: Making the Past Speak*. London: Pickering & Chatto, 2011.

Heyck, T. W. *The Transformation of Intellectual Life in Victorian England*. London: Groom Helm, 1982.

Hingley, Richard. "The Society, Its Council, the Membership and Publications, 1820–50." In *Visions of Antiquity: The Society of Antiquaries of London 1707–2007*, edited by Susan Pearce, 173–197. London: The Society of Antiquaries of London, 2007.

Howsam, Leslie. *Past into Print: The Publishing of History in Britain 1850–1950*. London: British Library and University of Toronto Press, 2009.

Jones, Anna Maria. "Victorian Literary Theory." In *The Cambridge Companion to Victorian Culture*, edited by Francis O'Gorman, 236–254. Cambridge: Cambridge University Press, 2010.

Karstens, Bart. "Bopp the Builder. Discipline Formation as Hybridization: The Case of Comparative Linguistics." In *The Making of the Humanities*, edited by Rens Bod, Jaap Maat, and Thijs Weststeijn, 103–127. Vol. 2. Amsterdam: Amsterdam University Press, 2012.

Kirby, James. "From Tractarian to Democrat: The Intellectual Formation of E. A. Freeman." In *Making History: Edward Augustus Freeman and Victorian Cultural Politics*, edited by G. A. Bremner and Jonathan Conlin, 31–45. London: Proceedings of the British Academy, 2015.

Langford, Paul. *Englishness Identified: Manners and Character 1650–1850*. Oxford: Oxford University Press, 2000.

Levine, Philippa. The Amateur and the Professional: Antiquarians, Historians and Archaeologists in Victorian England, 1838–1886. Cambridge: Cambridge University Press, 1986.

Maitzen, Rohan Amanda. *Gender, Genre, and Victorian Historical Writing*. New York: Garland Publishing, 1998.

Mays, Kelly J. "The Disease of Reading and Victorian Periodicals." In *Literature in the Marketplace: Nineteenth-Century British Publishing and Reading Practices*, edited by John O. Jordan and Robert L. Patten, 165–194. Cambridge: Cambridge University Press, 1995.

McKitterick, David. *A History of Cambridge University Press*. Vol. 3. Cambridge: Cambridge University Press, 2004.

Paul, Herman. "The Heroic Study of Records: The Contested Persona of the Archival Historian." *History of the Human Sciences* 26, No. 4 (2013): 67–83.

Paul, Herman. "Habits of Thought and Judgment: E. A. Freeman on Historical Methods." In *Making History: Edward Augustus Freeman and Victorian Cultural Politics*, edited by. G. A. Bremner and Jonathan Conlin, 273–289. London: Proceedings of the British Academy, 2015.

Peltz, Lucy and Martin Myrone. "Introduction: 'Mine Are the Subjects Rejected by the Historian': Antiquarianism, History and the Making of Modern Culture." In *Producing the Past: Aspects of Antiquarian Culture and Practice 1700–1850*, edited by Martin Myrone and Lucy Peltz, 1–13. Aldershot: Ashgate, 1999.

Readman, Paul. "The Place of the Past in English Culture c. 1890–1914." *Past and Present* 186 (2005): 147–199.

Schor, Naomi. *Reading in Detail: Aesthetics and the Feminine*. New York: Methuen, 1987.

Speight, S. J. "A Gentlemanly Pastime: Antiquarianism, Adult education and the Clergy in England, *c.* 1750–1960." *History of Education* 40, no. 2 (2011): 143–155.

Towsey, Mark. *Reading History in Britain and America, c.1750–c.1840*. Cambridge: Cambridge University Press, 2019.

White, Paul. "Darwin's Emotions: The Scientific Self and the Sentiment of Objectivity." *Isis* 100, no. 4 (2009): 811–826.

Yeo, Richard. *Science in the Public Sphere*. Aldershot: Ashgate, 2001.

PART II

Historians as Educators

Assuming different educational roles was part and parcel of historians' job description and Creighton declared with confidence that it mattered "a great deal" more who taught history at the universities than who was the head of the government.[1] Owing to this, pedagogical skills and knowledge were essential elements in the historian's persona. Importantly, historians realized that the educational sphere opened up numerous opportunities for spreading reliable historical knowledge and disseminating the new historic spirit and the persona it entailed. They understood the educational responsibilities broadly stretching from university lecterns to encounters with individuals curious about history. Following Creighton, "the work of culture is to be done as much by private as by public talk, as much by unauthoritative utterances at a dinner table, as by solemn ones in the pulpit or the newspaper."[2] As educators, historians reached out to a wide range of non-expert audiences and seized the opportunity to consolidate the persona among different readerships.

Historians' educational activities were vested with contradictions as they were compelled to square the conflicts between the scholarly virtues and the specific demands of different pedagogical tasks. Small educational histories looked anything but scholarly or virtuous, the performance of the inductive method in footnotes betrayed significant weaknesses in the epistemological foundations of historical knowledge, and histories with paratexts instilled with political lessons impaired the cherished virtue of impartiality. The following three chapters address different forms of educational writing and pivot on the one hand on the educational goals which historians pursued, and on the other the conflicts which the different formats, audiences, and aspirations created between the pedagogical and scholarly aspects of the persona.

Notes

1. Creighton to Acton, February 21, 1895, Add MSS 8119/1/C282, CUL.
2. Creighton to Louise Creighton, September 1871, in Creighton, *Life and Letters*, 82.

CHAPTER 5

Teachers with Scientific Credentials

William Stubbs discussed the "smaller works of a more distinctly educational character" in one of his statutory lectures in 1876. He connected the flourishing market in educational histories to the advancement of scientific history as he stressed how these histories "all have in common the mark of a purpose to secure skilled labour of the best sort." Hence, "boys are not to be taught any longer by book-makers." As proof of this change, he mentioned publishers' eagerness to establish high-quality educational series for history and to recruit the best historians to write and edit them. "Mr. Longman's Epochs, Mr. Freeman's Manuals, Mr. Green's Primers, Messrs. Rivington's Handbooks" were, according to Stubbs, prime examples of the new era of educational history.[1] Stubbs's observations were correct. The Education Acts of 1870 and 1880 exploded the need for cheap and accurate textbooks, and publishers responded to the demand by launching schoolbook series with ambitious scientific standards, with an eye to the educational policies and demands of the school boards, and with a unified form, style, and look that were easy and cheap to reproduce. The small histories also came to have such a wide appeal among adults that publishing houses created thematic series with content and design tailored to the tastes of the grownup middle-class reader. Owing to the publishers' inventiveness, Victorians had plenty of opportunities to cumulatively learn about history at every stage of their lives, as

E. Garritzen, *Reimagining the Historian in Victorian England*,
https://doi.org/10.1007/978-3-031-28461-8_5

nursery histories, illustrated children's histories, primers, textbooks, handbooks, popular histories, essays, and original studies flowed out from the printing presses.[2] The deluge of small histories was widely reported in the press. The *Athenaeum* wrote in 1875 how "One of the most striking of the publishing features ... is the quantity of school-books that pour from the press. Every firm ... has its series of school-books" and *The Times* wondered in 1888 whether there was perhaps an overproduction of the small and cheap volumes "of all manner of subjects."[3]

Historians recognized in the small histories a convenient channel for obtaining public acceptance for their persona and authority beyond learned circles. As the bulky multi-volume histories reached only a limited audience, historians needed novel forms to communicate with the public. The educational histories provided them with an excellent opportunity not just to disseminate accurate historical knowledge but also to perform their persona and to instruct readers to distinguish between good and bad history. The writing of educational histories came with its own cluster of skills, knowledge, and virtues which historians had to acquire. In this chapter, I explore how they incorporated the writing of educational histories into their persona and how they negotiated the contradictions between their epistemic virtues such as comprehensiveness and thoroughness and the limited number of pages available in schoolbooks. Despite the many practical challenges which historians faced when they experimented with educational writing, they embraced the opportunities the small histories created for them; they did not only promise visibility among the different types of readers but also gave hope of profit, which the big histories rarely brought. Edith Thompson captured the remunerative potential in 1879 when she, alluding to the ongoing economic stagnation, observed how "under our modern system people cannot dispense with education, even when the trade is bad."[4]

The prominence educational histories gained in periodicals and newspapers indicates that their quality was considered so important for English society that they demanded thorough consideration and exposure. The reviewers were not necessarily familiar with daily life in schoolrooms or specialists in pedagogical theories, despite their confidence in judging the textbooks. Nevertheless, it is safe to say that their proclamations shaped the public image of small histories and granted publicity to historians' educational endeavors. Regarding the persona project, one topic which resurfaced in the reviews was the belief that the existing history textbooks were antiquated, inaccurate, and unreliable. One reviewer summed up the popular sentiment by simply crying "How bad is the class!"[5] The general consent was that textbooks were so defective that if they did not altogether ruin the educational efforts in schools and homes, they certainly did not leave the students any wiser about history.[6] This dubious reputation of educational histories proved to be both a challenge and an opportunity for historians' persona project. It helped historians to garner generous support for their educational writing and for their assertion that educational histories should be written by the experts alone. The writer who labeled the entire class of history schoolbooks "bad" was indeed hopeful about the future, because a growing number of scientific historians were committing themselves to educational writing. Readers, the writer professed, should be grateful that men like Freeman were "doing something to introduce into schools an idea of scientific history." The *Athenaeum* agreed, and reminded how important it was that textbooks were written by capable people because they spread so widely due to their cheapness and conciseness. Hence, when executed poorly, their "influence may be very prejudicial."[7] This catered to historians' needs, as it gave them an opportunity to relate themselves as the saviors of the nation from the inadequate textbooks which the incompetent amateurs and hacks had been writing. This rendered the writing of educational histories a moral duty of the scientific historians and justified the integration of the authorship of short histories into the persona.

But to get entangled with a format that suffered from such a bad reputation had its own risks. To make the textbooks an eligible format for a scientific historian, the scholarly community set certain conditions for their educational pursuits to ensure that the small histories did not undermine the integrity of their persona. First, the writing of textbooks was acceptable only for those who had proven their skills with big histories. Powell instructed one young historian to forget "pot-boilers" until he had "*arrive*," that is, until he had written a big history to demonstrate his scholarly prowess.[8] This was considered a critical condition across the disciplinary lines, as scientists, too, set similar provisions for the popularization of scientific knowledge.[9] Second, it was crucial that educational histories conformed to scientific principles. It was not enough that historians applied the scientific method and embraced the epistemic and moral virtues; they had to communicate to the readers what exactly made these books accurate and scientific, as smallness implied lightness and low quality. Prefaces provided an appropriate stage to affirm that despite the conciseness, compression, and recycling of existing knowledge, the small books complied with scientific principles.

Prefaces can indeed exercise significant power over readers who regard them as authorial testimonies and interpretations of the ensuing text, its value, and purpose.[10] Even the brief prefaces that historians wrote for the textbooks functioned as assurances of scholarly value, as historians briefly described their methods and the epistemic virtues which underpinned the narrative, explained the subject matter and its historical significance, and defined the anticipated audience and the intended use of the book. Moreover, the prefaces added a new layer to the scholarly persona, as some historians invested them with deliberations about their pedagogical philosophies and visions. Although educational histories were crucial for establishing authority for scientific history, for historians with a pedagogical ethos they were also a chance to fulfill their educational mission. This extended the notion of what constituted a historian.

The support and encouragement from the publishers who recruited leading historians to contribute and edit the educational histories was vital for the success of the historians' experiments with different formats. Both historians and publishers benefitted from this collaboration. Alexander Macmillan, who relished a didactic ethos, was unable to understand why could "schoolbooks that pay best not be done by men who *know* what they are talking about."[11] Moreover, as Alexis Weedon has written, when genre, price, and format gained new importance in the segmented book market and cheapness provoked fears of low quality, the value of an author's name and a publisher's imprint for conferring prestige increased.[12] This was confirmed by the reviewers, who maintained that a "happy selection of the authors" was a genuine asset for educational publishing.[13] Names were also critical for the publishers, as they vouched for the reliability of the small histories, which were mostly issued without references. The *Athenaeum*, writing about Gardiner's *The Thirty Years' War*, which Longman published in the series "Epochs of Modern History," pointed out how the rule of the series was to omit all footnotes and that "There are comparatively few writers on whom we can rely implicitly, when, from the form in which their work is cast, we are forced to take upon trust all that they say; but Mr. Gardiner is one of those few."[14]

The production of educational histories was shaped by economic forces in the literary marketplace. However, the alliance between historians and the commercially commissioned short histories was scarcely discussed or contested. The *Pall Mall Gazette* complained in a rare critical appraisal about the commercialization of history, as the writer thought that Freeman's *William the Conqueror* in Macmillan's "Twelve English Statesmen" series bore "too much the look of work written to order."[15] Usually, however, the commercial dictates of the educational book market were quietly accepted.

Historians' eagerness to jump on the textbook bandwagon varied. Freeman, Gardiner, Creighton, and Green wrote and edited schoolbooks, concise biographies, and short histories. Another set of key contributors were the Oxbridge tutors. Although they had rarely authored "big" histories, they compensated the lack of scholarly credentials with their experience in teaching, and with knowledge about what did and did not work in textbooks. Educational histories were for them an important avenue to contribute to the scientification of history and to participate in the scholarly community. Seeley and Stubbs represented those historians who were uninterested in educational writing, although they agreed that textbooks were in dire need of a scientific facelift and that scientific historians should lead the way in this. For Stubbs, improvements in educational materials held the promise of a brighter future for history as schools, the "great feeders of University life," could prepare students better for their later studies in history.[16] Because of historians' relatively wide engagement in educational writing, textbooks came to be influential in spreading the awareness of scientific history and the adjoining scholarly persona. For many readers, the small histories were the only book-length acquaintance with these historians.

Popular, Small, or Something Else?

When historians talked about small histories, they meant textbooks, manuals, and other similar books that educated, not entertained, the reader. Because of this, I have chosen to use *educational* histories instead of popular histories, as there were imagined and actual differences between the two types of histories. "Popular" was a value-laden term in the Victorian literary and scholarly cultures. It denoted fashionable entertainment, ephemerality, and a courting of the pedestrian tastes of the multitude. It carried connotations of cheapness, low quality and cultural authority, and was linked to both the commercialization of books and the appearance of mass markets and audiences.[17] Educational histories, too, were "books of large circulation," but nonetheless had unique standards according to Mario Infelise's definition.[18] This description fits well the late-Victorian educational histories which historians strove to make more scientific by highlighting their scholarly qualities and purging the dramatization, emotions, and imaginary elements that had characterized earlier textbooks.

The emergence of popular history as a distinct form of history writing has been traced to the rise of the literary mass market and modern publishing industry in Europe and to the concurring professionalization of history which rendered historical knowledge too specialized for the public to indulge.[19] In England, the popularization of knowledge is chiefly associated with the sciences, as practicing scientists enhanced their status by promoting their knowledge to be too technical or complex for an ordinary reader. This created both demand and opportunities for a wide cast of popularizers who translated scientific discoveries into the vernacular. In history, the chronicling of facts came closest to this kind of technical knowledge, but it did not inspire a similar popularizing movement. The manuals which Stubbs's dry prose inspired were an exception, as amateur historians rather immersed themselves in original research than in making scientific histories accessible to a large audience. Nonetheless, historians and scientists were equally concerned about their diminishing presence in the public literary sphere and feared losing their status as producers of truthful and reliable scientific knowledge to non-expert authors.[20] The burgeoning textbook market brought relief to them all. Without surrendering to popularization, they could use small, scientifically accurate books to restore their authority among the public and protect the dignity of their persona.

As historians did not identify themselves with popularization, they categorized their educational books by foregrounding their small size. Small could mean to them and to their publishers anything between the number of pages, a book's physical dimensions, or a low price. Smallness could be a neutral or derogatory descriptive category and it therefore required qualification and contextualization. When Edward Armstrong reviewed Mary Robinson's biography of Margaret of Navarre, he noted that it was a "little book," but added that it nonetheless had "to be taken seriously."[21] Historians used "small" and its variants surprisingly casually in their correspondence. Susan Walton has proposed that "small" carried gendered meanings of femininity and low value, as Freeman employed attributes such as "little," "wee," or "less-than-little" to describe the primers that Margaret MacArthur, Edith Thompson, and

Charlotte Yonge wrote for the series "Historical Course for Schools" which he edited for Macmillan.[22] However, the femininization of small histories should not be overemphasized. Freeman, after all, applied similar vocabulary to his own books, calling them "less-than-little book," "less-than-least European History," "wee wee wee bit bookikie," and "prim, prim-er, prim'est."[23] Green, too, talked about "little" and "wee" histories, and Alexander Macmillan described Freeman's Norman Conquest primer as "your wee wee N.Q." without intending to denigrate its value.[24]

Although "small" did not necessarily mark femininity, histories for the youngest audience in particular were linked with womanliness and, accordingly, with low value, as they had often been authored by early- and mid-Victorian women. This association could have potentially engendered historians' manliness and persona, but their solution was not to expel women from educational writing. Instead, they encouraged women to adopt the scientific method and at least Creighton, Freeman, Gardiner, and Powell invited women to write for the series they oversaw. Women were regarded as appropriate writers for children for several reasons. Their presumed maternal qualities, it was believed, rendered them attentive to the needs of the young audience.[25] Furthermore, as educational histories were not premised on original research, even those who opposed women's intellectual ambitions agreed that women could be useful in reshaping original knowledge which men produced into a readable and compact format. This came close to the kind of popularization that took place in the sciences, and Emily Lawless, for instance, helped to disseminate Lecky's views about eighteenth-century Ireland in her *The Story of Ireland* (1887).[26] Of course, some remained skeptical. Bryce, after hearing that Freeman was recruiting young women to write for one of the series he was editing, declared that "these little things must be done by big people—that they are the most difficult things of all to do, and that till big people can find time to do them they had better wait."[27] Women, obviously, were not big enough people for Bryce, but perhaps even more important in his statement is the significance which he placed indirectly on educational histories while disregarding the agency of women historians.

Virtuosity of Educational Histories

James Gairdner wrote to the *Athenaeum* in 1879 to draw attention to the fact that the Education Act had done nothing to improve history schoolbooks. His daughter's school used *Outlines of the English History* by Henry Ince and James Gilbert, and he was dismayed by the authors' ignorance and the numerous embarrassing mistakes in the book which rendered it "utterly unfit" for educational purposes. It had, according to its 1877 title page, sold 515,000 copies but Gairdner found nothing in it that would have warranted its success. He could only ask in disbelief how

> Is it possible that over half a million of young persons have of late years been crammed with statements relating to English history of such a kind that it would really have been far better for them that they had never read a word of English history at all?[28]

Statements like this promoted the idea that there was a dire need for scientific historians to rescue children from the dangers of inaccurate history books. Historians accepted the invitation, but as the book's materiality did not make the scientific standards or persona evident, they furnished them with prefaces where they reminded readers about the new era in history writing. Originality, completeness, and industry were among the virtues which demanded most urgent prefatorial justification, because the writing of concise histories entailed skills that were almost antithetical to these normative ideals. Historians had to abandon their obsession for recounting everything they knew and instead learn to compress and simplify complex historical phenomena, identify significant and insignificant details, select facts, and recycle existing knowledge. As they had to learn to generalize and summarize, they needed self-discipline, emotional restraint, and courage to omit many encaptivating details and anecdotes. This went against their nature, and Freeman admitted that he suffered from a severe "temptation to enlarge" when he was preparing a short history of Rome for Macmillan's series "Historic Course for Schools."[29] Since the short format required significant adjustment from historians who cherished the big histories as embodiments of their persona, it was critical to maintain that, notwithstanding these alternate procedures, they remained faithful to the scientific ideals and that the small histories were incarnations of their persona too.

A striving for original knowledge was, as Stubbs emphasized, the engine that drove a scientific historian, and the reshuffling of "purely second-hand knowledge" was condemned as an inferior activity.[30] Educational histories made an exception to this, and the *Athenaeum* observed that all that was expected from an author of handbooks, was an acquaintance with secondhand authorities, familiarity with the latest historical knowledge, and the ability to adopt the most trustworthy views on each point. It would have been "out of question" to demand from such histories "either new ideas or new researchers."[31] Historians doubted whether the public truly grasped this, and they armed their prefaces with assurances of original ideas and fresh knowledge. John Mahaffy, professor in ancient history at Trinity College Dublin, warranted the originality of his *Social Life in Greece* (1874) with its novel approach. He departed from the traditional accounts of laws, religion, and lifestyle, and instead inquired how Greeks had felt about their life. Oman validated the freshness of his *History of Greece from the Earliest Times* (1890) with the new information it contained.[32] Macmillan promoted Mahaffy's book by quoting the *Athenaeum*, who praised how *Social Life in Greece* was "so fresh in its thought and so independent in its criticism." Oman, too, was complemented in reviews for showing "proof of independent judgment" instead of putting together "a mere summary" of larger histories.[33]

Sometimes historians stretched their claims for originality, compromising their adherence to honesty. Readers vigilantly exposed such inflated claims. While many of Freeman's educational titles were genuinely original undertakings, he shortened some of them from the five volumes of *The History of the Norman Conquest of England*. The title of *A Short History of the Norman Conquest of England* (1880) already gave away its true nature, and Freeman allowed that he was telling "in the shape of a primer" the same tale which he had already told "in five large volumes." As he knew very well that he needed something original in addition to the short format to induce readers, he added that he was not offering an abridgement, but "strictly the same [tale] told afresh."[34] Eight years later, he affirmed in the preface to *William the Conqueror* (1888) that it deviated from his previous publications because it did not narrate historical events over a long stretch of time, but painted a portrait of one man and his character.[35] These prefatorial avowals did not convince the critics. The *Athenaeum* decided that *A Short History of the Norman Conquest* was just a mutilated version of the larger one, and concluded that "as Mr. Freeman has perpetrated the sacrilege himself, no one has a right to complain."[36] The *Scottish Review* accepted that Freeman had added some fresh touches to give *William the Conqueror* "value of its own," but *The Times* disagreed and regretted that Freeman only repeated what he had already said elsewhere.[37]

The commitment to completeness was another delicate issue in educational histories where the generalization and selection of facts were unavoidable procedures. Historians resolved this contradiction by listing omissions in the prefaces to prove that they had not missed important historical details out of ignorance or carelessness. It was also popular to turn the vice of incompleteness into a virtue by presenting the compressions as a favor to the readers. The risk was that historians might appear arrogant as they used the readers' limited capacity to process historical details as an excuse for the absence of minutiae. Mahaffy wrote in *Social Life in Greece* how he had kept the reader in mind when he eliminated many details to save the narrative from becoming a mosaic of incoherent fragments instead of "a work fit for ordinary perusal." He even added philosophizing on how it was "generally true that no work is so disappointing as that which professes completeness."[38]

Another illustrative, but perhaps surprising, example of the same comes from Seeley's short history of Napoleon (1886). Seeley was better known as an advocate of scientific history than a proponent of popularization and his *A Short History of Napoleon the First* which Seeley & Co. published was the book that came closest to what could be defined as educational history writing. He furnished the book with a detailed preface where he excused the deletion of details for the benefit of the general reader. He had suppressed interesting events into one or two sentences and carefully shifted the bewildering number of facts pertaining to his subject so that he could present only the most significant incidents. With a tinge of superiority, he added how he could not ask from the ordinary reader the same kind of extraordinary attention and memory which he as a historian possessed and which he could not do without when processing the copious details in the material that he had gathered. This "attentiveness" toward the readers' limited cognitive abilities had encouraged Seeley to compromise on completeness, although he could not help wondering whether it was at all acceptable to blot out details in histories—regardless of their format.[39]

Finally, historians wanted to avoid an impression that short histories were fast and easy to write or that anyone could become an author of textbooks. Hence, Freeman energetically professed that a "shorter amount of space certainly does not represent a smaller amount of work."[40] Advertising industriousness in a preface was, however, a sensitive issue, because prefatorial self-praise was condemned as embarrassing egotism. To avoid such accusations, historians focused on their methodological procedures and research processes instead of their personal traits and accomplishments. Seeley boldly announced how much hard work and intellectual toil he had put into the writing of *A Short History of Napoleon the First*. He had not compromised his work ethic or scholarly standards even if the appearance of the book did not speak of remarkable industriousness or thoroughness. He verified this by claiming how his aim had been more ambitious than to offer a mere sketch of Napoleon; his book demonstrated how historical events had shaped Napoleon and how Napoleon had shaped his age. He had been able to reach this goal only by compiling everything that had been said about Napoleon and hence, instead of "Hasty investigations," he had condensed into the short volume the results of many years of research on the Napoleonic era. This, he explained, had demanded the exhaustive consultation of published and unpublished records and the examination of a vast body of scholarship which Napoleon had inspired. To further highlight the strenuousness of his scheme, he went on to state that he did not "flatter" himself that the project would have been "easy" to execute.[41] By foregrounding his consideration for the readers, the ambitiousness of his plan, and its diligent execution Seeley made a sustained attempt to reconcile the scholarly and educational sides of the scholarly persona.

The responses to historians' prefatorial pleas varied. The paradox was that while reviewers admitted that educational histories represented a distinct type of history, they were nonetheless tempted to collate small and big histories and evaluate the textbooks against the same repertoire of virtues which they expected from the big histories. Revealingly, when a textbook fulfilled the implicit expectations of originality and elaborateness, it was flagged as too valuable for its class. Bury's *A History of the Roman Empire from Its Foundations to the Death of Marcus Aurelius* was issued in Smith's "Student Manuals" and, according to one writer, would have deserved "a better audience than schoolboys."[42] Mary Dormer Harris's *Life in an Old English Town* (1898) was published in a series called "Social England" by Swan Sonnenschein and provoked a similar response. Jessopp wrote to Dormer Harris to express his astonishment that such a good work had appeared in a series.[43] If anything, the amazement of Jessopp and the anonymous reviewer indicate how the image of educational history series as outlets for second-rate work was firmly rooted in the minds of the readers. Historians, however, had more success in convincing reviewers that the writing of educational histories was the entitlement of qualified members of the scholarly community. One writer in the *Athenaeum* praised the "specialists" who gave their time and skills for writing proper educational histories. According to the writer, "Nothing, in fact, is harder to write than a primer, and the shorter it is harder the task" because an author of a textbook "ought to know everything … How else is he to choose, to distinguish the principal from the accessories, and give a general sketch which will be a faithful representation of reality?" An obvious consequence of this state of affairs was that "Specialists alone are capable of what has hitherto been too often left to compilers."[44] This kind of publicity was invaluable for historians. It provided external approval for their campaign to gain control over educational histories and also confirmed the virtuosity of their persona in the context of short histories.

Innovativeness and Its Limits

The educational book market was conservative, and an understanding of this was essential for forging the pedagogical persona. The content and structure of textbooks was influenced by the preferences of parents, teachers, and school boards. Generally, educational histories discouraged innovativeness. The emphasis was on accounts that confirmed the traditional view of English history, outlined the exceptional success of the English nation which had made it a world leader, instilled patriotic sentiments in the students, and cultivated their moral character. The political and institutional history of England were considered the most appropriate topics for teaching to young boys about duty, honor, and deference to the established institutions.[45] Instead of intellectually challenging themes, publishers wanted "periods" and chronological accounts, Seeley complained.[46] One book that shook the accepted canons of educational writing was John Richard Green's *A Short History of the English People* (1874). Although it was an immediate publishing sensation, its reception unmasks reservations that show how the writing of schoolbooks and manuals was embedded in conventional views of English history.

Green was not an ordinary kind of historian, and he enjoyed his role as a rebel whose philosophy of history, colorful style, and interest in social history set him apart from the scientific historians. The persona he cultivated, though, did not exclude him from the scholarly community, but he knew that he was accepted there more as a friend than a scholarly equal. He had confidently pronounced in 1870 that, "A short book need not be shallow, and a large book need not be big," but he knew that *Short History* was too small and popular to qualify him as an eminent historian and he tried to redeem himself later by writing "big" histories.[47] A certain ambiguity about the target audience of the book surrounded its preparation and reception. Green considered it as both a manual for schools and universities and an entertaining history for ordinary readers. Macmillan, by contrast, fathomed the book primarily as a schoolbook and adjusted its paratexts accordingly, creating confusion among readers, as Chapter 10 shows.

The process of making *Short History*, the book's unique literary qualities and ideological underpinnings have been explored in detail by earlier historians.[48] It is enough to mention here that the book's unforeseen success rested on its original approach to history, pictorial narrative, and affordable format. It was not just a triumphant story of England: it was a proto social history of the nation and as such, voguish, as the Second Reform Act in 1867 had inspired historical, theoretical, and polemical tracts that gave English people historical agency.[49] Green answered this demand more than well, as his understanding of history was grounded in the belief that the course of history was determined by the free will of individuals and the aspirations of the people.

Green dispensed with the stylistic, topical, and structural ideals of the standard educational histories and his defiance toward the prevailing norms provoked both enthusiasm and protest. He dismissed the matter-of-fact style that was the dominant mode in textbooks and argued that the scientific historians' uninteresting manner of presenting the past at least partially explained why youngsters found English history "hard" and "uninteresting."[50] Remembering his own childhood, he recalled how the dry raddling of facts had paralyzed his interest in history, and therefore he persuaded his fellow historians to enliven their narratives with stories and verbal portraits. Dryness should not have been "a *sine qua non* in educational books."[51] Green's vigorous use of language and colorful metaphors both in his histories and in his private correspondence have captivated later historians and somewhat overshadowed the fact that he did not enthrall all his contemporaries with his style and did not get significant imitators, as his elaborate use of language was judged too unorthodox.

Indeed, some considered Green's extravagant style an embarrassment and a sign of his inability to control his passions and subjectivity. He sacrificed the "higher qualities" of a historian—accuracy, scrupulousness, scientific reasoning, and self-restraint—to picturesqueness.[52] One reviewer confessed, "It makes us feel uneasy as we read of such incessant excitement and emotion as Mr. Green's characters always indulge in" and went on to add how they "do not talk nor walk like ordinary folk, but behave like heroes of melodrama on the stage." The excessive picturesqueness overpowered the story of historical progress and the philosophy of history which Green had developed, the same writer regretted.[53]

Short History did not inspire Green's contemporaries to make stylistic experiments. Neither did reviewers demand literary elegance from educational histories. Quite the contrary: the dryness which was branded as a vice in big histories was considered a virtue in textbooks. Gardiner, who was constantly reminded about the dullness of his bulky histories, was ranked as a first-rate author of textbooks.[54] If Stubbs was shorthand for scholarly virtuosity and Freeman for profuseness, the *Athenaeum* branded Gardiner as a paragon of excellence in educational writing: his books embodied accuracy, clearness of statement, sobriety of judgment, and "a due sense of historical proportion." His dry style gave no reason for complaints in this context because readers did not expect picturesqueness from textbooks: love of truth, impartiality, and good historical sense were more important than a love of color or a strong imagination. Indeed, Gardiner's dullness was insignificant because "brilliancy of style" did not belong in textbooks.[55] There is some irony in this: historians whose style was regarded too heavy for the long books which adults consumed were complemented for their dullness when they wrote educational books for English youth. For the historians, this was a great relief. By restricting the popularization ventures to educational histories, they could escape from making stylistic allowances or mimicking *Froudean* picturesqueness. They were more than satisfied with the narrative standards which dispensed with the drama, emotions, and colorful stories that were the general characteristics of popular histories and had been integral to the early-Victorian primers and textbooks which they loathed.[56]

Green deviated from the standard educational histories with his focus as he positioned himself in the preface against the "drum and trumpet" history that recounted wars, diplomatic maneuvers, and political intrigues. He questioned the importance of military campaigns and political conspiracies in shaping the course of history and was determined to prove that ordinary English people had been a genuine historical force that had influenced the fate of their nation. Accordingly, he zoomed in on social, intellectual, constitutional, religious, and industrial progress and introduced "figures little heeded in common history—the figures of missionary, poet, the printer, the merchant, and the philosopher."[57] He accentuated his novel approach by abandoning the traditional periodization according to regnal periods and instead structured his narrative thematically and chronologically.

Short History was received as a novelty, but it was not as original or radical as Green would have readers believe. Kings, politics, and battles had a prominent place in his book.[58] Also, there had been earlier attempts in England and elsewhere to add "the common man" into historical narratives and to democratize the national narrative by showing how people, not rulers, had been historical agents. Just like Green, these historians had contested the elitist regal periods as a received narrative structure or chronological order.[59] Nevertheless, Green's programmatic preface came to condition the book's reception, and almost without exception, the reviewers engaged with it, either approving or disapproving of Green's philosophy of history, his democratic views, and his dismissal of the Crown and the English Church, two staples in textbook histories. J. S. Brewer, for instance, connected Green's repudiation of the monarchy, aristocracy, and church, the foundations of English patriotism, to a dangerous attempt to infiltrate the gospel of democracy into schools. He cautioned parents to consider if this was the kind of history to "be put into the hand of the young and incautious." He did not think so.[60]

Green's perspective meant that many events and personalities that usually appeared in educational histories were missing from the pages of *Short History*. Green quickly learned how readers inconsistently expected brevity and comprehensiveness, simultaneously. Freeman, after reading the manuscript, was surprised that Green had excluded Thomas Seymour and suspected that Green had sacrificed historical facts to his "brilliant *talkee-talkee*." Green defended himself by explaining that he had omitted only such details which had no bearing "on the general current of our history," and Seymour's death, according to him, fell into this category. He also added that had he been writing "a general history in detail—say eight volumes or so—it might be fairly urged that, as one can hardly tell what facts will in the end turn out to be important, it is better to put in too many than too few."[61] Freeman did not question the need for omissions as such. He wrote elsewhere how there should be gradations in completeness between little books and books that wore "Stubbs's coat." Short textbooks begged generalization, and the "best rule" was to put only "so much as folk are likely to remember," while the big histories were the great works where "you put in a good deal [of details]" as it was the sum of the details that created "the general effect."[62] The difficulty was that readers thought differently on what should and should not be excluded from textbooks, as historical knowledge was epistemologically shaped by the choices which each historian made. The exchange of letters made Green gloomily predict that "nine readers out of ten ... will suppose I have forgotten him [Seymour]."[63]

It is hard to see why Green did not use the prefatorial opportunity to justify his omissions. He regretted this as soon as the reviews began to flow in. The book was received exceedingly well, but reviewers were disappointed in Green's brevity about more recent literary life and contemporary history. He admitted to Freeman that he should have added a prefatorial assurance that he was not ignorant of Dryden, Pope, Addison, or Wordsworth and an explanation that he had given less room to writers after 1660 because since then literature had lost much of its value in forming the national character.[64] Strikingly, Green did not explain himself in the later versions of his book either. He never thoroughly revised it or wrote a new preface to it. These he left to his widow, Alice Stopford Green, who finally replied to the critics in the first posthumous edition, which was released in 1888. Just as Green had written to Freeman, she now clarified in almost the exact same words that he had excluded many writers and artists because he had been convinced that "after 1660 Literature ceased to stand out in the fore-front of national characteristics, and that Science, Industry, and the like, played a much greater part."[65]

Although *Short History* enjoyed immense popularity, unique, original, or controversial educational histories were not unanimously approved. The scientific historians who contributed to the educational market were expected to observe the unwritten rules that defined the histories in this category because this made their books much more appealing to schools. James Franck Bright recognized this and used Green's celebrity status to shamelessly market his *A History of England* (1875) as a "safe" alternative to *Short History* and to show that he had adopted the "right" kind of scholarly persona. Bright represented a typical textbook author: he was a fellow and lecturer at Oxford and cultivated a persona which integrated scholarly and pedagogical virtues and skills. He highlighted his pedagogical insider knowledge by telling in the preface how he had attended a meeting of public school teachers where the weaknesses of existing history books had been discussed, and that he had done his best to attend to these shortcomings in his book.[66]

Bright positioned his *History of England* as an alternative to Green in terms of its content and structure. He first identified his book as useful "for school teaching" and then rationalized why he had come to give prominence to political and national events. His original plan had been more innovative, as he had aspired to merge social, constitutional, and political history. But he had quickly realized that this would have presupposed too much prior knowledge from readers of school age. Discussion of the growth of English society was therefore limited to a few chapters and passages, while the main storyline advanced chronologically from one historical event to another. This seemed to be the preferred mode for school histories: the chronological account of events was paired with brief sketches of manners, ideas, literature, and institutions so that the students learned both the facts and the "ideas of which those facts are the expression."[67]

Bright, furthermore, justified his minimal treatment of social history by noting how the topic had already been covered by Green "so adequately."[68] We can read this as a polite nod to Green or as an opportunity to mark the distinction between Bright's and Green's histories and as an attempt to peg the merits of the former to the recognized shortcomings in the latter. Green's friends had warned him that *Short History* demanded too much prior knowledge from its readers in schools and Ward had suspected that even university students would struggle with the book for the same reason.[69] Objections from teachers confirmed such doubts. The book was tried out in schools in England and North America. It was adopted as late as in 1895 for the upper grammar grades in public schools in Chicago due to its more judicious treatment of religious questions than was common for books in this class. However, it was quickly dropped, as it was too "voluminous" in a classroom use.[70] Bright must have realized this, as he emphasized vigorously how too demanding textbooks were useless in schools.

Bright also promised relief for another recognized pedagogical defect in Green's book: its unconventional structure. Immediately after acknowledging Green's achievement, Bright went on to state that he had retained the "old and well-known division into reigns." This immediacy, together with the curiosity which Green's unusual composition had attracted, ensured that readers understood what Bright was implying. He bolstered his message by maintaining how unwise it would be to "disturb the knowledge boys have already gained by the introduction of a new though more scientific division."[71]

Green's friends had pointed out this defect to him as well by remarking how the unusual order complicated the use of the book as a reference aid. A king was, Freeman explained, like "a finger-post that shows you the way," and he added how readers lost their way in Green's narrative without such familiar pointers.[72] There were also other reasons to prefer a conventional chronological or reginal arrangement in schoolbooks: their presumed neutrality. Bright, possibly out of consideration, did not refer to the ideological implications of the question, but the *Athenaeum* addressed the topic when reviewing another educational history, G. W. Kitchin's *A History of France*. Kitchin had grouped the reigns into periods according to their own "peculiar characteristics." The reviewer was unenthusiastic about this "surprising" disposition and hinted that such divisions were subjective and open to debate, while regnal periods were a neutral way of organizing past events and therefore more suitable for textbook histories.[73] Kitchin, perhaps anticipating criticism, offered in the preface an explanation for his reasons to prefer "natural epochs" over reginal periods whose historical importance, according to him, varied too greatly to form a structural foundation for a historical work.[74]

It seems safe to conclude that Bright offered his book as a conventional and practical alternative to Green for schools where accepted facts about English history were more valuable than novel perspectives and experimental storylines. The historians' persona was embedded in originality, but innovativeness was not a desirable quality in an author of educational histories. Deviations from the approved canons made the use of such histories difficult, and unorthodox approaches appeared to be suspicious attempts to infiltrate (radical) political or religious ideologies into the schoolroom. Merging an understanding of how textbooks were used in schools and the principles which informed history teaching with the scientific method and scholarly virtues should have been at the core of the historians' persona.

Pedagogical Visions

Scaling down history into smaller formats was not just a matter of adjusting historians' working patterns. Educational authorship placed more specific demands on the historians' persona as well. These encompassed, for instance, an ability to both envision the cognitive capabilities of different age groups and adapt the narrative accordingly. Historians also needed practical knowledge of the use of different paratextual components in educational histories and a pedagogical vision. Publishers hoped that historians contributed to those paratexts that helped readers to organize and digest historical knowledge. Schoolbooks could contain "an analysis," a chronological inventory of historical events, lists of sovereigns, genealogical tables, maps, and illustrations. Other standard features were analytical running heads, marginal notes, dates in the margin, bold typography, and an index. These were all paratexts that provided an authorial commentary on the text and indicated details the writers wanted to highlight. Ideally, well-composed paratexts supported the learning process. Moreover, when historians wrote books that were aimed at specific forms or examinations, they needed at least superficial familiarity with the regulations which the school boards had instigated. As they also understood the commercial value of this, they added prefatorial assurances about the book's compliancy with the official requirements. Oman opened the preface in *History of Greece* by noting how he had "kept in view" the requirements of upper forms in schools as well as the final pass examination at the universities and Powell informed how *Old Stories from British History* contained 57 lessons "in accordance with the instructions of Her Majesty's inspectors."[75]

Mostly historians wrote educational books without experience teaching schoolchildren other than what they gained from raising their own sons and daughters. Therefore, their educational writing was informed by general ideas about the usefulness of history in molding character and habits of mind such as the forming of conscientious judgments. Accordingly, Prothero praised Gardiner for inspiring the minds of older learners in *The Constitutional Documents of the Puritan Revolution* by setting them problems that involved shifting and comparing historical materials.[76] Pedagogically, too, Gardiner exemplified educational history writing. Another broader aim that historians set for their educational books was to spark curiosity about the past and induce even those boys who loved cricket and football to find pleasure and instructive lessons in history.[77]

Those historians who fostered a deep educational ethos wrote prefaces where they meditated about their pedagogical philosophy and informed parents, teachers, and governesses about their deeper pedagogical aims. Creighton was one of these historians. He was committed to improving history teaching at Cambridge and encouraged students to actively engage with the past and apply historical knowledge in practice. For him, a teacher was not like a performing monkey butler who poured knowledge into the heads of passive students. A good teacher made students think and become thirsty for learning more.[78] He found biography an ideal means to achieve these goals both in the classroom and in educational histories, because biographical accounts tended to awaken the historical imagination of even those who were not otherwise attracted to the subject. He applied this idea to the series "Historical Biographies" which he edited for Rivington and which was aimed at pupils who were not yet mature enough to read the "standard" histories of Hallam, Freeman, Macaulay, or Froude. The biographies, Creighton explained, opened a window onto history and helped children to get "accustomed to historical details, and to judging historical characters." Through the individual life stories, children could grasp the meaning of "the general principles of history."[79] In other words, the purpose of the biographies was not merely to list facts but to give parents and teachers tools for training children's cognitive and intellectual faculties and introduce them to the underlying historical forces.

Alice Gardner shared Creighton's enthusiasm for improving teaching, and they developed the history syllabus in Cambridge together. She specialized in history of philosophy, but also authored several histories for children, setting exceptionally demanding pedagogical targets for her young readers. *Rome. The Middle of the World* (1897) was aimed at children between six and seven and encouraged an emotional engagement with historical figures. Although historians endorsed detachment, self-restraint, and regulation of emotions, and criticized the earlier nursery histories for unnecessarily stirring the emotions, they did not necessarily abandon emotional alignment with the past as a pedagogical strategy. Instead, in accordance with Victorian ideals, they recommended a curated attachment to historical persons. The purpose of this was to bring the reader closer "to the mind and feelings" of past times, as Powell wrote in an introduction to the series "English History by Contemporary Writers."[80] Gardner pushed the method even further and advised parents to instruct their children to form a friendship with "a man who belongs to that time" and to discover "what he thinks about things around him" and "what he is *doing* to keep or hinder the progress or changes going on." Such an affinity would have created a bond with the historical ages and generated a realization of the sameness and strangeness between the present and the past. This "personal feeling towards people of bygone times" would have helped children to overcome "the feeling of strangeness" and to recognize the forerunners as "*makers of history*," Gardner explained. She assisted this imaginative and emotional learning process by choosing a chatty tone and snatching narrative "photographs" of historical events which parents could study with their children to stimulate each child to travel to the past using their imagination.[81]

While the chattiness echoed the early-Victorian nursery histories, Gardner used the front matter to distance herself from amateurism. She affirmed her scholarly authority by introducing herself on the title page as a "Historical Lecturer of Newnham College, Cambridge" and enforced the message in the preface by briefly summarizing her philosophy of history. Her historical understanding rested on the idea of the unity of history which had provided moral instruction throughout the ages. However, she was afraid that the eroding trust in Providence and human progress had propelled history into a fragmentary state "without assuredly becoming more scientific." Her ultimate pedagogical goal was to challenge this modern tendency and to steer children to see history "as one great whole."[82]

If Gardner was worried that the forces of modernization were causing a fragmented view of history, several others blamed the condensed histories altogether for producing a distorted image of the past. The unified look of the chronologically structured series plus the promises of a seamless narrative proceeding from one volume to another were considered misleading. The editors and publishers advertised the series with claims of a cohesive impression of history despite the multiple authorship. The multiple authorship model introduced modern efficiency, organization of work, and division of tasks into scholarly publishing. In France, such serialized histories were dubbed a "*revolution américaine*" and regarded as testimonies of modern ingenuity that extended to the "*travaux de l'esprit*" as well.[83] However, contrary to the hype and the marketing lingo, the critics complained that the small volumes created a disconnected picture of history, as the contributors interpreted the series' guidelines differently. Burrows, an author of educational histories himself, doubted whether little books "about great things" with "piecemeal" impressions of the past could ever "raise the level of historical knowledge" among the English.[84] Charles Plummer, Fellow at Corpus Christi College at Oxford, concurred. He wrote in the *English Historical Review* about the "flood of 'handbooks,' 'epochs,' 'primers,' &c." hoping that historians would abstain from writing too narrowly delineated small histories.[85] As many of the series failed to present a cohesive view of history, they undermined the most essential goal in history teaching: to demonstrate continuity and steady progress from the earliest times to the present. Instead of this, the incoherence was a reminder of the hubbub of the Victorian era and the constant changes that made the world appear fragmented, disconnected, and temporary—just like the past that the small histories evoked.

*

Francis Palgrave had famously pronounced that "no small book can ever really teach great things."[86] Most histories fell in between the extreme poles of "small" and "big," but the polarization nonetheless afforded a persuasive model for evaluating histories and historians. The small histories encapsulated several important aspects of the historians' scholarly persona and its entanglement with the literary marketplace as the historians strove to gain dominance over the educational histories to protect their right to define the standards for reliable historical knowledge and its producers. As they acknowledged that the segmentation of the readership was a reality, they developed and took advantage of the new channels to communicate their persona to the different reading communities. Thus, literary multitasking became a core component in their scholarly persona. Gardiner captured this neatly when he wrote in the preface to *The Thirty Years' War* how in this particular book he "must be content to be the retailer rather than the manufacturer of history, knowing that each kind of work has its use."[87]

The small histories helped to promote the persona only if they met the scientific standards and if their virtuosity was spelled out in the prefaces, as their modest size did not speak on behalf of comprehensiveness, thoroughness, or meticulousness. While listing the book's virtuous qualities, the historians insinuated that only qualified historians were competent to write scientifically accurate textbooks and that because educational authorship was now informed by the inductive method and epistemic virtues, immersion in the writing of small histories did not conflict or compete with the historians' scholarly persona. Although the historians emphasized their scientific status in the prefaces, their keenness to educate the public should not be underestimated, either. The pedagogical dimension was an actual factor in their persona.

Reviewers were crucial in popularizing the message that only specialists were eligible authors of educational histories. The *Pall Mall Gazette* endorsed this when eulogizing Freeman's *The Historical Geography in Europe* (1881). The reviewer recognized the immense labor which its preparation had demanded and stressed how its smallness should not misguide readers, as it contained "a mass of information altogether disproportionate to its size." The writer concluded on a high note:

> Hardly any living Englishman approaches Mr. Freeman in these respects; none surpasses him; and it is greatly to his honour that he should have bestowed so many years of toil on a work which ... cannot be expected to receive from the unskilled reader all the admiration it deserves.[88]

The review exposed the prevailing prejudices toward small histories and challenged many common misconceptions about their poor quality. At the same time, it made explicit that only experts were competent writers of high-quality educational histories. Public opinions like this were critical for historians who feared that the writing of small histories might threaten the integrity of their scholarly persona or the disciplinary status of history. Quite the contrary, carefully composed educational works like Freeman's *Historical Geography* embodied the persona in a concise package and popularized it among non-expert readers. According to the commentator in the *Pall Mall Gazette*, the problem was not the historians who wrote modest-looking histories, but the "unskilled" readers who were unqualified to judge the value of textbooks or grant their authors the admiration they deserved. The historians certainly could not have hoped for a better promotion for their educational endeavors and for the compatibility of their endeavors with their scholarly persona.

Notes

1. Stubbs, *Seventeen Lectures*, 45–47, 58.
2. Howsam, "Growing up with History," 55–56; Howsam, "Academic Discipline," 526; McKitterick, *History of Cambridge University Press*, 2:355, 3:66; Green, "Look of the Books," 273–274.
3. "Literary Gossip," *Athenaeum*, February 13, 1875, 230; "Twelve English Statesmen," *Times*, August 15, 1888, 3.
4. Thompson to Macmillan, January 10, 1879, Add MSS 55,078, BL.
5. "Compendium of European Geography and History," *Athenaeum*, October 19, 1872, 496.
6. "*English History*," *Athenaeum*, September 15, 1866, 333–334.
7. "School Manuals," *Athenaeum*, July 31, 1880, 142.
8. Powell to a young historian, 1900, in Elton, *Frederick York Powell*, 1:300.
9. Lightman, *Victorian Popularizers of Science*, 417–418.
10. Genette, *Paratexts*, 197–198, 222–223; Masson, "Marginalité de la préface," 13–15; Murray 2013, "Politics of the Preface," 50–51; Arroyo Redondo, "Aproximaciones teóricas al prólogo," 59.
11. Macmillan to Freeman, March 23, 1863, FA 1/7/495, JRL.
12. Weedon, *Victorian Publishing*, 89–90, 100, 107–108.
13. "William the Conqueror," *Western Daily Press*, March 31, 1888, 3.
14. "Thirty Years' War," *The Athenaeum*, July 25, 1874, 108.
15. "William the Conqueror," *PMG*, April 2, 1886.
16. Stubbs, *Seventeen Lectures*, 45–47, 58.
17. Jackson, *Those Who Write*, 81–83; Newey, "Popular Culture," 148–149.
18. Infelise, "Libri per tutti," 3.
19. Berger, "Professional and Popular, 16–18.
20. Lightman, *Victorian Popularizers of Science*, 9–1, 353–375, 418, 493–495.
21. Armstrong, "Review of Mary Robinson," 169–170.
22. Walton, "Charlotte M. Yonge," 226–255.
23. Freeman to Thompson, February 19, 1871, U DX/9/26, HHC; Freeman to Macmillan, May 21, 1871, Add MSS 55049, BL; Freeman to Thompson, June 6, 1874, U DX/9/48, HHC; Freeman to Green, September 13, 1874, FA 1/8/1–30, JRL; Freeman to Green, December 19, 1875, FA 1/8/31–60, JRL; Freeman to Green, September 6, 1876, FA 1/7/8/60a–60b, JRL.
24. Green to Olga von Glehn, August 2, 1873, in Stephen, *Letters*, 354–357; Green to Freeman, November 12, 1873, FA 1/7/279; Macmillan to Freeman, April 2, 1880, Add MSS 55410 (2), BL.
25. Mitchell, *Picturing the Past*, 59; Hesketh, *Science of History*, 125.
26. Smith, *"Manly Study,"* 21–24.
27. Green to Freeman, June 27, 1871, in Stephen, *Letters*, 305.
28. Gairdner, "How We Teach History Now," 79.

29. Freeman to Macmillan, 5 April 1881, Add. MSS 55052, BL.
30. [Roscoe], "Memoirs of the Verney Family," 414.
31. "History of France," *Athenaeum*, July 21, 1877, 76.
32. Mahaffy, *Social Life in Greece*, vii; Oman, *History of Greece*, v.
33. Macmillan's advertisement, *Athenaeum*, January 9, 1875, 43; Giles, "Review of C. W. C. Oman," 770.
34. Freeman, *Short History*, introduction.
35. Freeman, *William the Conqueror*, preface.
36. "School Books," *Athenaeum*, September 4, 1880, 305.
37. "Contemporary Literature," *Scottish Review*, July 1, 1888, 193; "Twelve English Statesmen," *Times*, August 15, 1888, 10.
38. Mahaffy, *Social Life in Greece*, vii.
39. Seeley, *Short History*, vi–vii.
40. Freeman, *NC*, 5:v.
41. Seeley, *Short History*, vi–ix.
42. "History of the Roman Empire," *Athenaeum*, December 2, 1893, 762–763.
43. Jessopp to Dormer Harris, May 22, 1898, CR3874/1/8, WCA.
44. "School Manuals," *Athenaeum*, July 31, 1880, 142.
45. Mitchell, *Picturing the Past*, 82.
46. Seeley to Browning, May 21, s.a. GBR/0272/OB/1/1455/, King's Cam.
47. Green to Freeman, [April 1870], in Stephen, *Letters*, 249.
48. Brundage, *People's Historian*, Chapters 6 and 7; Hesketh, *Science of History*, 119–124; Howsam, *Past into Print*, 37–40; Jann, *Art and Science*, 141–169; Garritzen, "Framing and Reframing," 181–185.
49. Mandler, *English National Character*, 71–72.
50. Green, *Readings*, iii.
51. Green to Freeman, June 27, 1871, and December 30, 1872, in Stephen, *Letters*, 303–305, 340.
52. "History of the English People," *Athenaeum*, August 7, 1880, 170.
53. "Mr. Green's English History," *Examiner*, December 15, 1877, 1584–1585.
54. "Thirty Years' War," *Athenaeum*, July 25, 1874, 108; "Longman's Ship Historical Readers," *Athenaeum*, June 9, 1894, 740.
55. "Student's History of England," *Athenaeum*, November 29, 1890, 732–733; "Student's History of England," *Athenaeum*, September 19, 1891, 383.
56. Of the characteristic features of popular history see Korte and Paletschek, "Popular History," 10.
57. Green, *Short History*, v–vii.
58. Himmelfarb, *New History*, 152–153.

59. Baár, *Historians and Nationalism*, 47–50; Mitchell, *Picturing the Past*, 15.
60. [Brewer], "Green's *History*," 322–323.
61. Green to Freeman, September 16, 1873, in Stephen, *Letters*, 357–358; Freeman to Green, September 21, 1873, FA 1/8/1–30, JRL.
62. Freeman to Macmillan, March 3, 1880, Add MSS 55052, BL; Freeman to Thompson, September 4, 1870, U DX9/21, HHC.
63. Green to Freeman, September 16, 1873, in Stephen, *Letters*, 357–358.
64. Green to Freeman, January 18, 1875, in Stephen, *Letters*, 407–408.
65. Stopford Green, "Introduction," xv.
66. Bright, *History of England*, v.
67. "A History of France," *Athenaeum*, July 21, 1877, 76.
68. Bright, *History of England*, v–vi.
69. Freeman to Green, May 16, 1875, FA 1/8/31–60, JRL.
70. Harper Brothers to Stopford Green, December 5, 1895, MS 15,124/3, NIL.
71. Bright, *History of England*, vi.
72. Freeman to Green, January 23, 1876, and February 13, 1876, FA 1/8/31–60, JRL.
73. "History of France," *Athenaeum*, July 21, 1877, 76.
74. Kitchin, *History of France*, v.
75. Oman, *History of Greece*, v; Powell, *Old Stories*, vi.
76. Prothero, "Review of S. R. Gardiner," 393–394.
77. "Readings from English History," *Athenaeum*, July 26, 1879, 111.
78. Creighton to a pupil, June 21, 1889, and October 14, 1890, in Creighton, *Life and Letters*, 294–295.
79. Creighton, *Life of Simon de Montfort*, vii–ix.
80. Powell's introduction, in Thompson, *Wars of York and Lancaster*.
81. Gardner, *Rome*, 2–4.
82. Gardner, *Rome*, ix–x.
83. *Revue universitaire* quoted in Leduc, *Ernest Lavisse*, 143.
84. Burrows, "Review of Edward E. Morris," 183.
85. Plummer, "Review of James Hamilton Wylie," 786.
86. Armitage, *Childhood*, vi.
87. Gardiner, *Thirty Years' War*, vi.
88. "Freeman's 'Historical Geography," *PMG*, July 14, 1881, 211–212.

References

Unpublished Primary Sources

British Library: The Macmillan Papers.

Cambridge University Library: Acton Papers.
Hull History Centre: Letters from Edward Augustus Freeman to Edith Thompson.
John Rylands Library, Manchester: E. A. Freeman Archive.
King's College Archive Centre, Cambridge: Oscar Browning Papers.
National Library of Ireland, Dublin: Alice Stopford Green Additional Papers.
Warwickshire County archive: Mary Dormer Harris Collection.

Printed Primary Sources

[Anon.]. "*English History from the Earliest Period.*" *Athenaeum*, September 15, 1866, 333–334.
[Anon.]. "A Compendium of European Geography and History." *Athenaeum*, October 19, 1872, 495–496.
[Anon.]. "The Thirty Years' War." *Athenaeum*, July 25, 1874, 108.
[Anon.]. "Literary Gossip." *The Athenaeum*, February 13, 1875, 230.
[Anon.]. "A History of France." *Athenaeum*, July 21, 1877, 76.
[Anon.]. "Mr. Green's English History." *Examiner*, December 15, 1877, 1584–1585.
[Anon.]. "Readings from English History." *Athenaeum*, July 26, 1879, 111.
[Anon.]. "School Manuals of French history." *Athenaeum*, July 31, 1880, 142–143.
[Anon.]. "History of the English People." *Athenaeum*, August 7, 1880, 167–170.
[Anon.]. "School Books." *Athenaeum*, September 4, 1880, 305.
[Anon.]. "Freeman's 'Historical Geography of Europe'." *Pall Mall Gazette*, July 14, 1881, 211–212.
[Anon.]. "William the Conqueror." *Pall Mall Gazette*, April 2, 1886.
[Anon.]. "William the Conqueror." *Western Daily Press*, March 31, 1888, 3.
[Anon.]. "Contemporary Literature." *The Scottish Review*, July 1, 1888, 193–194.
[Anon.]. "Twelve English Statesmen." *The Times*, August 15, 1888, 10.
[Anon.]. "A Student's History of England from the Earliest Times to 1855." *Athenaeum*, November 29, 1890, 732–733.
[Anon.]. "A Student's History of England from the oldest times to 1855." *Athenaeum*, September 19, 1891, 383–384.
[Anon.]. "A History of the Roman Empire from Its Foundation to the Death of Marcus Aurelius." *Athenaeum*, December 2, 1893, 762–763.
[Anon.]. "Longman's Ship Historical Readers." *Athenaeum*, June 9, 1894, 740.
Armitage, Ella S. *The Childhood of the English Nation, or the Beginnings of English History*. London: Longman, 1877.

Armstrong, E. "Review of Mary Robinson's *Margaret of Navarre*." *English Historical Review* 5, no. 5 (1887): 169–171.

[Brewer, J. S.]. "Green's *History of the English People*." *Quarterly Review*, April 1876, 285–323.

Bright, J. Franck. *A History of England. Period I: Mediæval Monarchy*. London: Rivingtons, 1877.

Burrows, Montagu. "Review of Edward E. Morris's *The Early Hanoverians*." *English Historical Review* 2, no. 5 (1887), 181–183.

Creighton, Louise. *Life and Letters of Mandell Creighton, Sometime Bishop of London*. Vol. 1. London: Longman, 1904.

Creighton, Mandell. *Life of Simon de Montfort. Earl of Leicester*. London: Rivington, 1876.

Elton, Oliver. *Frederick York Powell. A Life and a Selection from His Letters and Occasional Writings*. Vol. 1. Oxford: Clarendon Press, 1906.

Freeman, Edward, A. *The History of the Norman Conquest of England*. Vol. 5. Oxford: Clarendon Press, 1876.

Freeman, Edward A. *A Short History of the Norman Conquest of England*. Oxford: Clarendon Press, 1880.

Freeman, Edward A. *William the Conqueror*. London: Macmillan, 1888.

Gairdner, James. "How We Teach History Now." *Athenaeum*, July 19, 1879, 79.

Gardiner, Samuel Rawson. *The Thirty Years' War 1618–1648*. London: Longman, 1874.

Gardner, Alice. *Rome. The Middle of the World*. London: Edward Arnold, 1897.

Giles, P. "Review of C. W. C. Oman's *History of Greece from the Earliest Times to the Macedonian Conquest*." *English Historical Journal* 5, no. 20 (1890): 770–772.

Green, John Richard. *A Short History of the English People*. London: Macmillan, 1874.

Green, John Richard. *Readings from English History*. Vol. 1. London: Macmillan, 1879.

Kitchin, G. W. *A History of France down to the Year 1453*. Oxford: Clarendon Press, 1873.

Mahaffy, J. P. *Social Life in Greece from Homer to Menander*. London: Macmillan, 1874.

Oman, C. W. C. *History of Greece from the Earliest Times to the Mediterranean Conquest*. London: Rivington, 1890.

Plummer, Charles. "Review of James Hamilton Wylie's *History of England under Henry the Fourth*." *English Historical Review* 1, no. 4 (1886), 786–788.

Powell, Frederick York. *Old Stories from British History*. 3rd ed. London: Longman, 1885.

Prothero, G. W. "Review of S. R. Gardiner's *Constitutional Documents of the Puritan Revolution*." *English Historical Review* 6, no. 22 (1891): 393–399.
[Roscoe, E. S.]. "Memoirs of the Verney Family." *Edinburgh Review*, October 1892, 411–430.
Seeley, John Robert. *A Short History of Napoleon the First*. London: Seeley & Co., 1886.
Stephen, Leslie. *Letters of John Richard Green*. London: Macmillan, 1902.
Stopford Green, Alice. "Introduction." In *A Short History of the English People* by John Richard Green, v–xvi. Revised ed. London: Macmillan, 1888.
Stubbs, William. *Seventeen Lectures on the Study of Medieval and Modern History and Kindred Subjects*. Oxford: Clarendon Press, 1886.
Thompson, Edith. *The Wars of York and Lancaster 1450–1485*. London: David Nutt, 1892.

Secondary Sources

Arroyo Redondo, Susana. "Aproximaciones teóricas al prólogo: su papel en la narrative española reciente." *Revista de Literatura* 76, no. 151 (2014): 57–77.
Baár, Monika. *Historians and Nationalism: East-Central Europe in the Nineteenth Century*. Oxford: Oxford University Press, 2010.
Berger, Stefan. "Professional and Popular Historians 1800—1900—2000." In *Popular History Now and Then: International Perspectives*, edited by Barbara Korte and Sylvia Paletschek, 13–29. Bielefeld: Transcript, 2012.
Brundage, Anthony. *The People's Historian: John Richard Green and the Writing of History in Victorian England*. Westport: Greenwood Press, 1994.
Garritzen, Elise. "Framing and Reframing Meanings in History Books: The Original and Posthumous Paratexts in J. R. Green's *Short History of the English People*." *History of Humanities* 3, no. 1 (2018): 177–197.
Genette, Gérard. *Paratexts: Thresholds of Interpretation*. Cambridge: Cambridge University Press, 1997.
Green, Maureen. "The Look of the Books." In *The History of Oxford University Press*, edited by Simon Eliot, 227–274. Vol. 2. Oxford: Oxford University Press, 2013.
Hesketh, Ian. *The Science of History in Victorian Britain: Making the Past Speak*. London: Pickering & Chatto, 2011.
Himmelfarb, Gertrude. *The New History and the Old*. Cambridge: Cambridge University Press, 1987.
Howsam, Leslie. "Academic Discipline or Literary Genre?: The Establishment of Boundaries in Historical Writing." *Victorian Literature and Culture* 32, no. 2 (2004): 525–545.

Howsam, Leslie. "Growing up with History in the Victorian Periodical Press." In *Popular History Now and Then: International Perspectives*, edited by Barbara Korte and Sylvia Paletschek, 55–71. Bielefeld: Transcript, 2012.

Infelise, Mario. "Libri per tutti." In *Libri per tutti: Generi editoriali di larga circolazione tra antico regime ed età contemporanea*, edited by Lodovico Braida and Mario Infelise, 3–19. Turin: UTET, 2010.

Jackson, H. J. *Those Who Write for Immortality: Romantic Reputations and the Dream of Lasting Fame*. New Haven: Yale University Press, 2015.

Jann, Rosemary. *The Art and Science of Victorian History*. Columbus: Ohio State University Press, 1985.

Korte, Barbara and Sylvia Paletschek. "Popular History Now and Then: An Introduction." In *Popular History Now and Then: International Perspectives*, edited by Barbara Korte and Sylvia Paletschek, 7–11. Bielefeld: Transcript, 2012.

Leduc, Jean. *Ernest Lavisse: L'histoire au cœur*. Malakoff : Armand Colin, 2016.

Lightman, Bernard. *Victorian Popularizers of Science: Designing Nature for New Audience*. Chicago: University of Chicago Press, 2007.

Mandler, Peter. *The English National Character: The History of an Idea from Edmund Burke to Tony Blair*. New Haven: Yale University Press, 2006.

Masson, Pierre. "Marginalité de la préface autoriale." In *L'Art de la préface*, edited by Philippe Forest, 11–23. Nantes: Éditions Cécile Defaut, 2006.

McKitterick, David. *A History of Cambridge University Press*. Vol. 2. Cambridge: Cambridge University Press, 1998.

McKitterick, David. *A History of Cambridge University Press*. Vol. 3. Cambridge: Cambridge University Press, 2004.

Mitchell, Rosemary. *Picturing the Past: English History in Text and Image 1830–1870*. Oxford: Oxford University Press, 2000.

Murray, Jessica. "The Politics of the Preface: Lady Anne Barnard's Gendered Negotiations in a Liminal Textual Space." *English Studies in Africa* 56, no. 2 (2013): 49–59.

Newey, Katherine. "Popular Culture." In *The Cambridge Companion to English Literature 1830–1914*, edited by Joanne Shattock, 147–161. Cambridge: Cambridge University Press, 2010.

Smith, Nadia Clare. *A "Manly Study"? Irish Women Historians, 1868–1949*. Basingstoke: Palgrave Macmillan, 2006.

Walton, Susan. "Charlotte M. Yonge and the 'Historic Harem' of Edward Augustus Freeman." *Journal of Victorian Culture* 11, no. 2 (2006): 226–255.

Weedon, Alexis. *Victorian Publishing: The Economics of Book Production for a Mass Market, 1836–1916*. Farnham: Ashgate, 2003.

CHAPTER 6

Mentors of Scientific History

Historians had innumerable reasons to call Stubbs the personification of a scientific historian. Maitland topped the list with yet another one when he alleged how Stubbs's most valuable gift to English historians was the methodological model that he set out in his books. "No other Englishman has so completely displayed to the world the whole business of the historian from the winning of the raw material to the narrating and generalizing," Maitland explained, and went on to say how Stubbs had invited the readers to his "laboratory" and showed them "the unanalysed stuff, the retorts and test tubes" to allow them to witness "the organic growth of history in an historian's mind." Making the research process visible was a greater service to the discipline than any "abstract discourse on methodology" could have ever been.[1] Books were an indispensable educational arena for historians, but they did not only teach about the past itself. Just as Maitland asserted, they could also be instructive about the inductive method and its application. This chapter takes Maitland's words as its starting point and evaluates how historians relished the pedagogical potential of footnotes to teach readers how historians did their job and what it took to be a proper historian.

Jacqueline Labbe has suggested that footnotes allow writers to bolster their authorial selfhood.[2] I argue that the bottom of the page was ideal both for delivering practical lessons about historians' methodological habits and for performing the persona, a collective selfhood, to

E. Garritzen, *Reimagining the Historian in Victorian England*, https://doi.org/10.1007/978-3-031-28461-8_6

institutionalize scientific history and historians. The pedagogical and performative overlapped in footnotes as historians' educational techniques ranged from giving instructions to teaching by setting an example; that is, by performing specific methodological tasks, skills, or ethical precepts. Scientists' performances are usually geared toward gaining authority and recognition of expertise and they rely on theories, methods, and procedures to inspire confidence in their audiences.[3] Against this background, and considering how footnotes are mostly associated with historical evidence and scholarly digressions, historians' enactment of the persona contained some unexpected elements as they also charged footnotes with lessons about research ethics, scholarly conduct, and collegial respect. Out of the notes emerged a recognizable model of a scientific historian, and footnotes are therefore highly instructive in showing how important the performative dimension was for validating the persona and establishing history as a discipline.

To whom was this performance directed? The audiences varied, as historians dramatized their procedures in different types of histories starting from Freeman's *Old English History for Children* and ending with the five volumes of his *History of the Norman Conquest*. Historians themselves formed an important group of spectators, and in this context the enactment of scholarly procedures and codes of conduct confirmed the author's attendance to the persona and vouched for membership of the scholarly community. The reputation of footnotes as an outward expression of scholarly dryness and excessive erudition limited the market for the performance, and reviewers endorsed the public's aversion to notes with snarky comments on how "The honest traveller resents this constant plucking by the skirt, this incessant jolting off the track."[4] Even some of the first-rank historians such as Ranke and Fustel de Coulanges complained how the constant call to the bottom of the page disturbed the pleasure of the narrative and accepted notes only grudgingly. In England, Green confessed that he had "a sort of *physical* antipathy to notes."[5] Because of this, historians accepted that they were performing their persona to a limited audience. Freeman, when writing *History of Federal Government*, explained that he had composed the narrative to appeal to "any thoughtful reader," and the footnotes to satisfy the need "of the most exacting scholar." If he had "overdone" the notes, he was confident that "every real student will allow that it is a fault on the right side."[6] Footnotes were indeed like a club where scientific historians and initiated readers came together to discuss the technicalities of historical research.

Students were another important target group for the pedagogical and performative footnotes. The idea of books as conduits of methodological and disciplinary lessons was particularly popular in Oxford, where students rarely met the professors and where the set books and texts formed the core of the curricula. Because of this, Oman could assert in his inaugural lecture that students did not need any tutorial guidance in methods:

> If a man cannot pick up the art of weighing and comparing facts and theories from studying … his Maitland and his Stubbs, he will not pick it up from any lectures on method. If he can read all the prescribed books for his special subject without learning how to compare sources and evaluate their worth … then he is not a person about whom we need to bother our heads at all. He will never make a historian, though you drive "method" into him with a hammer.[7]

Oman surely exaggerated, as students read the great masters only sparsely, focusing on Stubbs's oeuvre and mostly consumed instead textbooks and manuals under the guidance of their tutors. But there are also traces of their exposure to Freeman and some other heavyweight historians. Ward, who was an examiner at Cambridge, reported already in 1871 how the students' answers showed that Freeman's "books, great and small, are becoming living influences."[8] What Oman's inaugural also indicated was that not everyone was predisposed to becoming a proper historian. The success of historians' educational pursuits was always conditional because according to a widely shared belief, only those with certain inherent, untrainable abilities could internalize the inductive method. Access to the persona was an exclusive privilege of a selected few who embodied the scientific spirit and innate qualities that constituted a proper historian.

The Bottom of the Page and Resolving Pedagogical Anxieties

A footnote was a versatile and flexible paratext in history books. Acton summed up the notes' most critical function by remarking, "The text contains one's own thought and researches. The notes contain the proofs and illustrations."[9] In similar fashion, footnotes assigned credibility to the historian through the enactment of the method, and Anthony Grafton has observed how notes persuade readers to believe that the historian "has done acceptable amount of work, enough to lie within the tolerance of the field."[10] Furthermore, footnotes stocked up supplementary explanations and details. Freeman reckoned that Herodotus would have been glad "of the chance of Appendix, Notes, Margins, to stow away their digressions."[11] Historians cherished this opportunity and lured readers to visit the notes with the promises of "pretty," "curious," or "romantic" historical sketches awaiting them at the bottom of the page. Referential and explanatory notes dominated in history books, but footnotes also served ethical, social, ideological, aesthetic, symbolic, and pedagogical purposes. The multiple voices which historians adopted in notes attested to the myriad functions of the footnote apparatus. They avoided the satire and irony which had made Gibbon's notes famous and instead appeared as self-consciously scholarly, argumentative, defensive, opinionated, playful, and even confessional and intimate as they shared personal experiences with the readers.[12] They also allowed the joys, disappointments, vexations, and other emotions which research stirred to shine through in the notes. The different discursive registers brought historians closer to the readers, introducing themselves as embodied individuals with opinions, feelings, and tastes. The various adaptations of footnotes encouraged readers to consider the scholarly, cultural, ideological, economic, and social contexts that had shaped the production of the book.

Considering footnotes as an educational space adds new light to their role in nineteenth-century history books. The bottom of the page differed from face-to-face teaching as it created an intimate setting for learning at the junction of the public and private sphere. As Shari Benstock has suggested, footnotes shift the limits of the textual universe, as they occasionally admit readers to the closed space of scholarly activity.[13] Footnotes were often compared to laboratories, workshops, and kitchens where readers were invited to watch how historians worked. Scientific performances are usually located in the public domain, but footnotes confused the conventional line between the front and back stage. This becomes evident when footnotes were collated with kitchens. Acton made this comparison in a review of Creighton's *A History of Papacy* in the *English Historical Review*. Creighton had been frugal with the notes, as he had written the book with an ordinary reader in mind. This irritated Acton, who complained how the "economy of evidence" might please those who longed for the pleasure of reading but did not serve scholars whose "souls are vexed with the insolubility of problems and who get their meals in the kitchen."[14] Acton was unfair: Creighton had added a decent number of references and footnotes with methodological meditations to his book. It is therefore the reference to a kitchen, not Creighton's footnoting habits, that makes Acton's comment important.

The kitchen as a metaphor resurfaced in reviews and other texts that discussed historical research. It had both positive and negative overtones, while food was used mostly to describe incompetent historians or unsound historical knowledge. If Acton considered the kitchen as the place to be for historians and educated readers, publishers saw the matter differently. Macmillan maintained that readers did not want to know how historians prepared their dishes. They cared about the dish itself and its easy digestion: "modern medicine recognizes that the way your food is prepared has a good deal to do, not only with the delight which accompanies its consumption, but also with the facility and efficacy of its digestion." Mental food, he added, did not deviate from this recipe.[15] If a kitchen was an ambivalent space, bad science was plain adulterated food that upset the body.[16] Carlyle, who recycled well-known pieces of history, was accused of serving "twice-cooked cabbage" to readers and Seeley, who filled his books with unprocessed facts, offered dough instead of bread. Oman, who wrote superficial histories on topics on which he was not an expert, was "a sausage machine" and those who contented themselves with printed materials available for everyone produced "the ordinary beverage drawn from the usual wells." Food was so common an allegory for poor scholarship that it was also used for criticizing the Oxbridge examination system, which allegedly "salted, soused, and pickled" the students.[17] The ordinariness of food made these metaphors compelling and the gendering of food management in middle-class families contributed to the allusions of weak scholarship. A kitchen was a more ambiguous metaphor, as illustrious historians were expected to cook their dishes under the prying eyes of their readers. Apart from Acton, historians avoided talking about kitchens and preferred distinctly masculine tropes. They compared research to hunting and battling or to conquering new territories and constructing edifices. Stubbs talked fondly about erecting buildings and superstructures using the materials which antiquaries supplied, and Freeman extended this idea to the bindings of a solid historical work whose cloth covers were "like a wooden roof."[18]

Footnotes as a pedagogical space saved historians from some dreaded aspects of lecturing and the humiliation of an empty or near-empty lecture hall, which was a common destiny of history professors at Oxford. In Victorian England, lectures were choreographed performances and a lecturer's success depended on their stage presence.[19] Many of the historians were painfully aware of this and of their shortcomings as engaging public speakers. The likes of Maitland were simply uninterested in the "full houses and the limelight."[20] Powell, who suffered from poor stage presence, complained how even professorial lectures demanded special skills which made teaching foreign to the scientific ideals he believed in. An inspirational lecturer, he noted bitterly, was not a scholar but a literary artist who excited the emotions of his audience and stirred the imagination of the attendants.[21] The professionalizing scientists had striven to cut their ties to the spectacle of public lectures to enhance their authority, but the audiences continued to expect entertainment and judged scientists' performances accordingly.[22] Because of this culture of sensation, *seeing* scientists and historians perform was an essential part of the experience and, according to one testimony, the Oxford audience was "agog to see the historian of the 'Norman Conquest'."[23] As historians could not enthrall the audience with awe-inspiring scientific experiments, all eyes were on their bodies, voices, and composure. Gardiner was devastated to discover how even some historians judged their colleagues' bodily practices and reported on them. One of these was the Belgian historian Paul Fredericq.

Fredericq toured English and Scottish universities in the early 1880s and published his observations in *De l'enseignement superieur de l'histoire en Eccose et en Angleterre*. His report reached the English-speaking audience when Johns Hopkins University published a translation of it in 1887. Fredericq assessed both the content of the teaching and the charisma of the teachers, and Gardiner apparently

> spoke simply and clearly, without attempt at eloquence, reminding me of a German professor, the resemblance being heightened, no doubt, by the somewhat Teutonic cast of the professor's features and his expression of learned candor and almost anxious good will. From time to time he readjusted his eyeglasses and polished his nose with his large colored handkerchief, like Droysen of Berlin.[24]

The comparison to an acclaimed German professor did not appease Gardiner, who was humiliated by the remarks about his handkerchief and fumbling with spectacles.[25] It was as if he had failed in the manly duty of commanding his own body.

Those historians, who were either performers against their will or discouraged by the poor attendance of their lectures, discovered in books an alternative channel for fulfilling their pedagogical obligations. Stubbs wisely advised his successor Freeman not to fret about lecturing and instead "to lead the way by writing books."[26] This idea of books as an extension of and even a surrogate for a classroom gave footnotes an important pedagogical role in demonstrating how historians did their work and what kind of persona a historian needed to cultivate to attain reliable knowledge.

Introducing the Persona to Children

One of the most remarkable performances of the persona in footnotes was in Freeman's *Old English History for Children*, which Macmillan published in 1869. The book answered Alexander Macmillan's aspiration to expose children to the foundation of historical research:

> It has long been an idea of mine that children from a very early age should be used to the idea of evidence: that the person who writes the book, or tells them the story was not present; but heard or read of it from some authentic source. I confess I don't see how this is to be introduced without being cumbersome. But it would be a great thing if it could.[27]

When Freeman approached Macmillan with his idea about a children's history, this was exactly what he envisioned achieving. He drew inspiration from the lessons he had given to his own children and wanted to offer a model of both scientifically accurate history and a historian to other children as well.[28]

Old English History for Children is an important landmark for the persona project because Freeman invested the book with a detailed demonstration of the persona. As Macmillan allowed him to furnish it with copious footnotes, he was able to show the young readers how historians used original authorities to produce truthful knowledge. This kind of ambitious footnote apparatus was extraordinary in a children's history book. Textbooks were mostly issued either with no footnotes at all or with a sporadic note here and there, as publishers wanted to ensure that the books did not appear too demanding for the aspired audiences. This no-note policy did not please everyone. The *Nottinghamshire Guardian*, referring to Longman's series "Epochs of Modern History," complained how "Foot-notes have been avoided—unnecessarily, we think—on the ground that 'they tend to interrupt the reader's interest in the narrative'."[29] Historians were not entirely happy with the situation either and tried to retain their scientific standards by listing their most important sources in a preface. Green and Freeman advocated an "intermediate way" with lists of key sources at the head of each chapter, as readers had a "right to ask whence the information comes," Freeman explained.[30] This practice had its critics as well, as readers were unable to infer which of the listed sources was used for verifying specific statements in the text.[31]

Freeman's particular goal was to teach children to distinguish good and bad history and he outlined his mission statement in the preface by explaining how his purpose was to

> show that clear, accurate, and scientific views of history … may be easily given to children from the very first. In truth the more rigidly accurate and scientific a statement is, the more easy it is for a child to take it in. The difficulty does not lie with the child, who has simply to learn, but with the teacher who often has to unlearn.[32]

He also specified the learning targets in the preface: to teach scientific exactitude, guide children to distinguish truthful history from legends, and to introduce historical authorities to them. He advanced these goals side by side in the text and footnotes. The notes imparted an ideal image of a historian who worked with sources to tease information out of them. He described the common characteristics of historical sources whenever he and his readers encountered them in the text and even informed children about the unique epistemic quality of historical knowledge. Freeman wanted children to understand that historians strove for facts and acquired them by piecing together different types of sources whose reliability varied. How the historian handled the sources distinguished good and bad history. In other words, children learned what constituted a historian, historical research, and historical knowledge. What they did not learn about were the disagreements between historians, as Freeman carefully concealed the controversies and competing interpretations from the young audience. Moreover, he chose a style that was suitable for addressing children, as he portrayed himself as a guide or a teacher repeating phrases like "you should remember" or "you should take care" to steer the children toward a scientific approach to history. His tone was chatty and he made his message digestible with concrete examples. Altogether, he created a nurturing atmosphere and sought to establish a confidential relationship with the children.

A theme that ran through the book was the difference between historical facts, stories, and legends. The reckless and uncritical use of stories and anecdotes in textbooks exasperated the scientific historians. Freeman wrote angrily to George Macmillan in 1889 how publishers, even Macmillan, still released histories with "childish romances" and old "hideous" myths whose authenticity historians had disputed.[33] Historians admitted the pedagogical value of stories but advocated a more conscientious method for introducing them to readers. When Louise Creighton published *Stories from English History* (1883), she did exactly what historians were expected to do to retain the dignity of small histories: she disconnected herself from unscientific conventions in the preface. Stories, she admitted, helped to enliven the past and awaken "an intelligent interest" in the young audience, but she had excluded "imaginary conversations" and "fanciful illustrations." All the stories were "either extracted from Chronicles or woven together from well-authenticated facts."[34] This technique made the use of stories acceptable for educational history and framed the book and its author as arbiters of the new, scientific educational history writing.

As Freeman included legends and stories in *Old English History for Children*, he too emphasized how they could have a place in histories only if they were treated critically. He separated the stories from his narrative with headings in Gothic type and prefixed them with methodological soundings both in the text and the footnotes as Fig. 6.1 shows. He told, for example, how "The Story of King Edwin" contained miraculous incidents and material that formed "an actual part of history," and showed how he had evaluated its truthfulness to reach a verdict that there was no reason to question its overall veracity even though some parts of it were doubtful.[35]

The footnotes were packed with talk about authorities to ensure that children understood that historical facts were derived from the sources, not from the historians themselves. Freeman wrote about the common characteristics of his sources and told how they could contain later additions, elements from different stories, and adaptations to different geographical locations. He made this comprehensible to children by demonstrating how one northern story had circulated in two versions and

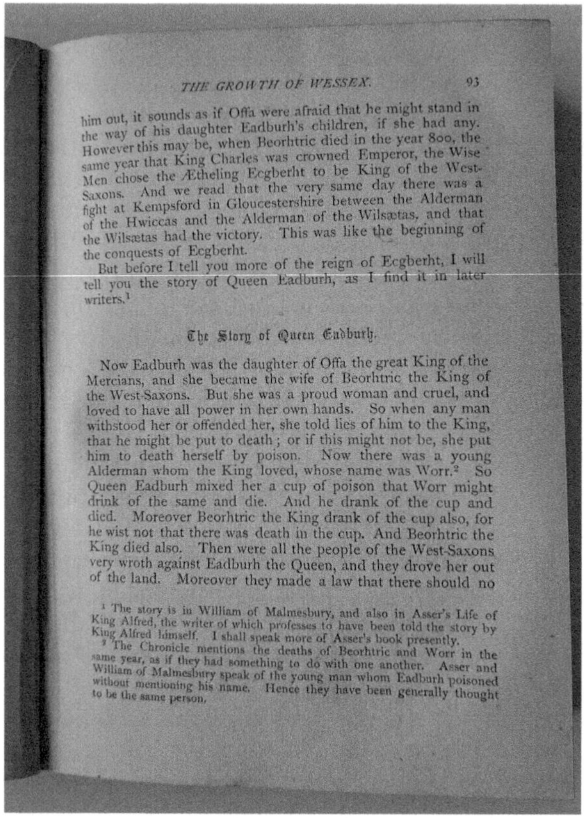

THE GROWTH OF WESSEX. 93

him out, it sounds as if Offa were afraid that he might stand in the way of his daughter Eadburh's children, if she had any. However this may be, when Beorhtric died in the year 800, the same year that King Charles was crowned Emperor, the Wise Men chose the Ætheling Ecgberht to be King of the West-Saxons. And we read that the very same day there was a fight at Kempsford in Gloucestershire between the Alderman of the Hwiccas and the Alderman of the Wilsætas, and that the Wilsætas had the victory. This was like the beginning of the conquests of Ecgberht.

But before I tell you more of the reign of Ecgberht, I will tell you the story of Queen Eadburh, as I find it in later writers.[1]

The Story of Queen Eadburh.

Now Eadburh was the daughter of Offa the great King of the Mercians, and she became the wife of Beorhtric the King of the West-Saxons. But she was a proud woman and cruel, and loved to have all power in her own hands. So when any man withstood her or offended her, she told lies of him to the King, that he might be put to death; or if this might not be, she put him to death herself by poison. Now there was a young Alderman whom the King loved, whose name was Worr.[2] So Queen Eadburh mixed her a cup of poison that Worr might drink of the same and die. And he drank of the cup and died. Moreover Beorhtric the King drank of the cup also, for he wist not that there was death in the cup. And Beorhtric the King died also. Then were all the people of the West-Saxons very wroth against Eadburh the Queen, and they drove her out of the land. Moreover they made a law that there should no

[1] The story is in William of Malmesbury, and also in Asser's Life of King Alfred, the writer of which professes to have been told the story by King Alfred himself. I shall speak more of Asser's book presently.

[2] The Chronicle mentions the deaths of Beorhtric and Worr in the same year, as if they had something to do with one another. Asser and William of Malmesbury speak of the young man whom Eadburh poisoned without mentioning his name. Hence they have been generally thought to be the same person.

Fig. 6.1 Freeman separated stories and historical facts with headings in Gothic type and with methodological observations in the adjoining footnotes (*Source* Author's collection)

how "One makes the man a merchant of York, the other a Thane called Beorn; one calls the King Ælla, the other Osberht; one makes him bring in the sons of Ragnar Lodbrog, the other makes him bring in Guthorm or Guthrum."[36] Furthermore, he tutored the children about research practices as he showed how to compare different authorities, piece together information from multiple sources, and apply historical knowledge to test their reliability. "The Story of Saint Kenelm the Little King" gave him an opportunity to exemplify these operations in practice as he explained why it was “a very unlikely story,” regardless of the fact that it was “hard to see what should have made anybody invent such a tale, if nothing of the kind had ever happened.” He analyzed the story against the knowledge he had about old English customs, which enabled him to identify the contradictory and unlikely elements in it.[37]

A proper historian, then, had to possess critical faculties and extensive knowledge about the past, but the footnotes also highlighted the cultivation of other epistemic virtues such as precision and accuracy. Freeman, renowned for his obsession with correct dates and historical spelling of names, introduced his pet theme immediately in the preface and in an ensuing essay “A Few Words on Old-English Words and Names.” He explicated how the meaning of geographical concepts such as “France” and “England” were historically contingent and therefore “Without perfect accuracy in these matters, no clear view of history can ever be gained.”[38] Throughout the footnotes, he kept reminding the young audience to remain attentive to the historicity of different concepts and place names. Altogether, Freeman’s footnotes reproduced an image of a historian that corresponded with the scholarly persona he embraced. Unsurprisingly, this historian championed the inductive method, was knowledgeable about history and historical authorities, and was accurate, precise, and persistent. This was the model which the prospective historians should emulate.

But the persona which Freeman advocated contained one unexpected feature: the ability to endure uncertainties. Indeed, one of the most important lessons which Freeman imparted was that historians were not unerring authorities since they had to rely on incomplete and inconsistent authorities. One example of this was the story about King Edwin. Freeman admitted that he was unable to draw any definite conclusions about its reliability as he lacked comparative evidence. Because of this, he passed the burden of judgment to his readers, telling how

> each man must settle for himself whether the marvellous part of the tale was a real miracle, or a dream, or a mere remarkable coincidence, or the misconception or invention of some one afterwards.[39]

This was a surprising shift of interpretive responsibility considering the average age of his readers. Freeman endorsed his message about inconclusive evidence by not overstepping the boundary between probability and fact. He cautiously talked about possibilities, probabilities, and events that *perhaps* had happened to remind his audience—children, parents, governesses—that sometimes historians could only draw inferences or make informed guesses. Attaining the truth was the moral goal of historians, but often they had to accept that the goal was unattainable.

Freeman succeeded in convincing at least one young reader with his method-talk, as he received fan mail from Miss A. V. Ponsonby of ten years of age. She urged him to write a sequel as, "It would be so much clearer, truer, and more interesting than what other people write."[40] The experiment also provoked approval among his friends. Green was impressed by Freeman's boldness of introducing to children "the whole criticism of authorities, etc." which constituted "the real originality and value of the book." There was no way of returning "to the old *Ipse dixi*" in children's books after this.[41] Yet Freeman failed to persuade those who doubted children's ability to comprehend the historical method or to process abstract information about source criticism. Green reported how one Oxford Fellow had described *Old English History for Children* as "a charming child's book, for children of twenty-four."[42] This was also what the *Athenaeum* had to say about it. The reviewer protested that the philosophizing, discussions about the unreliability of sources, and complex demonstrations of unfounded historical myths showed an overestimation of the "capacity of the youthful mind." Accordingly, "An abridgement, with all the hard unnecessary names, two-thirds of the discussions, and many of the minor facts expunged, would be more to the purpose."[43] There was too much originality in the book for this writer who longed for the conventional children's histories that concentrated on stirring imagination and serving Alfred's cakes, while keeping the door to the historian's kitchen shut.

Despite the criticism, *Old English History for Children* turned out to be one of Freeman's most successful books. It went through one revised edition and was then stereotyped and reprinted at regular intervals. This indicates that multiple generations of children became acquainted with the persona through Freeman's educational use of footnotes. The book was also exceptional in its scope and its undisguised attempt to proselytize the young readers. Freeman did not repeat the exercise in his later educational histories, nor did he inspire others to experiment with a similar endeavor. An obvious challenge was the publishers' no-note policy in educational histories. When textbooks contained notes, they were sparse and lacked the chattiness which was a critical component in Freeman's footnote pedagogy. A good example of this is Ella S. Armitage's *The*

Childhood of the English Nation, or the Beginnings of English History, which Longman published in 1877.

Armitage was one of the first five students at Newnham College in Cambridge, and the bibliography which she appended to *The Childhood of the English Nation* betrays her vast reading of original authorities and historical studies, ranging from Kemble, Palgrave, and Stubbs to foreign historians such as Michelet, Guizot, Michaud, Pauli, Schmid, Gneist, and Brentanno. Her style suggests that the book was targeted at an audience that was older than Freeman's, although she told in the preface that it was aimed at those who had no knowledge about English history as she hoped to induce "the unlearned" to study the "larger works."[44] She used footnotes thriftily and explained in the preface how copious references were undesirable in such a small book. Most of her notes were references or cross-references. The referential notes showed to the readers in a matter-of-fact style what kind of proofs historians used, and the occasional methodological remarks offered a brief peek into a historian's workshop. She followed Freeman's example as she emphasized the historicity of geographical names and admitted the uncertainty of historical knowledge.[45] The sparseness of the notes, however, meant that they gave only a meager idea of historical research and its practitioners.

A Model Persona in the Big Histories

Footnoting flourished in the "big" histories and notes were considered a visible testimony of their scientific character. Edwin Guest maintained in the 1860s how the use of footnotes had changed since the days when Palgrave had gotten away with scanty references. Now readers expected a historian "to show his voucher for every statement." It was very well so, as it brought history closer to truth, he added.[46] Notes and the constant allusions to sources gave history books a recognizable identity, and careful scrutiny of the footnote habits of Bryce, Burrows, Creighton, Freeman, Gardiner, Norgate, and Stubbs shows the interdependence between the primary sources, inductive method, and a historian who treats the material critically to acquire historical facts. They all used footnotes to perform a ritualized set of research procedures, just as Freeman did in *Old English History for Children*. However, the substantial footnote apparatus in the big histories opened possibilities for expanding the enactment of the persona from methods and practices to research ethics, conduct, and social relations in the scholarly community. These themes added depth to the persona, but an attentive reader could detect discrepancies between historians' words and deeds, between the staged virtuosity and reality.

If anything, footnotes showed that historians had a multidimensional relationship with their sources. This relationship was intellectual, epistemic, cognitive, passionate, and even fervently religious, as they worshipped original authorities in footnotes. The incessant talk about sources, however, recounted only a partial story of a document's journey from an archive to a finished book. Historians made no noise about the quest for manuscripts. Exhilarating adventure stories about archival research or the joys of discovery were found only in antiquarian or amateur histories. Gardiner's *The Fall of the Monarchy of Charles I.* (1882) offered a rare glimpse of what historians might encounter in archives, as Gardiner told how historians also needed luck and kindness from others next to impeccable research skills. He had wanted to consult the Act Books that were stored "in a room over the porch of the parish church at Colchester" which were under the charge of the registrar, who had fallen ill. Gardiner thanked "the Rev. Sir J. Hawkins, Bart., and F. T. Veley, Esq., for their kind assistance in helping me to see these books at a time when the illness of the late registrar made it difficult for me to procure access to them in the ordinary way."[47] This was an exceptional footnote, and historians seemed to trust that they did not need stories about dusty and damp archives or insupportable archive bureaucracy to prove their diligence, perseverance, and manly fortitude. Cataloging the heaps of manuscripts they had consulted in archives near and far should have been enough to convince the reader of their industriousness and thoroughness.

While historians were cursory about archival research, they were more verbose when they showed how to interpret geographical and physical traces of history. These notes invited readers to accompany them to exciting foreign locations and battlefields in Britain. According to one footnote in Bryce's *The Holy Roman Empire* (1866), he had visited six churches in Rome, Pisa, and Milan and now applied his observations to support the assertion that many churches in Italy carried only slight Gothic features.[48] Another note took the audience to two palaces in Würtemberg and Kaiserslautern which attested to the greatness of the Hohenstaufen family. As Bryce explained in the same note how the ruins could be reached by train and which were the closest stations, the note gained an air of an erudite Baedeker.[49] This was not incidental: footnotes could resemble a travelogue. Freeman was known for his enthusiasm for medieval architecture and the topographical method, and his notes mixed historical evidence with recollections of his journeys with Green in France and with his brother-in-law, the archeologist Arthur Evans, in Sicily. He told with great pride in one footnote how he and Evans had discovered a tomb whose whereabouts only the locals had known and which had not been noticed by any "traveller" before them.[50] The lesson of such notes was that historians were nothing like the frivolous tourists who followed a trodden path from one site to another. A historian, by contrast, always remained attentive and on the lookout for physical traces of the past. Freeman emphasized this uninterrupted dedication by ridiculing tourists who did not comprehend that "*travelling* is work."[51] Not everyone, however, mixed references to physical sources with travel accounts. Gardiner's take was purely methodological. He toured historical battlefields with his beloved tricycle but restricted the delight which these visits caused to his private correspondence.[52] Instead, he assumed a scholarly tone in the notes as he taught how the physical evidence complemented written testimonies and how the exploration of landscapes made the course of historical battles more comprehensible.[53]

If historians were haphazard about how they gathered their sources, the talk about authorities began in earnest when they submitted the material to their critical gaze. Just like Freeman in *Old English History for Children*, they took readers by the hand and taught them to organize and weigh evidence and to train skills that were pivotal to making conscientious judgments. As English historians drew more heavily on printed sources than many of their continental colleagues, they rarely talked about external source criticism or auxiliary sciences in their notes. Gardiner bore almost the sole responsibility for educating the public to authenticate documents. He demonstrated how to assess external qualities such as handwriting, and explained how important it was to ask whether the style of the document was appropriate to the office of its writer and whether its tone was in line with other documents from the same source.[54] More common advice which historians gave to their readers was encouragement to observe the tones, registers, moods, sentiments, and other hidden meanings in the records. These were just as important clues as the names, dates, and descriptions of events. Freeman exemplified this when he wrote about Rodulf of Ivry and the peasant revolt in 996. He cited William of Jumièges in the note and appended the quote with a recommendation to "Mark the brutal levity with which Rudolph's cruelties are dismissed."[55]

What footnotes then taught to readers was not how to track down, consult, and collect source material, but how to treat the gathered material. Footnotes invited those readers who cared to read them to witness how historians turned their stuff into facts, and the facts into a narrative. As readers turned the pages, they encountered a steady flow of scholarly parlance about the methodological procedures that ensured the acquisition of reliable historical knowledge. This pedagogical and scholarly performance of scientific rituals institutionalized the skills and virtues that a scientific historian needed. But this was only one act. Footnotes were intriguing as a performative space because they also conclusively showed the limitations of the scientific method in achieving firm historical facts. If the aim of a performance is, according to Goffman, to offer an audience a carefully curated end-product which conceals the disorganized production process, footnotes created an entirely opposite impression disclosing the messiness of historical research.[56]

Just like Freeman in *Old English History*, other historians, too, were frank about the incompleteness and inconsistency of historical records. They informed readers about gaps and contradictions, the one-sidedness of survived testimonies, unexplored documents and collections that might one day complement or alter the accepted truths, and about the risk that some gaps could never be closed. This invoked a bleak image of history as a scientific discipline and of historians' ability to produce firm facts. All this required patience and toleration of uncertainty from the historians and could have created a tension in the persona which was grounded in the ideal of completeness. But historians consented that this stoic acceptance of the epistemologically unstable character of history was imprinted in the persona alongside the virtues of completeness and definitiveness. Gardiner admitted that sometimes he simply had to wait until the collections in Simancas had been ransacked further before he could try to solve more historical mysteries and Norgate, struggling with inconsistent chronicles, allowed that she could only evaluate the probabilities of the different alternatives.[57] Historical knowledge was unfixed, probable, and vulnerable to misuse and speculation. Historians handled conjectures and inferences, and the lesson to learn was that it was a sign of mental fortitude and self-restraint to shrink from attempting to establish truths beyond all reasonable doubt when the material did not extend to such certitude.

Historians developed an intriguing pedagogical strategy for teaching about the implications the inconsistent records had for historical knowledge. They used open-ended questions to steer readers' cognitive processes in two important ways. First, the questions directed attention toward a desired viewpoint or interpretation. Second, they highlighted a historian's active role in interpreting the documents. There were no right nor wrong answers, only reasoned and justified probabilities. The open-ended questions encouraged a subjective reading of history and, by assigning the interpretive responsibility to the reader, historians declined the authority which they claimed elsewhere in their books. They seemed to be unaware of this effect which the open-ended questions produced.

Gardiner, who was complemented as an educational writer for his ability to make younger readers use their cognitive faculties, similarly encouraged more mature audiences to train their historical thinking with the questions he posed to them. For example, he inserted multiple queries into the footnotes which he tagged to the account of the events that led to the poisoning of Thomas Overbury in 1613. He recorded the developments in the text and provided in the notes the perplexing details that arose from the letters, reports, and other materials at his disposal. He punctured the confusing array of facts and counter-facts with a series of open-ended questions that induced readers to evaluate—even speculate on—the motives and intentions of the main protagonists: the King, Rochester, Northampton, Helwys, and Lady Essex.[58] This turned the footnote apparatus into an interactive space where historians and readers tried together to solve historical riddles when the sources failed to guide them on the journey toward historical truth.

Imperfections in the Persona

Historians rehearsed certain footnote habits that contradicted the image of a flawless historian and ran counter to their pedagogical use as they unmasked carelessness, lack of attention to detail, and non-scholarly commitments. The referential footnotes tested historians' dedication to precision. Footnoting was guided by two principles: notes should give enough information so that readers could easily trace the sources and the text should correspond to the sources. These were both critical, as the reliability of historical knowledge rested on a relationship between the text and an absent source which the concise reference represented. A fair and ethically sound handling of sources bestowed trustworthiness on historical knowledge, and therefore historians' footnotes advanced collectively the integrity of the discipline. But historians struggled to live up to these demands. While they obsessively established exact dates, chronologies, and nomenclature, they were unexpectedly negligent about shelf-marks and bibliographical details. When pedantry would have been desirable, they evaded it.

Lack of any accepted rules for formulating footnotes was an obvious challenge for everyone engaged in scholarly publishing. For example, there were no commonly agreed standards for documenting sources in late-Victorian scientific journals.[59] The editors of the *English Historical Review* seemed to welcome all possible configurations. There were, though, signs of a growing demand for unified standards toward the end of the century. Victor Langlois and Charles Seignobos wrote briefly about footnotes in the popular *Introduction to the Study of History*, stating how a reference should contain the title of the document and its shelf-mark and cautioning that "the apparatus of demonstration, while needing to be complete, ought to be reduced to what is strictly necessary." Excessiveness in footnoting was a sign of "bad taste, a kind of naïve vanity, sometimes of mental confusion."[60] This did not yet say much about how to compose a footnote. A slightly more serious attempt to streamline references came from the University of Chicago Press, which published its first *Manual of Style* in 1906. The manual allocated three pages for footnotes, specifying which bibliographical information was necessary when referring to different types of publications.[61] The instructions were generic, but the appearance of the style guide was an indication of a growing pressure for a more consistent style in academic publishing.

The late-Victorian historians, then, were not yet confined by any detailed rules on how to refer to different types of sources from manuscripts and coins to monographs, anonymously published articles in periodicals, and signed articles in learned journals. Their footnotes were a hodgepodge of styles, as the level of detail and uniformity varied not just from one book to another, but within a book from one page to another. Historians standardized the abbreviations of the regularly consulted collections and archives, but this covered only a fraction of their materials and they invented abbreviations as they went. As bibliographies were almost non-existent, they listed the non-standardized condensations in the prefaces. This meant that prefaces could expand to unwieldy proportions and become cumbersome reading for the uninitiated. Gardiner told in one preface that *Venice MSS* referred to the documents he had consulted in Venice while *Venetian Transcripts* meant the "admirable collection sent to the Public Record Office by Mr. Rawdon Brown." Prothero explained that the references to "*Rish. Chron.* are to the Chronicle of Rishanger, edited by Mr. Riley for the Master of the Rolls; those to *Rish. de Bellis, &c.* are to the other Chronicle attributed to the same author, edited by Mr. Halliwell for the Camden Society."[62] This put the footnotes and prefatorial discourse into a dialogue and the deciphering of footnotes could demand constant shifting between the notes and the preface—to readers' great annoyance.

Historians were more consistent and detailed with the references to primary material than when they recorded bibliographical information about secondary literature. The references to articles followed the general rules of anonymity, although historians began to identify the writers of signed articles with a delay after first getting used to the idea that the authority of a signed article derived from its writer, not from the journal. Footnotes recorded authors and titles, but the shortening of titles complicated the identification of sources. When it came to volumes and editions, they were less cautious. It was Doble from Oxford University Press who had to remind Freeman to specify which "edition of Stubbs" he was using or otherwise readers could not verify the quotations.[63]

Historians themselves grew impatient with obscure footnotes during the last quarter of the century. Stubbs, the paragon of precision, was perfunctory with footnotes. This frustrated Alice Stopford Green, who tried to trace "*Robertson's Historical Essays*" which Stubbs mentioned in a footnote without any additional bibliographical details.[64] Firth was similarly annoyed by James Hosmer's slapdash notes in *The Life of Young Sir Henry Vane*. The American professor's references such as "Green's *Calendar*" were useless, as Mary Anne Everett Green had edited dozens of Calendars. Even more absurd was the note "From a letter to Cromwell, in 1656, from Carisbrook Castle, bound up with the 'Healing Question' in an old volume in the British Museum." Firth humiliated Hosmer in the *English Historical Review* by offering the exact shelf-mark for the letter.[65] It was easy to laugh at amateur historians' ludicrous references such as "Mr. Green's account" or "an interesting essay by Lord Lytton," but the same vices from applauded historians became an embarrassment to the discipline as they contradicted the virtue of precision associated with the scientific historian.[66]

Another virtue which the footnotes contested was detachment. As the next chapter shows, impartiality was a cardinal virtue that required careful monitoring and restraint of passions and personal bias. Footnotes, however, were received as a liminal space between the narrative and the world beyond the book and historians took liberties to invest them with political, religious, and cultural comments. Assuming the competing voices of a detached expert, politically committed activist, or concerned citizen did not seem to make historians fear that the ideologically and emotionally informed notes would erode trust in their neutrality. For readers, these notes gave an unexpected commentary on the text, as they were fixed to a specific segment in it but simultaneously oriented both at it and away from it, mediating between the past and the present. The personal reflections, remarks, and emotional confessions made historians appear embodied individuals, and the footnotes with non-scholarly opinions or personal experiences opened a window onto their private side. In the age of emerging celebrity journalism, such notes satisfied at least part of the curiosity readers had about historians as living and breathing individuals.

Notes of this kind were certainly problematic in terms of historians' presumed impartiality. Burrows, for example, talked strongly against the "mischievous" campaign for the separation of State and Church. Freeman fulminated in footnotes about how careless restoration projects damaged the authentic character of medieval churches and how some wanton barbarians had allowed this mutilation of historical buildings.[67] His son-in-law Arthur Evans followed suit when he edited the fourth volume of Freeman's *The History of Sicily* after Freeman's death, and added to it a footnote to complain how economic interests endangered the survival of historical sights. In Taormina, he wrote in one footnote, "an English proprietress" had gained the ownership of important buildings and had since then "barred the access and warned off the civilized portion of mankind in four languages."[68] The modality of the notes and the text could differ significantly, and while historians condemned partiality in the narrative, footnotes enjoyed a different status. Occasional personal remarks were not considered harmful for the integrity or reliability of their persona. The next chapter shows that this was an overly optimistic assumption and applicable only to such notes that agreed with historians' own ideological positions.

An Ethical, Fair, and Polite Persona

Historians used footnotes to perform the morality of the persona as they advocated the use of ethically sound references and due regard to the accomplishments of earlier writers. Moral uprightness and civil manners were closely monitored because their infringements in individual histories eroded the dignity of the entire discipline. Footnotes, then, ideally attested to honesty, transparency, respect, and politeness, and since these were recognized moral habits of the cultured classes, the footnote pedagogy was geared to train the character as well. This gave additional sway to historians' persona performance at the bottom of the page.

Footnotes were susceptible to manipulation and their correct handling was of the utmost importance. Misuse was tempting, for readers rarely collated the text and its sources and the amassed documents persuaded them to accept the facts without considering how the plain references said nothing about the words, intentions, and spirit of the sources. Hallam had written already in 1818 to condemn the dishonest use of sources as "a glaring violation of historical integrity" that tended to render "the use of references, that great improvement of modern history, a sort of fraud upon the reader."[69] Next to misrepresenting the sources, it was common to pile up unnecessary sources. Mary Bateson censored this by explaining how the frequent references to manuscripts did not render a text original if the references were there only as embellishments to well-digested facts.[70] This was alarmingly common practice in the *Dictionary of National Biography* that was serving as a training ground for young historians. The *English Historical Review* drew attention to this anonymously by professing how some writers had multiplied their "references without due regard to their authority or importance." Closer scrutiny had revealed how "half of them repeat one another without adding to the original source from which all alike have drawn."[71] This was certainly unwelcome in a publication that catered for the acculturation of fresh talents.

Historians were expected to know and credit fairly all the relevant research. Uncredited "borrowing" was not tolerated.[72] Historians did their best to recognize their colleagues amidst the startlingly rapidly appearing studies in Britain and abroad. They were by rule generous with their references and acknowledgments and found in this respect the publishers' no-note policy problematic. Dean Milman explained in the preface to the new edition of his *The History of the Jews* (1863) how it had originally been issued only with a few explanatory footnotes due to its "popular form" and "limited scale." This publishing strategy, Milman lamented, had disposed him to the charges of "claiming for his own" the thoughts of others and of presuming the originality of views "which have long been maintained by accredited authors."[73] To Milman's relief, Murray had allowed the revised edition to appear with more complete footnote apparatus, but his words show how difficult it could be to reconcile historians' scholarly needs and publishers' commercial concerns. Historians were willing to make allowances for the ethical norms about references in educational histories to get them published, but acknowledging the scholarly debts only in brief prefaces hardly compensated the lack of more precise recognition of earlier scholars in footnotes. The extent of the exemplarity of historians' performance of the persona, then, varied according to the format of their books.

Footnotes were also informative about polite criticism. It was commonly agreed that controversies belonged to footnotes and appendices, but historians also advocated a "fair and generous" treatment of earlier historians.[74] Controversialists like Round were exceptions. Historians found the offensive footnotes of more distant writers bemusing. James Robertson reviewed the Vatican archivist Heinrich Denifle's *Die Universitäten des Mittelalters* and described with seeming delight how his "pages of text are fringed with the scalps of distinguished scholars."[75] But the need to criticize closer colleagues caused discomfort and historians drew on stock phrases to soften the blow. Expression like "I cannot understand," "I cannot see," or "I cannot infer" were popular, as they made the purpose clear while suggesting that the disagreement may be caused by the limitations of the annotator's own comprehension. The camaraderie and smallness of the scholarly community spurred historians to emphasize humility and indebtedness. The amateur historian Henry Howorth described this ethos in his customary lively style in *History of the Mongols*:

> If we see further than those who went before, it is because we are raised higher from the ground by their efforts, we in fact stand on their shoulders ... To throw stones, to cast jibes at them for their mistakes, is surely very likely parricide ... [We] are but poor creatures if we cannot gauge their work, the vast mass of new matter they brought together, without a perpetual snarl at their small mistakes, or as perpetual cackle over our own superior wisdom.[76]

The idea of standing on the shoulders of earlier giants was not new. Newton had been said to use the same metaphor as an admission of his humility, but Steven Shapin has questioned this interpretation and argued instead that Newton used it to place himself in the tradition of exceptional men.[77] Historians were not ignorant of this possibility; as we have seen, by acknowledging Stubbs in prefaces and footnotes, they aligned themselves with the heroic historian to boost their own authority.

There were three notable exceptions to the polite criticism. First, historians did not forgive methodological and ethical violations, as these spoiled the reputation of their persona. Norgate was ginger in her criticism when she disagreed with historians' explanations, but when she detected that someone did not "adduce his proofs," she made it plain that she had no obligation to believe his interpretations.[78] Second, historians erased criticism altogether when it could have undermined the unified front of the scholarly community. Joseph Bensman writes how acknowledging, citing, and excluding certain names in footnotes institute social boundaries and how this can encourage the manipulation of notes as names are deliberately omitted and unduly recognized.[79] Green admitted that he was going to leave out all the references to Freeman whenever he disagreed with him and explained to Freeman that he did this because "it takes away from one's notes that air of controversy and personal conflict which is so odious in itself and so likely to hinder the just consideration of historical facts."[80] So, Green assured that he had not ignored his friend intentionally, but out of consideration for what was best for the discipline.

Third, it was acceptable, even desirable, to censor certain names that represented rival personae. Froude was the most obvious target, but Freeman and Stubbs also singled out Francis Palgrave as a useful prop for their persona performance in footnotes. John Forster, for his part, fell victim to Gardiner's scholarly hostility. Forster is today better known as a literary critic, but he was also an amateur historian and biographer with a keen interest in the Stuart period. Gardiner disputed many of Forster's views but balanced his criticism in prefaces with apologies and acknowledgments as he thanked Forster for his kindness and for loaning for him the copies of John Eliot's notes even though Forster had been aware "that my view of many matters differed from his considerably."[81] In the footnotes, Gardiner was less courteous. He used historical records and his comprehensive knowledge of the seventeenth century to disclose Forster's many blunders. Here and there he even faltered in his civility as he talked about Forster's "imaginative narrative" or accused him of telling things that had only happened "in his own imagination."[82]

Footnotes were also a sensitive paratext for historians because the notes contained features that disclosed their networks, alliances, and status in the scholarly community. References attested to respect, recognition, and peer approval and it is not surprising that historians scanned footnotes to discover their names. Arthur Gissing parodied this in *New Grub Street*, as he made the slighted old erudite Mr. Yule's entire composure alter when he discovered that the young Mr. Hinks had praised his learning in one footnote. Yule read, eyes glinting, the "effusive eulogy … of 'Mr Alfred Yule's critical acumen, scholarly research, lucid style,' and sundry other distinguished merits." When Mrs. Yule suggested that this might benefit her husband, he replied with an admission that "If Hinks goes on, he'll establish my reputation."[83] Whereas Gissing made satire out of scholars' obsession for the acknowledgments, historians did not take the matter that lightly. They were severely insulted if they thought that they had not been credited fairly.[84] Freeman went so far as to send a letter to the German historian Wilhelm Ihne to complain how Ihne had slighted him unjustly in his notes. Ihne replied with an apology that was entangled with self-defense:

> I was struck with your remark that I had not quoted your Sulla. I am sure I summed up my highly appreciative opinion of your essay in a long footnote, but on looking over the book I cannot find it. I cannot think how it slipped out & am very sorry for it. However, I find a mention of your Sulla on page 374N.1. corresponding to the note in the German edition p. 394. In this edition p. 454 I have given my opinion of your & Jochaninde's works in the following terms … This note I meant to enlarge in the English Edition & I feel convinced I did it; but the papers on which it was written, must have slipped out somewhere.[85]

Freeman complained privately about the collegial misconduct, but historians' hurt pride also spilled from the private to the public sphere. This happened when Canon Dixon discovered that three Catholic historians, Francis Aidan Casquet, T. E. Bridgett, and Mr. Hendrike, had apparently acknowledged his work inadequately. He wrote an embittered review for the *English Historical Review* to teach the writers a lesson about "literary courtesy" and charged them of either unduly using his work without mentioning his name, ignoring him altogether, or referring to others when his name should have been invoked. This was not just insulting but had deeper consequences, as the writers had caused with their recklessness what Dixon called "secondary injustice"; reviewers who were unaware of Dixon's accomplishments did not ascribe originality to him but to those who had copied him without due acknowledgment.[86] Footnotes as a performance of ethical uprightness and polite manners enhanced the firm link between conduct, knowledge production, and scholarly sociability. As honesty and politeness belonged to the moral toolkit of a Victorian gentleman, courteous and morally sound footnotes also helped historians to win approbation for their social status and respectability.

Non-Teachable Virtues and the Sacred Band of Scientific Historians

If historians invested footnotes with lessons about methods, research procedures, skills, virtues, and ethical norms that constituted the persona, they nonetheless held an elitist conviction that some qualities and virtues fundamental to the persona were inherent and unteachable. In other words, not everyone had what it took to be a proper historian. According to Herman Paul, virtues can be learned through imitation and practice, but they are often regarded as character traits embedded in a historian's self.[87] This idea was widely accepted during the late-Victorian era, as Francis Galton's studies on hereditary genius gave a scientific explanation for the unequal distribution of intellectual abilities. Galton argued that inborn qualities could not be acquired through training, practice, or a tenacious exercise of will and pointed out that only by combining these abilities with zeal and intense intellectual engagement could individuals become true geniuses.[88] Although many of the leading historians avoided talking about genius, they agreed with the premise of inborn qualities. The persona was not available for everyone and its exclusiveness was based both on individuals' inherent abilities and on the "natural" qualities associated with specific social groups. This was a protest against the intensifying debates about democracy and equal rights and catered to historians' claims for a unique social position. The persona had an elitist overtone.

Oman was blunt about these ideas in his inaugural lecture. While he defended the liberal education ideal, he affiliated himself with the born-to-be historians:

> In short, the true historian … is born and not made. If he has the root of the matter in him, he gets precisely such a preliminary education from his schools as will enable him to work for himself when his schools are over … In short, it seems to me that zeal, insatiable curiosity, a ready mind to shape hypotheses, a sound judgment to test them, above all a dogged determination to work at all times and in all places, are the real requisites of the historian rather than any array of technical training … If he is worth his salt, he will teach himself 'method' in very short time. Nothing astonishes me more than the way in which the real born historian learns to get to the heart of the matter within a year or so of starting the work.[89]

Froude was an obvious and easy victim of this biologization of the persona, as the scientific community diagnosed him as suffering from an inherent condition that prevented him from telling the historical truth.[90] But this was not all. A renegade persona seemed to run through the hereditary lineage from Thomas Babington Macaulay to George Macaulay Trevelyan as well. John Laughton detected so much stylistic resemblance between George Otto Trevelyan and his famous uncle that he discerned "an inheritance." He got further confirmation of his discovery several decades later when he noticed how Macaulay's style had filtered through G. O. Trevelyan to his son George Macaulay Trevelyan and how the family suffered from a "hereditary protest against 'the dignity of history'."[91] For the supporters of scientific history, this particular type of family legacy was a burden because it predisposed them to a flawed persona.

Natural inborn qualities extended from individuals to social groups. In the case of nationality and ethnicity, this did not necessarily mean scholarly deficiency, but a limited comprehension of certain historical phenomena. Accordingly, historians alleged that foreigners could not fully comprehend English history. French historians, according to Freeman, were "unable to understand any English matter" and one obituary of Reinhold Pauli excused the little interest the English had shown in his histories by the absence of "the insular atmosphere which is so necessary to an Englishman's view of his own country" in Pauli's histories.[92] Even Ranke, Green argued, was unable to escape the cognitive limitations which his German nationality imposed on him.[93] This kind of inherent knowledge was curious because it was transferable in favorable conditions. Americans, Norwegians, and the Swiss were well equipped to conduct research on federal institutions because they had "daily experience of their working," according to Freeman.[94] W. A. B. Coolidge was articulate on this principle when reviewing Bernard Moses's *The Federal Government in Switzerland*. Moses was professor at the University of California and this gave Coolidge reason to pronounce how

> Mr. Moses has some special advantages for writing a book on Swiss matters. His title page tells us that he is professor of history and political economy in the university of California. He is thus as an American resident, and probably an American citizen, well versed in the actual working of federal institutions.[95]

A persona was a conglomerate of acquired and innate qualities and skills, and historians formed an exclusive community of men (and women) who possessed the necessary natural traits. But this did not mean that historians' pedagogical efforts in footnotes were without meaning; even those who could never become proper historians could use the lessons to train their intellectual and moral habits so that they were better prepared for the public duties that were more suitable for their nature and to distinguish good and bad history.

*

Footnotes were a many-sided paratext serving epistemic, methodological, ethical, social, pedagogical, and performative functions. For the scientific historians they were one of the most important spaces in a history book and Maitland wrote to Paul Vinogradoff, after reading his footnotes, how "they are enough to show me that this is a great book, destined in course of time to run the current of English and German learning."[96] The intense talk about the methods and sources created an impression of history as a distinctive discipline which demanded special skills and qualities from its practitioners. The notes, however, stored contradictions and paradoxes which revealed to a careful reader the uncertain epistemic grounds of historical knowledge and the evasiveness of historians' virtuosity. Historians, seemingly oblivious of the inconsistencies in their performance, sought to appropriate the bottom of the page to establish their persona as *the* model to emulate.

Notes

1. Maitland, "William Stubbs," 419.
2. Labbe, "Transplanted into More Congenial Soil," 72–73.
3. Vandendriessche, Peeters, and Wils, "Introduction," 2–4.

4. "Mr. Freeman's Norman Conquest," *PMG*, March 2, 1872, 851.
5. Grafton, *Footnote*, 70–71; Green to Freeman, March 2, 1867, in Stephen, *Letters*, 179.
6. Freeman, *History of the Federal Government*, xii.
7. Oman, *Inaugural*, 22–23.
8. Ward to Freeman, June 30, 1871, FA 1/7/778–801, JRL.
9. Acton to Simpson, March 3, 1863, in Altholtz, McElrath, and Holland, *Correspondence*, 3:90.
10. Grafton, *Footnote*, 22.
11. Freeman to Green, January 23, 1876, FA 1/8/31–60, JRL.
12. Of Gibbon's footnotes, see Palmeri, "Satiric Footnotes," 249–259.
13. Benstock, "At the Margin," 204.
14. Acton, "Review of M. Creighton," 571–572.
15. Macmillan to John Addington Symonds, January 24, 1871, in Novell-Smith, *Letters*, 136–137.
16. Secord, *Victorian Sensation*, 449.
17. [Dasent], "*Early Kings*," 232; Hesketh, *Victorian Jesus*, 161; Soffer, *Discipline and Power*, 112; "History of England," *Athenaeum*, May 21, 1887, 665; Fowler, *Oxford Correspondence*, 38.
18. Freeman to Thompson, January 10, 1892, U DX9/195, HHC.
19. Gooday, "Ethnicity, Expertise," 19–20, 24–27.
20. Maitland to A. W. Verrall, February 14, 1903, in Fifoot, *Letters*, 272.
21. Powell to Stopford Green, January 23, 1895, in Elton, *Frederick York Powell*, 1:181; Eliot and Stray, "History, Law," 567–568.
22. Morus, "Worlds of Wonder," 808–814.
23. "Ladies Column," *Evening Telegraph*, December 25, 1885, 5; "History of Cathedral Church of Wells," *BCWZ*, November 25, 1869, 8.
24. Fredericq, *Study of History in England*, 49.
25. Gardiner to Browning, October 17, 1885, GBR/0272/OB/619/A, King's Cam.
26. Stubbs to Freeman, February 20, 1884, FA ¼/60–88, JRL.
27. Macmillan to Wright, August 19, 1862, in Macmillan, *Letters*, 119.
28. Freeman to Macmillan, December 1, 1865, Add MSS 55049, BL.
29. "Epochs of History," *Nottinghamshire Guardian*, October 6, 1876, 6.
30. Freeman to Hunt, June 24, 1885, in Stephens, *Life and Letters*, 2:337.
31. Ropes, "Review of Arthur Parnell," 182.
32. Freeman, *OEH*, v.
33. Freeman to George Macmillan, [1889], Add MSS 55053, BL.
34. Creighton, *Stories*, preface.
35. Freeman, *OEH*, 50–51.
36. Freeman, *OEH*, 108.
37. Freeman, *OEH*, 87.
38. Freeman, *OEH*, vi.

39. Freeman, *OEH*, 50–51.
40. Ponsonby to Freeman, October 18, 1872, in Stephens, *Life and Letters*, 2:62.
41. Green to Freeman, 1869, in Stephen, *Letters*, 237.
42. Green to Freeman, November 1869, in Stephen, *Letters*, 235.
43. "Old English History for Children," *Athenaeum*, January 1, 1870, 21–22.
44. Armitage, *Childhood*, v.
45. Armitage, *Childhood*, vi, 80, 84, 94–95, 97, 111, 118, 123.
46. Guest to Freeman, March 2, [1867], FA 1/7/288, JRL.
47. Gardiner, *Fall of the Monarchy*, 287.
48. Bryce, *Holy Roman Empire*, 321–322.
49. Bryce, *Holy Roman Empire*, 1866, 182–183.
50. Freeman, *History of Sicily*, 3:402.
51. Freeman to Macmillan, September 5, 1886, Add MSS 55053, BL.
52. Gardiner to Browning, October 17, 1885, and July 25, 1886, GBR/0272/OB/619/A, King's Cam.
53. Gardiner, *Fall of the Monarchy*, 419; Gardiner, *History of the Great Civil War*, 1:161, 1:200, 1:246.
54. Gardiner, *Prince Charles*, 116, 342; Gardiner, *Fall of the Monarchy*, 402–404.
55. Freeman, *NC*, ed. 2, 1:257.
56. Goffman, *Presentation of Self*, 52.
57. Gardiner, *History of England from the Accession of James I. to the Disgrace*, 23; Norgate, *Angevin Kings*, 1:158, 1:214, 1:279.
58. Gardiner, *History of England*, 2:178–185.
59. Meadows, *Victorian Scientist*, 94–95.
60. Langlois and Seignobos, *Introduction to the Study of History*, 305–306.
61. *Manual of Style*, 71–73.
62. Gardiner, *History of England*, 3:v–vi; Prothero, *Life of Simon de Montfort*, ix.
63. Doble to Freeman, December 24, 1889, Letter Books 49, OUP.
64. Stopford Green to Macmillan, February 6, [1882], Add MSS 55095, BL.
65. Hosmer, *Life of Young Sir Henry Vane*, 298; Firth, "Review of J. K. Hosmer," 165.
66. "Review of G. D. Boyle," *EHR*, 1890, 810.
67. Burrows, *Constitutional Progress*, 124–125; Freeman, *History of the Cathedral*, 184, 189; Freeman, *History of Sicily*, 1:47.
68. Freeman, *History of Sicily*, 4:110.
69. Hallam, *View of the State*, 19–20.
70. Bateson, "Review of Henry Worseley," 769–770.
71. "Review of *The Dictionary*," *EHR*, 1890, 788.
72. Stubbs to Freeman, March 13, s.a. FA 1/3/21–41, JRL; Lord Carlingford to Freeman, October 11, 1868, FA 1/7/53–86, JRL.

73. Milman, *History of the Jews*, iii.
74. Gardner, "Review of Adolf Holm," 374; Freeman to Hook, December 4, 1867, in Stephens, *Life and Letters*, 1:395; Freeman to Max Müller, June 2, 1870, FA 1/8/109–126, JRL.
75. [Robertson], "Mediaeval Universities," 446.
76. Howorth, *History of the Mongols*, 1:xv.
77. Shapin, *Scientific Life*, 32–33.
78. Norgate, *Angevin Kings*, 2:482.
79. Bensman, "Aesthetics and Politics of Footnoting," 449–452.
80. Green to Freeman, December 21, 1876, in Stephen, *Letters*, 443.
81. Gardiner, *History of England under the Duke of Buckingham*, xi–xii; Gardiner, *Fall of the Monarchy*, xi.
82. Gardiner, *History of England under the Duke of Buckingham*, 18–19, 68, 168, 227, 355; Gardiner, *Fall of the Monarchy*, 89; Gardiner, *History of England*, 10:71.
83. Gissing, *New Grub Street*, 121–122.
84. Stubbs to Freeman, October 14, [1864], MS. Eng. Misc. e. 148, Bodleian.
85. Ihne to Freeman, October 3, 1882, FA 1/7/389–423, JRL.
86. Dixon, "Review of T. E. Bridgett," 776–777.
87. Paul, "What Is a Scholarly Persona," 358–359.
88. Browne, "Inspiration to Perspiration," 78; Dahlberg, "Gifts of Nature," 182–189.
89. Oman, *Inaugural*, 23–25.
90. Langlois and Seignobos, *Introduction to the Study of History*, 125.
91. [Laughton], "Trevelyan's *Early History of Fox*," 540, and "Peasants' Rising of 1381," 77.
92. Freeman, *Reign of William Rufus*, 175; "Dr. Reinhold Pauli," *Athenaeum*, June 10, 1882, 733; Cornish, "Review of Moritz Brosch," 801; Oman, "Review of Fritz Hoenig," 571–573.
93. Green to Freeman, September 30, 1878, in Stephen, *Letters*, 474–476.
94. Freeman to Ihne, November 16, 1882, in Stephens, *Life and Letters*, 2:262–263; Freeman, *Methods*, 288–294.
95. Coolidge, "Review of Bernard Moses," 799.
96. Maitland to Vinogradoff, June 12, 1887, in Fifoot, *Letters*, 32.

References

Unpublished Primary Sources

Bodleian Library, Oxford: Stubbs Papers.
British Library: The Macmillan Papers.

Hull History Centre: Letters from Edward Augustus Freeman to Edith Thompson.
John Rylands Library, Manchester: E. A. Freeman Archive.
King's College Archive Centre, Cambridge: Oscar Browning Papers.
Oxford University Press Archive: Letter Books.

Printed Primary Sources

Acton, Lord. "Review of M. Creighton's *A History of the Papacy during the Period of the Reformation.*" *English Historical Review* 2, no. 7 (1887): 571–581.
Altholz, Josef L., Damian McElrath, and James C. Holland. *The Correspondence of Lord Acton and Richard Simpson*. Vol. 3. Cambridge: Cambridge University Press, 1975.
[Anon.]. "History of the Cathedral Church of Wells, By Mr E. A. Freeman." *Bath Chronicle and Weekly Gazette*, November 25, 1869, 8.
[Anon.]. "Old English History for Children." *Athenaeum*, January 1, 1870, 21–22.
[Anon.]. "Mr. Freeman's Norman Conquest—Vol. IV." *Pall Mall Gazette*, March 2, 1872, 851–852.
[Anon.]. "Epochs of History." *Nottinghamshire Guardian*, October 6, 1876, 6.
[Anon.]. "Dr. Reinhold Pauli." *Athenaeum*, June 10, 1882, 733.
[Anon.]. "Ladies Column. Notes in Passing." *Evening Telegraph*, December 25, 1885, 5.
[Anon.]. "History of England in the Eighteenth Century." *Athenaeum*, May 21, 1887, 665–666.
[Anon]. "Review of *The Dictionary of National Biography*." *English Historical Review* 5, no. 20 (1890): 783–788.
[Anon.]. "Review of G. D. Boyle's *Characters and Episodes of the Great Rebellions*." *English Historical Review* 5, no. 20 (1890): 809–810.
Armitage, Ella S. *The Childhood of the English Nation, or the Beginnings of English History*. London: Longman, 1877.
Bateson, Mary. "Review of Henry Worseley's *The Dawn of the English Reformation*." *English Historical Review* 6, no. 24 (1891): 769–770.
Bryce, James. *The Holy Roman Empire*. New ed. London: Macmillan, 1866.
Burrows, Montagu. *Constitutional Progress*. London: John Murray, 1869.
Coolidge, W. A. B. "Review of Bernard Moses's *The Federal Government of Switzerland*." *English Historical Review* 5, no. 20 (1890): 799–800.
Cornish, F. W. "Review of Moritz Brosch's *Oliver Cromwell und die puritanische Revolution*." *English Historical Review* 2, no. 8 (1887): 800–804.
Creighton, Louise. *Stories from English History*. London: Rivington, 1883.

[Dasent, G. W.]. "*Early Kings of Norway.*" *Edinburgh Review*, July 1875, 203–232.

Dixon, Richard Watson. "Review of T. E. Bridgett's *Life of Blessed John Fisher.*" *English Historical Review* 4, no. 16 (1889): 775–777.

Elton, Oliver. *Frederick York Powell. A Life and a Selection from His Letters and Occasional Writings*. Vol. 1. Oxford: Clarendon Press, 1906.

Fifoot, C. H. S. *The Letters of Frederic William Maitland*. London: Selden Society, 1965.

Fowler, W. Warde. *An Oxford Correspondence of 1903*. Oxford: Blackwell, 1904.

Firth, C. H. "Review of J. K. Hosmer's *The Life of Young Sir Henry Vane.*" *English Historical Review* 5, no. 17 (1890): 164–165.

Fredericq, Paul. *The Study of History in England and Scotland*. Baltimore: Johns Hopkins University, 1887.

Freeman, Edward A. *History of Federal Government from the Foundation of the Achaian League to the Disruption of the United States*. London: Macmillan, 1863.

Freeman, Edward A. *Old English History for Children*. London: Macmillan, 1869.

Freeman, Edward A. *History of the Cathedral Church of Wells as Illustrating the History of the Cathedral Churches of the Old Foundation*. London: Macmillan, 1870.

Freeman, Edward A. *The History of the Norman Conquest of England, Its Causes and Results*. 2nd ed. Vol. 1. Oxford: Clarendon Press, 1870.

Freeman, Edward A. *The Reign of William Rufus and the Accession of Henry the First*. Vol. 1. Oxford: Clarendon Press, 1882.

Freeman, Edward A. *The Methods of Historical Study*. London: Macmillan, 1886.

Freeman, Edward A. *The History of Sicily from the Earliest Times*. Vol. 1. Oxford: Clarendon Press, 1891.

Freeman, Edward A. *The History of Sicily from the Earliest Times*. Vol. 3. Oxford: Clarendon Press, 1892.

Freeman, Edward A. *The History of Sicily from the Earliest Times*. Edited by Arthur J. Evans. Vol. 4. Oxford: Clarendon Press, 1894.

Gardiner, Samuel Rawson. *History of England from the Accession of James I. to the Disgrace of Chief-Justice Coke*. Vol. 1. London: Hurst and Blackett, 1863.

Gardiner, Samuel Rawson. *Prince Charles and the Spanish Marriage. 1617–1623*. Vol. 1. London: Hurst and Blackett, 1869.

Gardiner, Samuel Rawson. *A History of England Under the Duke of Buckingham and Charles I. 1624–1628*. Vol. 1. London: Longman, 1875.

Gardiner, Samuel Rawson. *The Fall of the Monarchy of Charles I. 1637–1649*. Vol. 1. London: Longman, 1882.

Gardiner, Samuel Rawson. *History of England from the Accession of James I. to the Outbreak of the Civil War. 1603–1642*. Vol. 2. London: Longman, 1883.

Gardiner, Samuel Rawson. *History of England from the Accession of James I. to the Outbreak of the Civil War. 1603–1642*. Vol. 3. London: Longman, 1883.
Gardiner, Samuel Rawson. *History of England from the Accession of James I. to the Outbreak of the Civil War. 1603–1642*. Vol. 10. London: Longman, 1884.
Gardiner, Samuel Rawson. *History of the Great Civil War. 1642–1649*. Vol. 1. London: Longman, 1886.
Gardner, Alice. "Review of Adolf Holm's *Geschichte Grechenlands von ihrem Ursprung bis zum Untergange des Selbständigkeit des griechischen Volkes*." *English Historical Review* 6, no. 22 (1891): 374–375.
Gissing, George. *New Grub Street*. London: Penguin Books, 1985.
Hallam, Henry. *View of the State of Europe during the Middle Ages*. Vol. 2. London: John Murray, 1818.
Hosmer, James K. *The Life of Young Sir Henry Vane*. Boston and New York: Houghton, Mifflin and Company, 1888.
Howorth, Henry H. *History of the Mongols from the 9th to the 19th Century*. Vol. 1. London: Longman, 1876.
Langlois, Ch. V. and Ch. Seignobos. *Introduction to the Study of History*. New York: Henry Holt, 1898.
[Laughton, John Knox]. "Trevelyan's *Early History of Fox*." *Edinburgh Review*, October 1880, 540–577.
[Laughton, John Knox]. "The Peasants' Rising of 1381." *Edinburgh Review*, January 1900, 76–105.
Macmillan, George A. *Letters of Alexander Macmillan*. Glasgow: printed for private circulation, 1908.
Maitland, F. W. "William Stubbs, Bishop of Oxford." *English Historical Review* 16, no. 63 (1901): 417–426.
Manual of Style. Chicago: University of Chicago Press, 1906.
Milman, Henry Hart. *The History of the Jews. From the Earliest Period Down to Modern Times*. 3rd ed. Revised. Vol. 1. London: John Murray, 1863.
Norgate, Kate. *England under the Angevin Kings*. Vol. 1. London: Macmillan, 1887.
Norgate, Kate. *England under the Angevin Kings*. Vol. 2. London: Macmillan, 1887.
Novell-Smith, Simon. *Letters to Macmillan*. London: Macmillan, 1967.
Oman C. "Review of Fritz Hoenig's *Oliver Cromwell*." *English Historical Review* 4, no. 15 (1889): 571–573.
Oman, Charles. *Inaugural Lecture on the Study of History*. Oxford: Clarendon Press, 1906.
Prothero, George Walter. *The Life of Simon de Montfort Earl of Leicester with Special Reference to the Parliamentary History of His Time*. London: Longman,1877.

[Robertson, James C.]. "Mediaeval Universities." *Quarterly Review*, April 1896, 445–472.

Ropes, Arthur. "Review of Arthur Parnell's *The War of Succession in Spain during the Reign of Queen Anne*." *English Historical Review* 4, no. 13 (1889): 179–183.

Stephen, Leslie. *Letters of John Richard Green*. London: Macmillan, 1902.

Stephens, W. R. W. *The Life and Letters of Edward A. Freeman D.C.L., LL.D.* Vol. 1. London: Macmillan, 1895.

Stephens, W. R. W. *The Life and Letters of Edward A. Freeman D.C.L., LL.D.* Vol. 2. London: Macmillan, 1895.

Secondary Sources

Bensman, Joseph. "The Aesthetics and Politics of Footnoting." *International Journal of Politics, Culture, and Society* 1, no. 3 (1988): 443–470.

Benstock, Shari. "At the Margin of Discourse: Footnotes in the Fictional Text." *PMLA* 92, no. 2 (1983): 204–225.

Browne, Janet. "Inspiration to Perspiration: Francis Galton's *Hereditary Genius* in Victorian Context." In *Genealogies of Genius*, edited by Joyce E. Chaplin and Darrin M. McMahon, 77–95. Basingstoke: Palgrave Macmillan, 2016.

Dahlberg, Julia. "Gifts of Nature? Inborn Personal Qualities and Their Relation to Personae." In *Gender, Embodiment, and the History of the Scholarly Persona: Incarnations and Contestations*, edited by Kirsti Niskanen and Michael J. Barany, 181–214. Cham: Palgrave Macmillan, 2021.

Eliot, Simon and Christopher Stray. "History, Law, and Literature." In *The History of Oxford University Press*, edited by Simon Eliot, 559–599. Vol. 2. Oxford: Oxford University Press, 2013.

Goffman, Erving. *The Presentation of Self in Everyday Life*. London: Penguin Books, 1990.

Gooday, Graeme. "Ethnicity, Expertise and Authority: The Cases of Lewis Howard Latimer, William Preece and John Tyndall." In *Scientists' Expertise as Performance*, edited by Joris Vandendriessche, Evert Peeters, and Kaat Wils, 15–29. London: Routledge, 2020.

Grafton, Anthony. *The Footnote: A Curious History*. Kent: Faber and Faber, 2003.

Hesketh, Ian. *Victorian Jesus: J.R. Seeley, Religion, and the Cultural Significance of Anonymity*. Toronto: University of Toronto Press, 2017.

Labbe, Jacqueline "'Transplanted into More Congenial Soil': Footnoting the Self in the Poetry of Charlotte Smith." In *Ma(r)king the Text: The Presentation of Meaning on the Literary Page*, edited by Joe Bray, Miriam Handley, and Anne C. Henry, 71–86. Aldershot: Ashgate, 2000.

Meadows, Jack. *The Victorian Scientist: The Growth of a Profession*. London: The British Library, 2004.

Morus, Iwan Rhys. "Worlds of Wonder: Sensation and the Victorian Scientific Performance." *Isis* 101, no. 4 (2010): 806–816.
Palmeri, Frank. "The Satiric Footnotes of Swift and Gibbon." *The Eighteenth Century* 31, no. 3 (1990): 245–262.
Paul, Herman. "What Is a Scholarly Persona? Ten Theses on Virtues, Skills, and Desires." *History and Theory* 53 (2014): 348–371.
Secord, James A. *Victorian Sensation: The Extraordinary Publication, Reception, and Secret Authorship of* Vestiges of the Natural History of Creation. Chicago: University of Chicago Press, 2000.
Shapin, Steven. *The Scientific Life: A Moral History of Late Modern Vocation.* Chicago: University of Chicago Press, 2010.
Soffer, Reba N. *Discipline and Power: The University, History, and the Making of an English Elite, 1870–1930.* Stanford: Stanford University Press, 1994.
Vandendriessche, Joris, Evert Peeters, and Kaat Wils. "Introduction: Performing Expertise." In *Scientists' Expertise as Performance*, edited by Joris Vandendriessche, Evert Peeters, and Kaat Wils, 1–13. London: Routledge, 2020.

CHAPTER 7

From Public Intellectuals to Radicalized Historians

John Laughton saw a need to defend history as a useful subject in 1883 because many, even those "charged with the conduct of affairs," thought that history was a subject "which ought to be confined to girls' boarding-schools." He declared this to be untrue, and instead professed that history was "the proper study of the politician and statesman."[1] Some historians have argued that history began to lose its political role in England in the 1880s and that the professionalization of the discipline enhanced this development.[2] This might be so with the upcoming generations in academia, but otherwise it is hard to detect any strong signs of historians' declining confidence in the usefulness of history for current political debates—despite Laughton's pessimism. The public was not willing to abandon the idea of history as a guide to the present and future, and that is why so many were disappointed with what they saw as the lack of instructiveness in the bulky histories. As we have seen, historians ignored such complaints and, as Seeley phrased it, they conceived history as a scientific and a practical endeavor that profited both historians and society at large.[3] It was critical for historians to stress the benefits of history because it confirmed their cultural authority and legitimized their work. This made them reluctant to substitute their status of an engaged intellectual with that of a recluse ivory-tower scholar. The conviction that history served immediate political purposes was inscribed in the paratexts.

E. Garritzen, *Reimagining the Historian in Victorian England*,
https://doi.org/10.1007/978-3-031-28461-8_7

Negotiating the postures of a dispassionate scholar, public intellectual, and radical propagandist was a difficult exercise for the scientific historians whose scholarly persona was grounded on a disinterested commitment to facts and self-restraint. Any charges of partisanship could have undermined the integrity of the persona and endangered the usefulness of history as resting on both truthfulness and impartiality. Historians' public engagements were accepted with the condition that the past was not employed for raising support for any particular party but for explaining the roots of the current political questions. To uphold their image of impartiality, the scientific historians distanced themselves from the dogmatism of Macaulay and Carlyle and did their best to contest any claims about their own biases. This was not easy, for history enjoyed a reputation of rousing prejudices, generating controversies, and serving as a political and religious weapon. Cambridge University Press had declined the proposal to become the publisher for the *English Historical Review* precisely because history, according to the Syndicates, "was connected to politics, and the Press ought to be neutral."[4]

Despite the lofty proclamations of impartiality, historians continued to overstep the line between immediate scholarly and political considerations. As Rosemary Jann has stressed, the break between a scientific and a Romantic historian was not in this sense as dramatic as historians imagined it to be.[5] The Reform Acts of 1867 and 1884, Home Rule agitation, the Eastern Question, jingoism and anti-imperialism, the dwindling economy, plus the rise of radical socialism and labor unions had direct bearing on the uses of history, and historians' political, religious, and social beliefs were imprinted in their publications. It was common knowledge that Seeley and Burrows were advocates of the Empire, Acton a liberal Catholic, Green a liberal social reformist, and Freeman a Gladstonian liberal who famously declared that history is past politics and politics is present history. Stubbs was identified as a Conservative and Gardiner as a Liberal, though they escaped the allegations of biases. *John Bull* congratulated Stubbs for his rare "freedom" of political prejudices and Gardiner was bracketed as "not a party man." Obviously, such images of neutrality were myths.[6] The rival political allegiances shaped historians' relations as well. Home rule drove Seeley to take distance from Browning and to write with resentment to his friend how his

> favourite notion of making politics a matter of teaching seems to me to suffer a humiliating *reductio ad absurdum*, when two men who united in

advocating it are led by their historical studies to adopt views of politics so extremely opposite.[7]

Historians, then, trod a fine line between their scholarly and political obligations. Stefan Collini, Stuart Jones, and many others have discussed historians as public intellectuals and portrayed Freeman and Seeley as the embodiments of a late-Victorian historian who assumed the role of public commentator. According to Collini, public intellectuals used historical knowledge to offer a perspective on issues with societal relevance, presented themselves as historians, addressed the general audience, not just academic peers, and both needed and had access to a non-expert media. The public intellectuals gained authority from the audience who recognized their expertise and cultural authority.[8] This chapter, however, shifts the gaze from the iconic figures to the edges of the scholarly community where historians assumed more radical stances, cultivating personae that contested that of a scientific historian and a mainstream public intellectual. The ideologically driven personae defy easy categorization as one persona shaded off into another, but whereas the Freemans and Seeleys openly promoted political ideas in historical essays, lectures, and journalism and at least tried to keep their advocacy out of their scholarly works, the markedly radical historians also invested monographs and document editions with political advocacy.

Two radical historians discussed in this chapter, Mary Hickson and Alice Stopford Green, represent a strand of Irish women historians who mixed political engagement and historical research.[9] The rising tide of cultural and political nationalism in Europe and the growing tensions between London and Dublin formed the backdrop to their historical pursuits. Their scholarly and political views differed, but their goal was the same: to use history to change the course of present-day politics. Hickson, who was inspired by political nationalism, advocated a unionist reading of history. She was convinced of her impartiality and restricted her political participation mostly to the histories which she published. Stopford Green was the younger of the two and belonged to the post-famine generation and the Protestant establishment. Her thinking was shaped by cultural revivalism. She contributed to history with a nationalist bent, rejected neutrality altogether, and blended the personae of a public intellectual and a political activist. They were both bracketed as partisans and gained enough public attention that the scientific historians could not afford to dismiss them as "harmless." Instead, the scholarly establishment mobilized the *English Historical Review* and leading periodicals and quarterlies to expose how Hickson and Stopford Green manipulated the inductive method and historical sources, to ensure that the two would not be falsely counted as representatives of the scientific history. The public reproof allowed the scientific historians to protect the integrity of their persona, but the exposure of the vices of the contending persona also permitted them to underscore the cluster of virtues that warranted the desired detachment of a scientific historian.

For Hickson and Stopford Green, paratexts were a sphere where they fabricated an ideological context for their books and linked their texts to the ongoing historical and political debates. As Joep Leerssen has observed about paratexts in nineteenth-century Irish historical novels and histories, "these frayed edges of the text" engaged in significant ways with their "social and ideological environment."[10] Just like footnotes, other paratexts too, could be liminal spaces separate enough from the narrative to accommodate ideas that were not immediately linked to the historical content of the texts. At the same time, paratexts and texts were closely enough connected so that the non-historical interpretive clues which Hickson and Stopford Green planted in their prefaces, footnotes, titles, and running heads guided readers to draw appropriate political lessons from their books. The politicization of their paratextual sphere attested to the fine line between historians' scholarly and partisan goals and the accompanying persona.

While unraveling the links between history, politics, paratexts, and personae, this chapter highlights women as authors of politically charged histories. Political history was, according to Seeley, the only kind of history that could have practical utility. When done correctly, he explained, it was free from ornamentation and those chapters about people and their habits, literature, and science that impaired the standard national histories. In the Rankean fashion, he argued that history ought to be a "biography of states."[11] This view gained support in the *Athenaeum* where Lecky was accused of spoiling honest political history with sketches of dress, amusements, theater, art, gambling, dueling, and cock fighting. "History will never be regarded with the serious attention which is due to it unless the several branches of history are carefully separated from each other, and political history occupies a position of pre-eminence," the writer declared.[12] Meanwhile, women's historical pursuits were associated precisely with the lively descriptions of habits, material culture, and daily life. Andrew Lang mocked women's fascination for social history by talking about "our young women" who "peruse medieval washing-bills."[13] However, just as women subverted the gender restrictions and created opportunities to engage with the public sphere, they also invented ways to interact with the political master narratives. Mary Spongberg has argued that early-Victorian women, regardless the topics of their histories, reacted to the current political debates through their publications.[14] The late-Victorian women continued to write histories with public relevance—also in the masculinist sphere of political history—but their experiences were by no means identical as a result of their gender. Whereas Hickson chose political topics to push her agenda, Stopford Green preferred social history for advocating her ideas.

From Impartial Knowledge to Political Propaganda

Impartiality was so critical for the truthfulness of historical knowledge that most of the historians went to great lengths to defend their detachment and to specify what impartiality meant in the context of historical research. They offered multiple definitions for it and women, too, were eager to pitch in. Mary Robinson discussed detachment in the *English Historical Review* and came to question whether it was at all a possible or even a desirable quality in a historian. She was unable to comprehend how a historian could be neutral. "The great actions and questions of the world are not neutral. Life is not neutral, but vari-colored, multiform, and disconcerting." As the past was "dead beyond resuscitation," no "freedom from prejudice" could reanimate it, and therefore it was unwise to demand these "difficult qualities" from a historian. This, however, did not mean that a historian should become a "party pamphleteer." No, a historian ought to be "an honest student of facts, with such and such principles, convictions, and limitations."[15] As impartiality and the methodological strictures that it entailed were evoked so frequently when Hickson's and Stopford Green's radicalism was condemned, it is best to start with a brief overview of the meanings which historians ascribed to impartiality and of the procedures they developed to curb their biases.

If one thing was sure, historians rarely believed naively in objectivity, unlike Hayden White and a host of other postmodern theorists have argued.[16] On the contrary, they admitted the elusiveness of impartiality and developed epistemic and cognitive procedures to keep their biases in check. Moreover, they hardly ever used the term "objectivity." Instead, they talked about impartiality, detachment, and neutrality, or about biases, prejudices, and partisanship to denote qualities that compromised truthfulness. The popular late-Victorian understanding of impartiality resembled that which Ranke had formulated and his disciples had developed further. Objectivity was reserved in historians' lexicons for the epistemic goals that were restricted to the production of facts. "Impartiality," by contrast, was a more comprehensive idea. It had emerged as a concept in the seventeenth century and since then had been a coveted yet unattainable goal that merged epistemic and moral qualities.[17] Historians concurred with the dual character and emphasized how it was their moral obligation to recognize their own biases and rise above the political, religious, social, and economic divisions in Victorian society.[18] Detachment alone, however, did not make anyone a great historian. Mary Bateson reminded her colleagues that impartiality was arguably an important quality, but "does not make up for an absence of learning—the impartiality that comes of ignorance is worthless."[19] Only when the different epistemic and cognitive actions were merged could historians claim to produce truthful knowledge.

Historians cherished an idea that the impartiality they cultivated distinguished history from the other disciplines. Lorraine Daston and Peter Galison have argued how a mechanical form of objectivity emerged in the sciences in the mid-nineteenth century and identified the willful self of a scientist as a major obstacle to knowledge. Mechanical objectivity, too, fused epistemology and morality, as its supporters advocated the suppression of self and the disciplining of will as ways to resist the impulse toward personal intervention.[20] Neutrality was for scientists—just as it was for historians—a cultural force that advanced their status as experts, and the insistence on a science free from political and metaphysical shackles strengthened their authority. However, mechanical objectivity does not adequately describe what historians understood by impartiality. Seeley argued that impartiality had different meanings in history and science. In science, prejudices disqualified the practitioners from ascertaining truths, while in history impartiality was a virtue, but "it is impartiality in a secondary and very modified sense."[21] To appreciate Seeley's vision of historians' impartiality, it is worth quoting at length from his essay "History and Politics" from 1879:

> It is the impartiality of one who can acknowledge faults in his own side, and admire the virtues of an antagonist. It is the impartiality of one who controls his inclination by a violent effort. It is not that more complete impartiality which the Germans call objectivity. It is not the cool indifference of a judge who does not form any opinions at all until the investigation is finished, and who, if he detected in himself any initial bias toward either side, would desire to withdraw from the decision of the case. In a historian impartiality of this kind would seem almost monstrous. What! When he narrates some war in which his countrymen have been engaged, is he not to betray the smallest personal interest in the cause of the conduct of his countrymen, no inclination to believe their cause just, no wish to find their valour heroic? To expect this of him would surely be to require him to divest himself of this humanity.[22]

According to Seeley, historians' impartiality made an allowance for an emotional attachment to the historical figures who occupied a place on the historical trajectory that connected the present of the historians and the past of their histories. There was something patriotic in Seeley's view as he encouraged his fellow historians to rejoice in the greatness of the national historical actors. This resonated with the objectivity which Heinrich von Treitschke and his followers in Germany endorsed as an alternative to the "lifeless" objectivity of Ranke, Waitz, and the other *Fachmänner*. The model which Treitschke introduced blended head and heart and consisted of dispassionate rationalism, patriotism, manly pride, empathy, and sensitivity.[23] Allowing sentimental patriotism to enter historical inquiry and integrate it into impartiality illustrates well why Seeley could claim that historians and scientists thought differently of objectivity. This also came close to the stance which Stopford Green adopted. But it did not mean that Seeley or the other scientific historians accepted unchecked subjectivity. They agreed that there were methodological procedures that aided historians in their aspirations for impartiality.

Historians proposed four methodological steps for attaining neutrality. Most importantly, they should let historical actors speak for themselves through the "actual words" of the sources, as William Holden Hutton explained it.[24] Second, they should treat historical figures fairly. This required empathy, fairness, and willingness to hear the different sides, treat them justly, and give them equal attention. Third, conscientiousness and impartiality demanded an active process of recognizing and overcoming prejudices, passions, and subjective views. This put historians' moral character to the test, as detachment required self-disciplining which avowedly weak minds were unable to master. Fourth, self-reflection entailed awareness of the differences between the past and the present. Historians should ward off anachronisms and analogies by becoming instinctively familiar with the historical periods in which they specialized so that they would not confuse their own assumptions with those of historical figures. Seeley warned that projecting current party divisions onto history was simply ahistorical.[25] There was still one additional qualification that was essential for acquiring historical truth: intellectual honesty. Historians recognized a boundary between honesty and dishonesty but were careful with allegations of intentional deceptiveness. Even Froude was not considered dishonest, but a victim of an inherent quality that prevented him from telling the truth.[26] Historians disguised the charges of dishonesty as methodological failures. R. Garnett maintained that suppression of facts was "more subtly destructive" than the "direct misrepresentation" of history.[27] Unveiling the meanings of such statements, however, was not too difficult for the contemporaries.

It was important to emphasize detachment, intellectual honesty, and methodological rigor because only truthful and impartial knowledge could benefit society. Historians across the party lines agreed: they could help politicians only by making them understand the origins of the current political issues with histories that were not entangled in party politics.[28] Histories which rose above mere "party watchwords" legitimized the societal value of history even for historians like Gardiner who otherwise disregarded the persona of a public intellectual, abstained from publicly making political statements, and appealed to Ranke's authority when speaking against an anachronistic use of history in politics.[29]

How useful was history then considered, as the bulkiness and chronicling of facts tended to compromise the benefits? A host of critics not only complained about the dry pedantry that had replaced the moral judgments and reduced the instructiveness of history but also maintained that historians had become too wrapped up in the past to understand what was going on in the present. Under such circumstances, historians produced histories that lacked any practical value; they were too tedious for politicians and too colored to please scholars. One writer explained how Freeman's attempts to mix his historical expertise and political opinions failed to produce anything useful because

> He treats modern politics like an archeologist; and while he is rightly warning us not to allow the modern maps to confuse our notions of medieval history, he allows the medieval map to shape his ideas of modern history. This is a pity, for Mr. Freeman's prejudices lead not merely to an erratic view of the events of the present day, but to some extent falsify his picture of the past times.[30]

The newspaper reviewers grew suspicious about the value of historians' political participation. *John Bull* bluntly declared, "Professors have seldom achieved any distinct success in English politics."[31] The reason was that historians and politicians cultivated personae that were incompatible; the *Manchester Courier* reminded how a historian was expected to nurture calmness and restraint whereas a political commentator must "inevitably tinge his productions with his own sentiments."[32] The public, moreover, yearned for guidance from the historians but had reservations about their political engagements because they were uncertain whether historians could compartmentalize their scholarly and political dispositions. This confusion of overlapping personae was used against historians when they got embroiled in political controversies, as Freeman so often did. His political adversaries were only too delighted to confound the personae of a historian and a politician to sow the seeds of mistrust in Freeman's overall reliability. When Freeman ferociously attacked Lord Derby over the Eastern Question, his political opponents seized the opportunity to discredit his scholarly and political authority, professing how his "Lord Derby is a creature of his imagination ... If his Harold or William be as fictitious as his Lord Derby, his history of the Norman Conquest is the longest and most labored, as it would certainly be the bulkiest, romance in the English language."[33] Such comments deliberately misrepresented Freeman, but they also betrayed the reputational threats that were embedded in political interventions. If *John Bull* was able to identify Freeman's different roles when it pronounced how "Against Mr. Freeman the politician we have had to protest again and again, for Mr. Freeman the historian we have an unbounded admiration" many others failed either intentionally or unintentionally to do so.[34]

Freeman's political outbursts both irritated and amused his colleagues, but mostly his politically vested histories were considered somewhat comical—especially as there were signs during the last quarter of the century how a more shamelessly outright manipulation of the past was spreading. Aggressively political nationalism exposed the intersection of history, fanaticism, and propaganda in Europe. The Franco-Prussian War was one of the turning points in nationalist history writing because it showed how the Germans, on the one hand, had used history to justify the war and to motivate the soldiers by strengthening their German sentiment and, on the other, how history had helped Germany to achieve its military goals. Stubbs had clearly detected some nervousness in the air as he explored the motives of German historians in one of his Oxford lectures in 1876. He assured his audience that those Germans who investigated the history of England were doing so with good and honest intentions. Unlike von Moltke and his "agents," these scholars were not collecting maps and models of fortresses to train German soldiers or to scheme in any other way to conquer England. They only worked "to increase human knowledge" and "to perfect the instruments of historic study."[35]

The Franco-Prussian War made the connection between history, nationalism, and political antagonism evident and paved the way for ever more ingenious schemes to use the past as a resource for current gains. As Eric Hobsbawm and a host of other historians have demonstrated, the decades around 1900 were a heyday for nationalist history writing. The increasing competition between the European great powers, the aspirations of the small nations for independence, and the entangled national histories provoked ever more aggressive responses from historians who considered it either their sacred duty or an obligation as public servants at state-funded universities to join the ranks of the demagogues.[36] Bury and Powell were alarmed about the development. Bury touched on the topic in his inaugural lecture in 1902 and Powell warned his Manchester audience in the same year about the consequences of historical research being taken over by political fanatics. Their message was that nationalism had made history vulnerable to falsification, but that it had also made visible how relevant the past truly was.

Powell invited his audience to take a tour of European nationalism. He started dramatically with the dark side of nationalism, professing how "Bulgarians would not be blowing up Greeks with dynamite, or Greeks joining Turks to cut the throats of Bulgarians and keep Servans out of Macedonia to-day, but for history, written history." He crossed then to the brighter side of European nationalism and visited many small nations to show how history had helped them to keep their "patriotism alive." He was explicit that Britons had no reason to rest content and believe that the evils of nationalist history writing happened only elsewhere. No, it was taking place in Britain as well. He painted a bleak image of the current state of Irish history, professing how the Irish Catholics, Protestants, unionists, and nationalists all fabricated histories of their own and refused to accept "the unpalatable but salutary truth" about their common past. Powell closed his speech with a simple message: false histories and patriotic lies fed excessive nationalism, race hate, and political unrest and this might have unforeseen repercussions for Europe as they obstructed the development of democracy.[37]

Powell and Bury offered as a cure an increase in the endowment of historical research because conscientious histories could reduce the impact of historical dogmatism. Bury appealed to the growing importance of history in schools and emphasized how only truthful history had pedagogical value. He did not spare his words when he showed how poorly historians were financed in England, while "every little people in Europe devotes sums it can far less afford to the investigation of its particular history." Just like English scientists, Bury, too, referred to the national embarrassment which the low research budgets caused as nations raced against each other in the field of scientific discovery. He accused politicians and private donors of ignorance since they failed to recognize the importance of the immaterial gains which history produced. He also blamed the universities for not doing their job and urged them to "preach more loudly and assiduously that the advancement of research in history, as in other sciences, is not a luxury … but is a pressing need, a matter of inestimable concern to the nation and the world."[38] As the extent of historical manipulation was unfolding, historians like Bury and Powell were preoccupied about the integrity of their discipline; historical truth was too easily sacrificed for political gains.

Just as Powell asserted, the history of Ireland was a contentious topic. It had stirred discontent already in the eighteenth century when the Irish Patriots of the 1760s and 1770s had tried to repudiate the Anglophone histories that presented Ireland as a barbarous wasteland.[39] When the idea of the distinctiveness of Ireland crystallized during the nineteenth century, its uniqueness was traced to its Gaelic roots with a hope that the Gaelic past would enable the Irish to transcend sectarian lines and embrace a shared idea of Irishness. The strong connection between the Gaelic past and Catholicism undid such aspirations, causing only additional resentment among Irish Protestants.[40] Home rule, the Land War, the spreading radicalism, and the expanding political nationalism escalated the use of history in Irish politics as both unionists and nationalists searched history to find precedents to consolidate their current claims and stimulate the loyalty of their supporters. It was nearly impossible to write history that satisfied the different factions, as these were in constant movement in response to the shifting political and cultural circumstances. What they all agreed, though, was that a nation must have a past and a future and that the present politics were justifiable with historical examples. These negotiated representations of history were at the core of the political, social, and cultural changes in Ireland before the First World War.[41]

The historical community was aware of the tensions that eroded trust in Irish history and historians. English historians and reviewers, imbued with superiority and unionist agendas, reinforced this image by regretting time and again how the history of Ireland had not invited impartial and passionless judges. As Sydney Lee remarked in the *English Historical Review*, "A judicial tone is the last characteristic that one would expect to meet with in a work of an Irish historian."[42] It was common to underline the apparently despicable state of Irish history by hoping that younger Irish historians would be disposed to change the course of historical studies in Ireland. The good intentions of this were curbed by the habit of recording alongside all the imagined and real vices that had characterized earlier history writing in Ireland. Lecky, whose moderate unionist histories garnered approval across party lines, did exactly this when he reflected in 1898 on the state of historical scholarship in his native country. He was confident about the young historians who had accepted impartiality and the fairness of judgment as their "first duty," realizing

> that the falsest of all traitors are those whose statements in themselves are mainly true, but who made it their business to pick out of the annals of the past the misdeeds of one side, and to conceal the misdeeds of the other, and in the interests of a party or a creed habitually to suppress palliations on one side and provocations on the other.[43]

Characteristically, Lecky balanced optimism with rhetoric that evoked the conventional image of Irish history as exceptionally susceptible to manipulation. Considering the respect which he enjoyed in England and Ireland, his words mattered.

The mistrust in Irish historians, the rival notions of Irishness, a charged political climate, and the deep-seated conviction that the current problems carried historical baggage gave history high relevance in the

Irish question. As Joep Leerssen has put it, bygones were "anything but bygones" to Irish history writers.[44] This explosive mix of political, cultural, and scholarly forces informed Hickson's and Stopford Green's interventions in Irish history. Their attempts to use the past to legitimize their brand of Irishness were etched in their paratexts and scholarly personae.

Mary Hickson and Insistence on Impartiality

> Our ancestors were guilty of abominable and atrocious crimes, to which the present generation, thank God, looks back with all the horror and indignation they desire.

This quote from Reverend Charles O'Conor (1764–1828) on the title page of Mary Hickson's *Ireland in the Seventeenth Century or the Irish Massacre of 1641–2* (1884) identified her compellingly as a unionist and Protestant. O'Conor, a Roman Catholic priest and historian, had become famous for his Anglo-Irish sympathies and objectionable opinions about the Catholic Church that had outraged the Irish Catholics and led to the suppression of his priestly functions. By the last quarter of the century, he was known as a second-rate scholar and controversialist whose sanity was seriously questioned.[45] If O'Connor's name on the title page was not enough to impose a specific political reading on the book, an accompanying note about a preface by James Anthony Froude abolished any remaining doubts. The title page and Froude's allographic preface, together with Hickson's introduction and footnotes made her partisanship evident, but she nonetheless launched a spirited campaign to defend her impartiality after *Ireland in the Seventeenth Century* had been published and her motives and detachment challenged. She was unwilling to resign the persona of an impartial historian, although her book proved the contrary.

Hickson was from the Catholic-dominated County Kerry and saw the past in the terms of the present. She fostered a deep mistrust toward Irish Catholics and explored history to prove that Irish Catholics, who were according to her full of sectarian hatred and willingness to stir unrest, threatened Anglo-Irish Protestants. She drew on the Protestant narrative of victimization when composing the two-volume *Ireland in the Seventeenth Century or the Irish Massacre of 1641–2* (1884) which Longman published. The book was a collection of depositions about the 1641 Rebellion in Ulster. The rebellion as a historical event and the depositions as testimony of Catholic atrocities had been highly explosive in historians' hands. Catholic and Protestant historians disagreed about the causes and consequences of the rebellion as well as the extent of Catholic violence. As John Gibney notes in his study on the place of 1641 in Irish historical memory, for the Protestants it was a rebellion and for the Catholics a rising against the Protestant oppressors.[46] The year 1641 epitomized the divisiveness of Irish history, and for Hickson its symbolical weight was increased by the fact that the Protestants traced sectarian hatred to 1641 and used it to cast the blame for the current tensions on Irish Catholics. They also used it to oppose home rule, arguing that the depositions proved how Catholics had misbehaved under an earlier home rule experiment.[47] As T. C. D. wrote in the *British Quarterly Review* when introducing Hickson's book, "worse than all, the events of 1641 still act with a living force upon the political relations of Irishmen, and are the true explanation of the deep and apparently incurable animosity which animates alike the Nationalists and the Orangemen of the country."[48]

Hence, the 1641 rebellion was ideal for Hickson's purposes. She interpreted it as a turning point in Irish history and concrete evidence of the despicable character of Irish Catholics. She was determined to prove wrong the Catholics who questioned what she saw as a near extermination of Ulster Protestants in 1641.[49] Her mission gained additional acuteness from the latest historiographical trends. The unionist narrative was increasingly being questioned by nationalist and professional historians. More alarmingly, even the new generation of Protestant and unionist historians approached Catholic historians in their interpretations of 1641. Hickson had no understanding for this. She wanted to make clear that the historical guilt laid on the Catholic side and the historiographical falsification of the events had derived from that same quarter, and that the rebellion was a testimony of the true nature of Irish Catholics.[50] *Ireland in the Seventeenth Century* was as much a manifestation of her political convictions and historical understanding as an attempt to justify these two with primary sources.

The depositions were the main source on the 1641 rebellion. They contained eyewitness testimonies and hearsay about the killing of several thousand Protestant settlers by Irish Catholic insurgents. The depositions were kept at Trinity College in Dublin and bound in 31 volumes consisting of approximately 8000 statements from the Protestant survivors. They were an appalling source: the documents were arranged in a confusing manner and the spelling was erratic, if not incomprehensible. Importantly, most of the depositions said nothing about killings, but before Hickson's volumes, only some 200 carefully selected specimens had been circulating among historians, supplying one-sided evidence to the Protestant narrative. To gain a more comprehensive view of the depositions, John T. Gilbert, the librarian of the Royal Irish Academy, suggested in 1881 that they should be made available to the public, but to no avail. The plan for an extensive edition of the depositions resurfaced in the 1930s, but again brought no results.[51] Hickson, then, should be credited for her pioneering work with the depositions, and when rumors about her project spread, historians were eager to see the results. But as her critics quickly discovered, her biased selection of the documents impaired the value of her contribution.

The form of Hickson's work, a collection of depositions and an accompanying historical introduction, was important for her aims, as the original documents lent an air of authenticity to it. Producing document editions was a great nineteenth-century trend. States, academies, learned societies, and private donors funded the publishing of historical records all over Europe. In England, the Public Record Office oversaw the production of *Calendars of the State Papers* and the *Rolls Series*, and the Royal Commission of Historical Manuscripts was established in 1869 and the British Record Society in 1888. In addition to these, the Camden Society, the Selden Society, the Navy Record Society, and many others made records available to the public and had notable historians on their committees. The myriad projects were an important source of livelihood for many historians and opened up academic careers for junior scholars.[52] Despite the popularity of these ventures, they have attracted surprisingly little scholarly attention. Apart from Mary Anne Everett Green, even less attention has been paid to women editors, who seemed to be particularly active in England.[53]

Everett Green, whom Christine L. Krueger appropriately calls the first professional woman historian in England, made a career at the PRO, where she edited 41 volumes of *Calendars* and perfected the methods of abstracting and extracting the documents. Her other accomplishment was to convince the deputy keeper, Lord Romilly, that historical introductions, where the editors contextualized the ensuing documents, would make the publications more accessible to readers. But she was an exception. When the PRO introduced formal requirements and expected an academic degree from recruits, women were edged out from its ranks.[54] The learned societies continued to enlist women, and Lucy Toulmin Smith, Mary Bateson, Mary Dormer Harris, and others seized these opportunities to engage in historical work. Furthermore, gentlewomen used their spare time to publish records from family papers, local and regional collections, and parish records. A few women editors were also privately sponsored, as was Hickson. She had approached Froude with her financial troubles asking for help and he had reached out to his friend Lord Carnarvon to subsidize her work. Referring to the "most excellent service" Hickson had done for Ireland as a historian, he had persuaded Carnarvon to fund her.[55]

Joan Thirsk explains women's success in editing local documents with their "indefatigable" energy in "assembling a mass of documents."[56] Editing was indeed often portrayed as a mechanical task, and in central and northern Europe it was therefore allocated to historians whose talents were considered insufficient for creating original knowledge.[57] In England, it would have been inconceivable to discredit editing when the PRO and the societies enrolled such eminent historians as Stubbs, Gardiner, and Maitland. If editing was in general considered a valuable pursuit, editing and writing monographs were, nevertheless, considered distinct endeavors which required unique sets of skills and methods. Meticulousness and assiduousness alone did not suffice for editors. This is where Hickson ran into trouble, as she plunged into the work without earlier comprehensive experience of making document editions. Gardiner, who reviewed *Ireland in the Seventeenth Century* for the *Academy*, was unimpressed with her results and observed how the volumes betrayed "a novice in the art of editing."[58]

Document editions were received as impartial and truthful representations of the past, and this false conception benefited Hickson. The authenticity of document editions was eulogized time and again. They were the true embodiment of scientific history as they let the past speak without authorial intervention. J. S. Brewer praised Marquise Campana de Cavelli's *Les derniers Stuarts à Saint-Germaine-en-Laye* because she had corrected many biases by being "content to let her documents tell their own story, without putting a word into their mouths ... whether they seemed to make for or against her cherished [Tory] opinions."[59] The enthusiasm which record editions caused blinded many to the fact that they were always selections. Editors, the Keeper of the Records, publishers, and the committees of learned societies made decisions about which documents to include or exclude and which ones to calendar, extract, or print *in toto*. Yet, for many readers the editions appeared as unmediated historical reality. Historians such as Gardiner knew better, and therefore he was disappointed that Hickson failed to say anything about the methods she had applied to select and edit the depositions in her introduction.

When readers opened Hickson's book, they first encountered Froude's preface and then her own introduction. The preface was important in many ways: it confirmed Hickson's authority, legitimized her claims about the depositions, and informed readers about the ideological context of her work. Inviting an outsider to eulogize a text is a common method for boosting the sales and reputation of lesser-known writers. Genette, calling a preface that is solicited from an outsider an allographic preface, notes that it is an indication that someone other than the writer has considered the text worthy of their name.[60] Allographic prefaces are, as Jeri English suggests, usually written by renowned authors or cultural icons who have enough symbolic or cultural capital to recommend a text from an unknown writer. "The name of the preface writer," English explains, "should have enough resonance with the public," because the shaping of textual reception entails that readers "recognize the preface writer's name and acknowledge his or her cultural authority."[61] The allographic preface in *Ireland in the Seventeenth Century* ticked all the right boxes. Hickson was a relatively unknown Irish woman who previously had worked on local and family history and participated in antiquarian activities. Froude, in contrast, was a genuine factual paratext loaded with contradictory meanings.

Allographic prefaces are not any less neutral than other paratexts, as their writers may use them to gain extra-textual benefits. Referring to an allographic preface by Simone de Beauvoir, English notes that "the greater the symbolic capital of the preface writer, the greater the ease with which she can use her rhetorical abilities and discursive authority to appropriate the text in question for her own purposes."[62] This is exactly what Froude envisioned when he offered to compose the preface. Supposedly, he feared that Hickson's name would not have evoked the sort of authority he thought was needed to establish the depositions as a genuine historical source. Lending his name to the book was easy for him because Hickson had done exactly what he had hoped her to do with the depositions: she had made them appear as a reliable source and lampooned those who had questioned their veracity. He commented later with great delight how she had "with considerable humour" identified the mistakes "into which the impugners of their authenticity have blundered in their haste to condemn them."[63]

Just like Hickson, Froude subscribed to the idea of the 1641 rebellion as a turning point on which later tensions in Ireland hinged. He had been keenly interested in the depositions already for some time because he read them as a confirmation of his views. He had reinstated the depositions in *The English in Ireland in the Eighteenth Century* (1872), which was a violent vindication of the colonization of Ireland and an attack against Irish Catholics. He had manipulated historical evidence to prove that the Irish were racially inferior and thoroughly unfit for self-government.[64] The defenders of the empire found in Froude's book justification for their anti-Irish and anti-Catholic opinions. His theories and rhetoric, however, were so divisive that they provoked strong objections in both Ireland and England. The scientific historians hastened to point out that Froude had sacrificed the truth to his Tory biases and had suppressed material that was inconsistent with his views.[65] Powell described Froude's method in a brutally ruthless obituary, claiming Froude had handled his authorities "as a willful baby does her dolls." By alluding to girliness, Powell undermined Froude's manliness and scholarliness in one blow. He regretted that Froude had chosen "the enchanted forest of politics" instead of "the rugged acres of history."[66] However, Powell failed to see that Froude had chosen his course deliberately. According to Chiaran Brady, for Froude "historical research always remained a mere means to greater purposes rather than an end in itself" and he saw himself first and foremost as a moralist, and only then a historian.[67]

Hickson was just as convinced of the historical value of the depositions as Froude was, and one of the great aims of *Ireland in the Seventeenth Century* was to prove the critics wrong and establish them as reliable sources for once and all. To achieve this, it was essential that Froude awarded her the persona of a proper historian by framing her in the preface as an able scholar and impartial historian who had mastered the technical skills of an editor. Before doing this, however, he repeated his well-known notions on present-day Ireland and its history. He claimed that "irresponsible agitators" and "reverend and grave historians" had misguided the Irish to believe that the massacre and atrocities committed by Irish Catholics were a mere fabrication, "cowardly lies," which seriously insulted "English honour." The now-published documents, he continued, finally allowed everyone to judge the matter for themselves—although he made it perfectly clear what conclusion everyone should draw. After this, he moved on to explain why Hickson had been the best possible person to transcribe and annotate the depositions. According to him, what served as compelling proof of her abilities was the fact that she had convincingly resolved the controversy over the authenticity of the depositions and proved their reliability as historical evidence. Moreover, it was important to accentuate her "fairness of mind" and the "love of justice" that rendered her perfect for the task. She had no keen English prejudices and "on some points she is in full sympathy with Irish nationalism." Therefore, she was able to clarify "better than any previous writer the causes which drove them [the Irish] into fury," Froude explained.[68]

This was quite an overstatement of Hickson's detachment, but for the most ardent defenders of the union even the slightest hint of Hickson's nationalist sympathies was too much. Thomas Croskery, a renowned unionist and professor in theology at Magee College in Derry, protested in the *Edinburgh Review* against Froude's attempt to depict Hickson "as something of a Nationalist," but agreed with him about her impartiality. She was "singularly free from prejudice; being about equally severe on her strictures upon Protestants and Catholics," he assured.[69] It was critical for Froude to exaggerate Hickson's neutrality, and portraying her both as a unionist and a nationalist was therefore necessary. Readers, however, were not convinced, as Froude's name alone was enough to establish Hickson's allegiances.

Once *Ireland in the Seventeenth Century* was published, the *Pall Mall Gazette* declared it to be an "honour" for Hickson to have a preface from Froude.[70] The publisher saw the matter similarly. Froude was a best-selling Longman author and his name was used lavishly in the pre-release announcements and advertisements to promote *Ireland in the Seventeenth Century*.[71] This added to the paratextual impact which the allographic preface had for the contextualization of Hickson's book. The preface gained additional visibility as it was commented on in many reviews. As the *Scottish Review* put it, Froude's preface was "strongly worded, and somewhat pugnacious" and added "nothing to the value of the volumes, though it is certainly of value as affording additional evidence, if such were wanted, of the direction in which its author's sympathies lie."[72] Subsequently, Hickson was perceived as a Froude's assistant or an ally, a Froudean "she devil," as Prendergast who advocated a conciliatory approach to Irish history called her.[73]

This was taken to the extreme in a controversy that broke out between Hickson and Robert Dunlop in the *English Historical Review*. Dunlop specialized in seventeenth-century history and was affiliated with Manchester University. This gave him formal authority to teach Hickson a lesson about the veracity and historical significance of the depositions. He began by explaining how historians treated the depositions either mostly as fabrications, as mainly reliable, or as taken under too suspicious circumstances to qualify them as trustworthy evidence. Dunlop supported the last approach and placed Hickson and Froude in the group who uncritically believed in them. He did not mention Froude's preface, but he lumped Hickson and Froude together in a way that implied Froude's presence in *Ireland in the Seventeenth Century*.[74] Hickson replied to Dunlop in the following issue, insisting that he had misunderstood her arguments about the depositions. She also unexpectedly mentioned the allographic preface, as she explained how Froude had made "considerable alterations" to it at her request, because the first version had tended "to make the volumes appear as if written for a political purpose."[75] This comment drew Dunlop's attention to the preface and he brought it up in his answer to Hickson. Writing with obvious derision, he remarked, "She says she cannot be accountable of Mr. Froude, that she even does not know what his opinion is, and yet, strange to say, she asked him to write a preface to her book."[76] Hickson defended herself once more. The question of the preface had to have been important to her, as she addressed it even in her brief final reply, revealing only now that the preface had in fact been Froude's idea. "Mr. Froude offered, rather to my surprise," she explained, "to write the preface, and I very thankfully accepted his kind offer, on condition that there was to be nothing in it which seemed to connect the volume with present politics."[77]

It is unclear how much Froude had changed the original version of the preface, but the published one was not as apolitical as Hickson professed. It was explicit enough about Froude's opinions and for many it was irrelevant what he really wrote in it; his name was enough to indicate the intentions of Hickson's book. The positive effect of the endorsement on Hickson's scholarly credibility was limited by Froude's controversial reputation, as it was common that readers discerned a transmission of qualities from the writer of an allographic preface to the actual author. An anonymous reviewer complained in the *English Historical Review* how Freeman's preface in Louis Leger's *History of Austro-Hungary from the Earliest Times to the Year 1889* reasserted his "Slavonic prepossessions" and risked impairing the book's value to those who preferred "history without any infusion of modern politics."[78]

The allographic preface and Froude's name, however, were not the only paratexts that linked Hickson to radical unionist politics. Her historical introduction and the notes that accompanied the depositions were explicit about her partisanship. She used the footnotes conventionally to introduce the sources, state their current repositories, and on a few occasions to point out the unreliability of a specific deposition or a passage in it. This enabled her to argue that she comprehended that not *every* deposition was a truthful account of the events. Her rare critical reflections did not convince Dunlop or the others who thought that she overstated the accuracy of the statements in the depositions. The notes, moreover, served her ideological ambitions, as she scattered them with remarks about the unreliability of the Catholic histories and with her pet arguments, such as Cromwell having been a true blessing to Ireland. Every "wise and true-hearted Irishman" realized that it was not Cromwell who had been Ireland's worst enemy, she professed in one note.[79]

Yet, it is the introduction which deserves more attention here because it played a leading role in a confrontation between Hickson and Gardiner. Once the *Calendars* began to appear with historical introductions, such prefatorial attachments became a staple element of document editions in England. They hovered between texts and paratexts: they contained historical accounts that were more characteristic of texts than of paratexts, but also offered typical paratextual information that guided the reading of the "texts," that is, the historical documents. They contained descriptions of the editing process, the method of selecting the documents, and the principles of transcription, extraction, and orthography. Introductions also stated the significance of the documents and contained acknowledgments of helpful colleagues and archivists. This emphasized their paratextual quality. Hickson reserved much space for enthusiastic praise of the historical value of the depositions but said nothing about her method of selecting the depositions. This caught Gardiner's attention.

Gardiner had to have been disappointed with what Hickson had produced. He had been aware of her ongoing work and had even advertised it in a footnote in the new edition of his *History of England* (1884) by writing how "the celebrated *Depositions*" were being edited by Miss Mary Hickson and would soon be accessible in print.[80] When he got a copy of the work, he was appalled by its poor quality. He reviewed it in the *Academy*, where he connected Hickson's silence about her methods to her partisanship and implied that her detachment was impaired:

> If a selection of documents is made, everything depends on the spirit in which it is made. Not only is it impossible to give an opinion on this important point without a prolonged examination of the original MSS., but Miss Hickson herself does not give us any clue to the system which she has adopted ... She does not affix to the depositions any references to volume and folio to assist those who wish to test her work.[81]

If this only hinted partiality, Gardiner expressed his suspicions more explicitly when he noted that the introduction "conveys the impression of being the work of a writer in whom a sense of fairness and a desire to arrive at the truth are contending with the impetuosity of one who has formed strong opinions of her own." Hickson was offended and wrote a series of replies to Gardiner.

In her first rejoinder, Hickson explained that she had excluded the shelf-marks because the documents were so well marked in Trinity College that "no such references are necessary to assist really industrious and impartial students, the only persons whose opinions on the depositions would be valuable." This could be read as a bold dismissal of Gardiner's virtuosity or just the expression of a common disregard for detailed references. The latter was something entirely foreign to Gardiner and his meticulous method of recording his sources. Even more surprising was that she ignored Gardiner's complaints about the missing information on her editorial policies or about her partiality. Perhaps to divert attention away from her biases, she launched a spirited attack against Gardiner, Lecky, and everyone else who had written about the 1641 rebellion without consulting the depositions at the Trinity College. According to her,

> we have a real "Irish grievance" to complain of in the fact that eminent English writers, as well as eminent Irish ones, like Mr. Lecky, of whom we are all justly proud, have undertaken to write about our country without examining a leaf of documents which are essential to the right understanding of Irish history, and also of some of the most important Irish questions of the day.[82]

One of Hickson's claims to fame was that she was one of the first, if not the very first, historian, who had gone to the sources. Of course she was not entirely mistaken, but the loudness of her self-praise suggests that she calculated that the pitching of originality would increase the value of and interest in her work—and indirectly the support for unionist historiography.

Gardiner wrote a reply to clarify that it had not been his intention to accuse Hickson of partiality. Yet, he undermined his own words by emphasizing once more how document editions were always selections by the editors and could give only a partial image of history.[83] Gardiner was not explicit about Hickson's biases, but the meaning of his words was evident enough. This spurred Hickson to compose one more reply where she reacted to the insinuation about her partisanship. She had explained in one of her earlier rejoinders that a discussion of her methods was unnecessary because the chosen depositions themselves attested to the soundness of her editorial principles. The two volumes, she wrote, contained a fair sample of the material as they had a small selection of documents that revealed acts of kindness from Catholics and a few unreliable depositions to show that some of the material was untrustworthy. She restated this in the closure to her confrontation with Gardiner.[84] Whether this convinced her critics is unlikely. After all, most of the documents which she had selected focused on exposing Catholic brutalities.

Since Hickson mentioned Lecky as one of those who had written about the 1641 rebellion without examining the original depositions, she also received a reply from him via the *Academy*. Lecky, clearly not amused by the attention, defended himself dryly: he had written only briefly about the events because they fell outside of the scope of his *History of England in the Eighteenth Century*. It would also have been superfluous to consult the depositions in manuscript because they were not as unknown as Hickson maintained. They "have been seen and used by several writers, and the worst stories in them have been made very familiar by Temple, Borlase, and Carte," he added.[85] Surely Lecky knew that these earlier historians had consulted only a fraction, and a highly biased fraction, of the existing material. He obviously wanted to downplay Hickson, and this must have worked as an incentive for her to submit one more reply to the *Academy*. She admitted that the 1641 massacre was not at the core of Lecky's research, but nonetheless he had discussed it in his *History*. Therefore, she insisted that any attempt to write about the events without consulting the unpublished depositions was "like playing Hamlet with the part of Hamlet left out." She had used the same Hamlet metaphor in her Introduction, but this time it did not seem to be enough for her to drive her point home, as she passionately continued,

> Mr. Lecky's sketch of 1641–49 and the Cromwellian Settlement has been described in the *Northern Whig* as a 'masterpiece of historical criticism.' And it is so, as far as the very imperfect data on which it is based goes; but for want of a full examination of the College MSS. it fails. All histories or sketches of the same period by persons who have not carefully and impartially examined the MSS. in the College equally fail.[86]

Considering Lecky's eminence, it is somewhat startling that Hickson chose to oppose him so openly. He had been her adviser when she was composing *Ireland in the Seventeenth Century* and had instructed her to accept only the evil which the friends of the accused reported or the good which the friends of the victims admitted.[87] She had obviously disregarded his advice. It is likely that she tried to benefit from the meanings assigned to Lecky's name to accentuate the originality and significance of her work. By playing with the vested symbolism of proper names, she contrasted herself favorably with a renowned historian and elevated herself to the category of a proper historian.

Hickson's public appearances in the controversies had very little impact on how she and her book were perceived. The paratexts were obvious in their partisanship and the topic so deeply invested with prejudices that any appeals to impartiality were futile. Different groups of readers evaluated her against their own political opinions and religious beliefs. For fervent unionists like Thomas Croskery she was an incarnation of an impartial historian who gave great service to their cause by proving how the massacres of 1641 were one of the "disgraceful travesties of history." Croskery thanked Hickson for forcing Catholics to accept the full guilt of the events with her compelling evidence.[88] Meanwhile, for Catholics she was the personification of a doctrinarian historian, and Sylvester Malone took the opportunity to declare in the *Dublin Review* how she had filled her pages with offensive falsehoods, absurdities, and occasionally simply with "wild and unfair writing!".[89]

Alice Stopford Green and the Persona of a Partisan Historian

"Noblest defender of Éire." This was the title which the Irish medievalist Henry Egan Kenny gave to Alice Stopford Green when she was writing *The Making of Ireland and Its Undoing* (1908).[90] Her book became an immediate hallmark in cultural revivalism in Ireland, as it on the one hand turned the gaze to medieval Ireland with an intention of restoring the pride of the Irish in their Gaelic past and on the other aimed at balancing the Anglophone colonial narrative. She comprehended the utility which history had in defining Irishness and sanctioning her political claims. For the Irish, she explained, "the far past and the far future are part of the eternal present, the very condition of thought, the furniture without which the mind is bare."[91] This connection to the past, she argued, had been denied the Irish by the English, who had forcefully fed them the narrative of barbarism. It had not only distorted Irish history but had normalized the idea that the Irish were "a strange and childish race" and that this primitiveness justified their colonization.[92] She pitched herself confidently between historical scholarship and political propaganda, dispensing with impartiality as an illusion. She was greatly influenced by the strong tide of cultural nationalism in Ireland around 1900 and the cultural and political overlapped in her thinking. She did not question for a moment whether her book was a political vehicle: she presented copies of it to Theodore Roosevelt, Winston Churchill, and the U.S. Supreme Court Justice Oliver Wendell Holmes to advance the cause of Irish nationalists. Tellingly, the Library of the Royal Dublin Society banned *The Making of Ireland* because of its subversiveness.[93]

Hickson is today little known, but Stopford Green has garnered growing scholarly attention. Historians have been intrigued by her overlapping identities as a historian, social reformist, Irish nationalist, and philanthropist, and by her extensive networks, which exemplify women's indirect and soft political influence around 1900.[94] She was from an Anglo-Irish family and first became known as the wife and then the widow of John Richard Green, who instilled in her a passion for social history, taught her the historian's craft, and introduced her to the Macmillan family, to his fellow historians, and the circles of social reformers. This was a significant cultural, social, and ideological heritage for a woman who became a widow at the age of 35. She used the lessons to her benefit when Macmillan invited her first to finish Green's uncompleted *The Conquest of*

England and then to revise *A Short History of the English People*.[95] Her first major contribution to history was *Town Life in the Fifteenth Century* (1894) which established her as an expert in social and medieval history.

Stopford Green began to fashion herself as an advocate of Irish cultural nationalism in the 1890s and fused in it her anti-imperialism and social reformism. She got a push toward Irish history from the American publisher Harper Brothers, who induced her to abridge Green's *Short History* for the American schools where the children of Irish immigrants constituted a large market. She was unwilling to spoil the monument which Green's *Short History* was for her, so she instead proposed to Macmillan to write another book for the American schools with "a distinct centre of Irish history." [96] She never completed the plan, but it sparked in her a curiosity about Irish history. Some of her colleagues were astonished about her sudden interest in Irish history. When she asked Powell's opinion about her plan to investigate medieval Ireland, she received a reply full of warnings and preconceived ideas about Irish historians' "blind party spirit." Powell advised her to brace herself "for abuse in plenty, whichever side you take" and for a hostile audience because "The Irish hate the plain unvarnished truth about anything they really care for, and those that care for plain unvarnished truth hate the Irish usually." Powell admitted that a "history of Ireland in three volumes, cheap and clear, and as unprejudiced as it could be, would be a boon (but it would never pay, there would be fierce attacks and worry without an end)." Perhaps realizing how discouraging he was, he added that medieval Ireland was a relatively safe topic and "less a battle-ground than any other part of Irish history."[97] If anything, Powell's prejudices should have convinced her about the urgency of her project, as his letter documented the hostilities which the English historians fostered about Irish history and historians.

Stopford Green embraced Irish cultural nationalism, which gained prominence in the 1890s when the radicalization of political nationalism and the rigidity of English colonialism created room for a cultural movement that searched for the common denominators in Irishness. This encouraged Anglo-Irish Protestants such as Stopford Green to craft for themselves a new Irish identity located in the Gaelic heritage, language, and literature. She was convinced about the common Irish past and *The Making of Ireland* was a response to the Irish-Ireland movement as it both unearthed suppressed Irish history and encouraged the Irish to claim ownership of their past. She repeated her message in the paratexts, in lectures, and in essays in periodicals, hoping that the Irish would realize that

> The reform of Irish history must begin in our own country, among our own people ... we ourselves must be the keepers of our fame and the makers of our history. Let us in Ireland therefore remember that we have an ancestry on which there is no need for us to cry shame.[98]

Stopford Green appropriated multiple arenas to promote her ideas, and this strengthened the political dimension in her persona. She participated in associations and committees and provided financial support to the nationalist movement. She assumed a position as a mediator between British radical liberals and nationalists in Ireland, but paradoxically remained a Home Ruler. She joined other women in radical political action, was treasurer of the Howth gun-running of 1914, offered shelter for Irish volunteers in Dublin and London, and hosted meetings of leading politicians before the establishment of the Irish Free state in 1922. Accordingly, the officials of British state security frequently raided her property and confiscated her papers. Her work did not go unrewarded: she was one of the four women elected to the newly formed Irish Senate in 1922.[99]

The Making of Ireland was a manifestation of Stopford Green's ideological ambitions, and its paratexts portrayed her as a historian who cultivated a persona that blended historical methods and undisguised political intentions. She wrote the book at a time when English historians were becoming more pronounced about their professional status and standards. She, too, armed her book with a mass of primary sources but abandoned impartiality and distanced herself from the younger generation of scientific historians who advocated neutrality. Her blunt statement of how "We usually think people impartial who hold our own opinions" encapsulates her disillusionment with historians' detachment.[100] She also found intellectual impartiality too restrictive and professed how historians needed the sensibility of heart because it was impossible to understand human life without "great affectation of the soul." An approach that took emotions into account led to a more balanced view of the different historical parties, and she recommended to historians "fewer orthodox predilections of the head" and more "illumination from the heart."[101]

There is a curious echo of Treitschke and his advocacy of methodological empathy in Stopford Green's words. Otherwise, the pacifist Stopford Green and the militant Treitschke could not have stood further apart.

In addition to this, she also discarded impartiality because the recording of documentary information spoiled the narrative character of history. For the scientific historians, the principle of letting the documents tell their story was a cure for dramatized narratives as it kept their imagination in check, but Stopford Green maintained that it led only to pedantry and dry chronicling which put off ordinary readers. She preferred picturesqueness and stories that made history appealing to readers with different socioeconomic and educational backgrounds.[102] Kenny was impressed by her style. He read the manuscript and gave a lively description of her manner of writing: "When you push aside your authorities and write with their ideas in your head you rise to heights of eloquence ... Then short graphic iron pen of yours becomes a thing of life. It reproduces Life."[103] This suggests empathy and assimilation, which contested the methodological stipulations for distance and self-restraint. For Stopford Green, liveliness was also important for strategic reasons as it helped her to appeal to the large readership beyond learned circles.

The paratexts in *The Making of Ireland* flagged the book's ideological objectives and betrayed the political, scholarly, and commercial targets which Stopford Green had set for it. The material paratexts were designed to ensure wide circulation, dissemination of her historical views, and profit. She identified three separate markets for her book—England, Ireland, and North America—and proposed to Macmillan that the materiality of the book should be customized to each of these reading communities with distinct cultural, educational, political, and socioeconomic characteristics. Macmillan agreed, and at least separate English and Irish versions were produced.

Stopford Green did not expect large sales in England and the English version was tailored to appeal to educated readers. It was printed on thick paper, had a firm binding, and was pricier than the Irish version. Stopford Green and Macmillan anticipated a high demand in Ireland and designed the paratexts so that the book would be accessible to less affluent readers. It was printed on thin paper and had its own title page that accentuated its difference to the English one.[104] The book indeed became popular in Ireland, and Stopford Green reported to Macmillan in early 1909 how booksellers in Dublin were complaining that they did not get enough copies of it to satisfy the demand. She also sold copies from her home and witnessed steady incoming orders.[105] One reason for this success was printed on the book's spine: instead of a publisher, it stated "Gaelic League."

The Gaelic League was a hallmark of the Irish cultural revival. It had been founded in 1893 and provided an important intellectual home for Stopford Green. The League encouraged plurality in membership, and although its apolitical and non-sectarian outlook was an illusion and Catholic middle-class men made up the majority of its membership, it also accommodated as members Protestants, the lower middle classes, the well-educated, and women. The League was instrumental for women who were excluded from the official political parties as it provided them with a training ground for more openly political and radical action, of which the League disapproved.[106] Stopford Green had close ties with the League and one of its co-founders, Eoin MacNeill, who shared her scholarly interests in Irish heritage and history. The League, then, formed an important supply channel for *The Making of Ireland*, and Stopford Green's later publications were similarly distributed in Ireland via the League.

It seems that the North American version was never realized although the book was sold there. Stopford Green's plans for the American market are nonetheless informative about the overlapping roles which historians assumed as scholar–educators, political agitators, and commercial agents. When she introduced the idea of an American version to Macmillan, she painted a picture of large Irish immigrant communities and explained how there was "a great movement being made to get Irish reading books into the schools." She was convinced that *The Making of Ireland* could fill that void with certain revisions because it was not only "historical but will be used politically" and this double function "may give it a sale which would never attach to a historical book." She obviously had no scruples admitting the ideological character of her work. But to make the book appealing to this audience, it had to be cheap and lose its scholarly appearance, and so she planned to remove the footnotes and appendices to make it "seem less heavy."[107] In other words, to boost sales, she was willing to suppress the scholarly apparatus that validated her controversial claims. She probably calculated that her American audience would appreciate her tribute to Irish history without fretting about sources and evidence. She was so hopeful that she even proposed to Macmillan to copyright the book in America even though she knew that usually it was not worthwhile copyrighting English books there. Now the large Irish audience made a difference and she feared that otherwise the book might spread as a pirated copy and bring no profit to her.[108]

Stopford Green was explicit about her ambitions in the textual paratexts, and the preface confirmed that her aim was to offer what Angus Mitchell calls a "radical reimagining of the late medieval Gaelic world."[109] If the main function of a preface is to ensure that the book is read properly, to give clues about what to expect of it, and to provide the author's interpretation of the text, she nailed this initial framing of her text.[110] The preface revealed that her express goal was to show that the English had distorted Irish history and to give voice to the silenced people of Ireland. Drawing on her expertise in social and medieval history, she traced the progress of Ireland "in industry, in wealth, and in learning" and painted a picture of a flourishing Gaelic society whose "national life" the English invasion had brutally ruined. Her interpretation deviated sharply from the English narrative of backwardness and her "pious duty" was to recover the voices of the "noble men, Irish and Anglo-Irish, who built up the civilization that once adorned their country" and had since then been "buried by the false hands of strangers in the deep pit of contempt, reproach, and forgetfulness." She was convinced that regaining ownership of Irish history would have a tremendous impact on the Irish and lessen the suffering which the distorted image of a shameful past had caused. With great passion she wrote how Irish hearts had been depressed as they had been told how "they have inherited the failings of their race, and by the verdict of the ages have been proclaimed incapable of success in their own land, or of building up there an ordered society, trade, or culture." Confidently, she pronounced, "It is in the study of their history alone that Irishmen will find this just pride restored, and their courage assured."[111]

While the preface was a passionate testimony of Stopford Green's historical and ideological convictions, she ushered readers out of the book with similar force. The last chapter demonstrated how the English had deliberately manipulated Irish history to gain control of the locals and how this had eroded the historical consciousness in Ireland. She adopted the edges of the pages to drive this message into the minds of her readers. She called the chapter tellingly "The Political Myth and its Consequences" and the analytical running heads were equally revealing. Running heads, typically, indicated a subject matter or an abstract idea, but as *The Making of Ireland* shows, they were not neutral descriptions of the narrative. Stopford Green loaded them with direct and indirect messages as she used running heads such as "The effacing of history,"

"Origin of the fables," "The picture of barbarism," "Legend in modern history," "The barbarian legend," and "The uses of history." The final two echoed her wish to restore the pride in Irish history: "The Irish hope" and "Foundations of esteem."

The footnotes, too, were infused with anti-English sentiments and highlighted the liminality of notes between scholarly and non-scholarly utterances. Stopford Green talked in the final chapter, for instance, about school histories and their biases and extended the discussion to the accompanying notes. Just like Bury, she grasped how history was used in schools to generate patriotic sentiments. She was worried about what kind of lessons Irish children were taught to draw from the past. She returned to this topic on several occasions. She complained in one of her public lectures how history was taught in Irish schools from an English point of view. Accordingly, the history of Ireland began only in 1066, the periodization derived from the changes in England, not in Ireland, and the "centuries of Ireland's history as a free and independent country are blotted out."[112] In her book, she protested the dominance of English narrative as children were told how poverty, idleness, and barbarism had caused the "Irish misfortunes" throughout the ages. She provided evidence for these claims in a footnote where she interestingly relied on her own eyewitness testimony: "A few weeks ago," she wrote, "I saw children in an Irish school instructed from a reading book beginning 'The southern part of the island in which we live was not always called England.' The map of Ireland was rolled up behind the door."[113]

In another note, she pointed out how the handbooks that were used in Cambridge and Oxford were spoiled by English prejudices. She did not hesitate to censor the venerable seats of learning for fostering biases, and she enhanced her message by stating how even an eminent historian such as Stubbs had recommended to students books like *The Annals of England; an Epitome of English History* that contained falsehoods about Irish history. Questioning Stubbs was both daring and ingenious, as it underlined how ignorant or indifferent even the scholarly elite in England was about Irish history. Her grim conclusion was that nothing had changed since the days of Stubbs. The universities now used C. R. L. Fletcher's *An Introductory History of England* which, according to her, was hostile toward Ireland. She also mentioned in the note that Fletcher was a fellow at Magdalen College, to show how "ignorance and bigotry" continued to live in Oxford. She concluded by sharply adding how the fact "That such gross errors and fantastic absurdities could be printed in a 'historical' work, without fear of awakening any protest, is a curious illustration of the state of historical study so far as Ireland is concerned."[114]

Stopford Green was so vexed about the history teaching in schools that she returned to the topic in the preface she wrote for the revised edition which Macmillan issued in 1909. She appealed to her readers with a short anecdote about an "Irish scholar" who shared her opinion about the wretchedness of the textbook histories in English schools:

> The suggestion was made to me by an Irish scholar that in the interests of goodwill I should omit some reflections on modern English versions of Irish history; the suggestion was voluntarily withdrawn after a visit to England where this gentleman happened to see the effects produced in schools by teaching from such books.[115]

She borrowed here authority from an anonymous Irish scholar to confirm her claim that there was an urgent need for textbooks that were free from English prejudices. She closed the new preface by reminding her readers once more about the devastating impact which biased histories could have on children's sense of Irishness: "Those who have seen the results, both in schools and among general readers, must deplore the present state of things."[116] While Stopford Green argued that history teaching was politicized in schools and the ancient universities, she of course hoped that readers would consider her book as an alternative to the dominant Anglophone narrative in schools.

The paratexts framed *The Making of Ireland* as a specimen of radical history writing and, unsurprisingly, Stopford Green's polemical arguments and her unconventional manner of using Irish sources such as folklore engendered controversy. Among her like-minded readers she was celebrated for her innovativeness and mastery of primary sources and associated with scientific history. She was complimented for filling a void in Irish history and it was predicted that her discoveries might come as an unpleasant revelation to English readers accustomed to the English narrative.[117] Her brand of Irish history irritated the scientific historians in England and those who subscribed to the unionist narrative. They dismissed her arguments about the Irishness of medieval Ireland as an exaggeration and falsification of historical facts.[118] But it was not just her message that they found dangerous. It was the mixture of scholarliness and luminous style that made them worried. Kenny had predicted the book "to vindicate honor and renown of our beloved country to the world" precisely because of this reason. "We have had a few brilliant writers of historical monographs ... but no patriotic who was at once a scholar and a stylist," he wrote to her adding with great admiration: "You stand alone."[119]

There was, then, good reason for her critics to be anxious. They teamed up and presented themselves as representatives of the scholarly elite whose reputation they were protecting from a renegade historian who endangered their dignity with her radicalism. One by one her antagonists condemned her partisanship and undermined her scholarly competence by listing her apparent methodological distortions. They talked about ingenious explanations, wild assumptions, selective use of sources, and of bending facts to suit the hypothesis. They also made clear that all this was incompatible with the virtues of a scientific historian. The *Academy* reminded how "the main business of the historian is with the accurate presentation of facts."[120] The *Saturday Review* seconded this and excluded her from the class of scientific historians:

> It is no doubt difficult to treat the subject without trenching in some degree on topics that are even now contentious; but it ought surely to be within the competence of a skilled historian to touch on controversial subjects without being controversial.[121]

This thread was picked up by Robert Dunlop, who had earlier refuted Hickson's *Ireland in the Seventeenth Century*. Now he showed no appreciation for Stopford Green's history. Both women were too radical for his moderate unionist taste. Dunlop's fierce assault was published in the *Quarterly Review* in January 1909 with the revealing title, "Truth and Fiction in Irish History." He joined the ranks of those who feared that historians like Stopford Green posed a genuine threat to the discipline. Her book was exceptionally alarming because she had earned with *Town Life in the Fifteenth Century* (1894) a reputation "as an historical writer" and her public image together with "the evidence of wide reading displayed in her book" and "the array of references" which accompanied her statements mislead less-informed readers to trust her arguments.[122] Dunlop, then, had a duty to protect the unsuspecting reader, and he did his utmost to discredit her. Just like others before him, he presented abundant proofs of her incompetence and intentional mishandling of historical evidence, rounding off the review sternly:

> We deeply regret that Mrs Green has written this book. No doubt it will ensure her a certain popularity in circles where history is treated as the slave of politics; but it will be at the expense of forfeiting the respect of those who regard history as a serious subject, and the office of historian as one not lightly to be assumed.[123]

The criticism did not surprise Stopford Green. She had anticipated it and rightly calculated that the polemics would serve as unpaid advertisements for her book.[124] She replied to Dunlop in the *Nineteenth Century* because his assault was so excruciating and humiliating, but the others she answered collectively in the preface to the revised edition. Confidently she pronounced that although she had corrected "various errors of detail," she had limited the alterations "to minor points … [that] do not affect my general line of argument or my conclusions, which I have seen no reason to change."[125] To convince her opponents of the veracity of her arguments, she added to the new edition an appendix that contained more evidence to support her claims. As such, these attempts were futile: no amount of archival material could have changed the minds of her opponents. The deep-seated mistrust of the different political and sectarian parties, long-standing prejudices about Irish historians, and the habit of pinning current political questions to past precedents ensured that it was practically impossible to reconcile the rival notions of Irish history.

The scholarly establishment defended the authority and integrity of history by distancing itself from radical history writing. The copiousness of the remarks on the historian's office, methodological commitments, and ethical obligations indicate that they could not ignore historians such as Stopford Green whom the public associated with proper history writing. Although the historians who fostered a rebellious persona irritated the scientific historians, the alternate models were not entirely without value for them. They allowed the scholarly inner circle to demonstrate the rights and wrongs in history writing and to project themselves as the guardians of virtuousness.

The scientific historians were aware that impartiality was an unattainable goal, yet they agreed with Stubbs who professed how even "an honest partisan" could write "an honest book" if he only recognized his prejudices and kept them at bay.[126] This kind of talk was part of the persona project as historians tried to convince the public of the scientific quality of historical knowledge. Since detachment was at the core of the scholarly persona, they could not afford to ignore the politically charged "pamphlets" that the ideologically inspired historians wrote. The rival persona of a radical historian evoked strong opposition because it exposed how effortlessly history could be employed to advance political agendas. Paul Ferdinand Willert, Fellow and tutor at Exeter College in Oxford, wrote about this compellingly in the *English Historical Review* in 1886:

> Facts, whatever they once were, have long ceased to be stubborn. We have become adepts in the art of manipulating and bending them to suit our theories. So may the worm be fitted to any hook. But facts, like the worm, should be treated with some little consideration, some affectation of gentleness, if not of love.[127]

As nationalism swept through Europe, self-reflection and distancing between the past and the present proved to be empty words. Historians dived into foregone times to unearth material on which to build their national identities and political campaigns. This made the scientific historians' quest for impartiality look like a utopian endeavor and the "public intellectual" appear quite an innocent figure.

Notes

1. [Laughton], "Frederick II.," 422.
2. Mandler, *English National Character*, 116.
3. Seeley, "History and Politics," 298–299.
4. Seeley to Browning, February 2, s.a., GBR/0272/OB/1/1455, King's Cam.
5. Jann, *Art and Science*, 205–206, 223–224.
6. "Stubbs's Constitutional History," *John Bull*, July 20, 1878, 462–463; "History of the Great Civil War," *Athenaeum*, April 20, 1885, 495–496; Adamson, "Eminent Victorians," 641–642.
7. Seeley to Browning, April 6, [1887–1888], GBR/0272/OB/1/1455/A, King's Cam.
8. Collini, *Absent Minds*, 46–52; Jones, "Historical Mindedness," 299–303.
9. Smith, "*Manly Study*," 3.
10. Leerssen, *Remembrance*, 7.
11. Seeley, "History and Politics," 297–299.
12. "History of England," *Athenaeum*, May 21, 1887, 665.
13. Lang, "History as She Ought to Be," 266.
14. Spongberg, *Women and the Nation*, 1–8; Malečková, "Where are Women," 178–182.

15. Robinson, "Review of J. Dufresne de Beaucourt," 161–162.
16. Paul, "Distance and Self-Distanciation," 104–107.
17. Murphy and Traninger, "Introduction," 1–10.
18. Hesketh, *Science of History*, 6–7, 97–99.
19. Bateson, "Review of George W. Child," 382.
20. Daston and Galison, *Objectivity*, 36–37, 42–43.
21. Seeley, "History and Politics [II]," 374; Turner, *Contesting Cultural Authority*, 288.
22. Seeley, "History and Politics [II]," 374.
23. Paul, "Whole Man," 262–269.
24. Hutton, *Sir Thomas More*, viii.
25. Seeley, "History and Politics [III]," 449, 458.
26. Freeman to Thompson, May 18, 1878, U DX9/77, HHC.
27. Garnett, "Review of Ludwig von Pastor," 782.
28. Burrows, *Imperial England*, iii–iv; Gardiner, *History of England*, 10:vi–ix; Lecky, *Historical and Political*, 40–41.
29. Gardiner to Browning, March 7, 1881, GBR/0272/OB/619/A, King's Cam; Gardiner to Edward Westermarck, May 13, 1899, Edward Westermarck Papers X, Åbo Akademi; Gardiner, *History of England*, 10:v–vii.
30. "Historical Essays," *Athenaeum*, October 14, 1871, 493.
31. "Mr. E. A. Freeman," *John Bull*, July 25, 1891, 472–473.
32. "Mr. Freeman on Turkey," *Manchester Courier*, May 9, 1877, 3.
33. "Notes of the Week," *BCWG*, August 9, 1877, 4.
34. "The Regius Professor of Modern History," *John Bull*, December 27, 1884, 850.
35. Stubbs, *Seventeen Lectures*, 63.
36. Hobsbawm, "Mass-Producing Traditions," 263. For an overview of nationalist history writing in different national contexts, see *The Contested Nation: Ethnicity, Class, Religion and Gender in National Histories*, edited by Stefan Berger and Chris Lorenz (European Science Foundation, 2008); *Political Uses of the Past. The Recent Mediterranean Experience*, edited by Jacques Revel and Giovanni Levi (Franc Cass., 2002); *Writing National Histories: Western Europe Since 1800*, edited by Stefan Berger, Mark Donovan, and Kevin Passmore (Routledge, 1999).
37. Powell, "General Survey of Modern History," in Elton, *Frederick York Powell*, 2:8–13.
38. Bury, "Science of History," 13.
39. Towsey, *Reading History*, 152–155.
40. Leerssen, *Remembrance*, 4, 11–12, 101; Hachey and McCaffrey, *Irish Experience*, 114–115; Lane, "Ireland," 230–231.
41. O'Day, "Perspectives on Irish Identity," 59; Graham, "Ireland and Irishness," 2.
42. Lee, "Review of Richard Bagwell," 378.

43. *Memoir*, 1909, 321.
44. Leerssen, *Remembrance*, 9.
45. O'Brien and Lunney, "O'Connor"; O'Sullivan, "O'Connor," *ODNB*.
46. Gibney, *Shadow of a Year*, 8.
47. Smith, "*Manly Study*," 17; McCartney, *W. E. H. Lecky*, 71.
48. T.C.D. "*The Irish Rebellion*," 364.
49. Smith, "*Manly Study*," 15–21.
50. Hickson, *Ireland*, 1:155–167.
51. Gibney, *Shadow of a Year*, 9–10, 133–138. The depositions have been accessible online since 2010 via http://1641.tcd.ie.
52. Maitland to Poole, April 22, 1902, GBR/0012/MS Add. 7474/32, CUL.
53. Some rare exceptions to the absence of systematic scholarship in document editing in the nineteenth century include Saxer, "Monumental Undertakings," 47–64 and Poncet, *Les entreprises éditoriales*, 1–52. One of the few continental women editors was the Danish Anna Hude. Another exception was the Finnish Liisi Karttunen. Manniche, "Umaettelig Kundskabstorst," 152–153; Garritzen, "International Historical Institutes," 38.
54. Krueger, "Why She Lived at PRO," 72, 87–88.
55. Froude to Lady Carnarvon, August 18, [1885], and September 5, [1885], Carnarvon MSS, Add MS 60799B, BL.
56. Thirsk, "Women Local and Family Historians."
57. Garritzen, "International Historical Institutes," 52; Paul, "Heroic Study of Records," 76–77.
58. Gardiner, "Ireland," July 26, 1884, 53.
59. [Brewer], "Stuarts," 169.
60. Genette, *Paratexts*, 263–275.
61. English, "Literary Patronage," 127–128.
62. English, "Literary Patronage," 130–132.
63. Dunn, *James Anthony Froude*, xii.
64. Brady, *James Anthony Froude*, 281–296.
65. McCartney, *W. E. H. Lecky*, 86–87.
66. Elton, *Frederick York Powell*, 1:169.
67. Brady, *James Anthony Froude*, 22–27.
68. J.A.F., "Preface," v–xii; Garritzen, "Women Historians," 659–660.
69. Croskery, "Irish Massacres," 495.
70. "Ireland in the Seventeenth Century," *PMG*, July 26, 1884, 4.
71. "Literary Notes," *PMG*, February 7, 1884, 4. See also *Freeman's Journal*, February 8, 1884, 5; *Leeds Mercury*, February 11, 1884, 2; *Western Daily Press*, February 11, 1884, 7; *Derby Daily Telegraph*, February 12, 1884, 4; *Bath Chronicle*, February 14, 1884, 6; *Leicester Chronicle*, February 16, 1884, 4.

72. "*Ireland in the Seventeenth Century*," *Scottish Review*, October 1884, 359.
73. Prendergast quoted in Gibney, *Shadow of a Year*, 137.
74. Dunlop, "Depositions," *EHR* 1, no. 4, 740–744; Garritzen, "Women Historians," 660–661.
75. Hickson, "Depositions," *EHR* 2, no. 5, 133–137.
76. Dunlop, "Depositions," *EHR* 2, no. 6, 339.
77. Hickson, "Depositions," *EHR* 2, no. 7, 527.
78. "Review of Louis Leger," *EHR* 5, no. 18, 396.
79. Hickson, *Ireland* 1:156, 2:180.
80. Gardiner, *History of England*, 10:68.
81. Gardiner, "Ireland," *Academy*, July 26, 1884, 53.
82. Hickson, "Ireland," *Academy*, August 9, 1884, 95.
83. Gardiner, "Ireland," *Academy*, September 13, 1884, 169.
84. Hickson, "Ireland," *Academy*, September 6, 1884, 153–154, and September 27, 1884, 203.
85. Lecky, "Ireland," 121–122.
86. Hickson, "Ireland," *Academy*, September 6, 1884, 153–154.
87. McCartney, *W. E. H. Lecky*, 74.
88. Croskery, "Irish Massacres," 491–500, 523–524.
89. Malone, "Ireland," 475–477.
90. Kenny to Alice Stopford Green, July 7, 1905, MS 15,082/1/5, NLI.
91. Stopford Green, *Old Irish World*, 1–2.
92. Stopford Green, *Old Irish World*, 18, 22–23.
93. Mitchell, "Historical Revisit," 351, 360–362.
94. Mitchell, "Historical Revisit," *Irish Historical Studies* (2020); Mitchell and Ní Bheacháin, "Scholar-Diplomats," *Women's History Review* (2021); Kingstone, "Feminism, Nationalism, Separatism?," *Journal of Victorian Culture* (2014); Holton, "Gender Difference," *History Workshop Journal* (2002).
95. Garritzen, "Framing and Reframing," 185–186.
96. Harper Brothers to Stopford Green, December 5, 1895, Stopford Green to Harper Brothers, August 8, 1896, Stopford Green to George Brett, June 1898, MS 15,124/3, NIL.
97. Powell to Stopford Green, January 23, 1895, in Elton, *Frederick York Powell*, 1:182–183.
98. Stopford Green, *Old Irish World*, 61.
99. Mitchell, "Historical Revisit," 352–354; Mitchell and Ní Bheacháin, "Scholar-Diplomats," 204–219; Kingstone, "Feminism, Nationalism, Separatism?," 443; Holton, "Gender Difference," 120; Kirkland, *Irish London*, 154, 178.
100. Stopford Green, *Old Irish World*, 188.
101. Stopford Green, *Old Irish World*, 36–37.

102. Holton, "Gender Difference," 124.
103. Kenny to Stopford Green, December 3, 1906, MS 15,082/2/7, NIL.
104. Stopford Green to Macmillan, February 12, 1908, May 22, 1908, Add MSS 55059, BL.
105. Stopford Green to Macmillan, January 1, 1909, February 5, 1909, Add MSS 55060, BL.
106. McMahon, "All Creeds," 134–140, 153–154; Leerssen, *Remembrance*, 157–159; Pašeta, *Irish Nationalist Women*, 19–27.
107. Stopford Green to Macmillan, February 14, 1908, Add MSS 55059, BL.
108. Stopford Green to Macmillan, February 14, 1908, Add MSS 55059, BL.
109. Mitchell, "Historical Revisit," 349.
110. Genette, *Paratexts*, 197, 221.
111. Stopford Green, *Making of Ireland*, x–xi.
112. Stopford Green, *Old Irish World*, 45–47.
113. Stopford Green, *Making of Ireland*, 490.
114. Stopford Green, *Making of Ireland*, 486.
115. Stopford Green, *Making of Ireland*, revised ed., xv.
116. Stopford Green, *Making of Ireland*, revised ed., xv.
117. "An Epoch," *Academy*, July 25, 1908, 79–80; Barry, "Ireland's Case," 221–222.
118. Orpen, "Making of Ireland," 134; "Plantagenet Ireland," *SR*, July 11, 1908, 50–51.
119. Kenny to Stopford Green, April 27, 1907, MS 15,082/2/3, NLI.
120. "An Epoch," *Academy*, July 25, 1908, 80; Orpen, "Making of Ireland," 131–134.
121. "Plantagenet Ireland," *SR*, July 11, 1908, 51.
122. Dunlop, "Truth and Fiction," 259.
123. Dunlop, "Truth and Fiction," 275.
124. Stopford Green to Macmillan, February 12, 1908, July 28, 1908, Add MSS 55059, BL.
125. Stopford Green, *Making of Ireland*, revised ed., xv.
126. Stubbs, *Seventeen Lectures*, 109.
127. Willert, "Review of W. S. Lilly," 558.

References

Unpublished Primary Sources

British Library: Carnarvon Manuscripts.
British Library: The Macmillan Papers.
Cambridge University Library: Reginald Lane Poole Papers.

Hull History Centre: Letters from Edward Augustus Freeman to Edith Thompson.
King's College Archive Centre, Cambridge: Oscar Browning Papers.
National Library of Ireland, Dublin: Alice Stopford Green Additional Papers.
Åbo Akademi, Turku: Edward Westermarck Papers.

Printed Primary Sources

[Anon.]. "Historical Essays." *Athenaeum*, October 14, 1871, 493.
[Anon.]. "Mr. Freeman on Turkey and the Eastern Question." *Manchester Courier and Lancashire General Advertiser*, May 9, 1877, 3.
[Anon.]. "Notes of the Week." *Bath Chronicle and Weekly Gazette*, August 9, 1877, 4.
[Anon.]. "Stubbs's Constitutional History." *John Bull*, July 20, 1878, 462–463.
[Anon.]. "Literary Notes." *Pall Mall Gazette*, February 7, 1884, 4.
[Anon.]. "Ireland in the Seventeenth Century." *Pall Mall Gazette*, July 26, 1884, 4–5.
[Anon.]. "*Ireland in the Seventeenth Century.*" *The Scottish Review*, October 1884, 358–361.
[Anon.]. "The Regius Professor or Modern History at Oxford." *John Bull*, December 27, 1884, 850.
[Anon.]. "History of the Great Civil War." *Athenaeum*, April 20, 1885, 495–496.
[Anon.]. "History of England in the Eighteenth Century." *Athenaeum*, May 21, 1887, 665–666.
[Anon.]. "Review of Louis Leger's *History of Austro-Hungary*." *English Historical Review* 5, no. 18 (1890): 396.
[Anon.]. "Mr. E. A. Freeman." *John Bull*, July 25, 1891, 472–473.
[Anon.]. "Plantagenet Ireland." *Saturday Review*, July 11, 1908, 50–51.
[Anon.]. "An Epoch of Irish History." *Academy*, July 25, 1908, 79–80.
Barry, William. "Ireland's Case Stated." *Bookman*, September 1908, 221–222.
Bateson, Mary. "Review of Gilbert W. Child's *Church and State under the Tudors*." *English Historical Review* 6, no. 22 (1891): 381–383.
[Brewer, J. S.]. "The Stuarts." *Quarterly Review*, July 1872, 167–199.
Burrows, Montagu. *Imperial England*. London: Cassell, 1880.
Bury, J. B. "The Science of History." In *Selected Essays of J. B. Bury*, edited by Harold Temperley, 3–22. Cambridge: Cambridge University Press, 1930.
[Croskery, Thomas]. "The Irish Massacres of 1641." *Edinburgh Review*, October 1884, 490–524.
Dunlop, R. "The Depositions Relating to the Irish Massacres of 1641." *English Historical Review* 1, no. 4 (1886): 740–744.

Dunlop, R. "The Depositions Relating to the Irish Massacre of 1641." *English Historical Review* 2, no. 6 (1887): 338–340.
Dunlop, Robert. "Truth and Fiction in Irish History." *Quarterly Review*, January 1909, 254–275.
Elton, Oliver. *Frederick York Powell. A Life and a Selection from His Letters and Occasional Writings*. Vol. 1. Oxford: Clarendon Press, 1906.
Elton, Oliver. *Frederick York Powell. A Life and a Selection from His Letters and Occasional Writings*. Vol. 2. Oxford: Clarendon Press, 1906.
Gardiner, Samuel Rawson. *History of England from the Accession of James I. to the Outbreak of the Civil War, 1603–1642*. Vol. 10. London: Longman, 1884.
Gardiner, Samuel R. "Ireland in the Seventeenth Century." *Academy*, July 26, 1884, 53.
Gardiner, Samuel R. "Ireland in the Seventeenth Century." *Academy*, September 13, 1884, 169.
Garnett, R. "Review of Ludwig von Pastor's *Geschichte der Päpste*." *English Historical Review* 5, no. 20 (1890): 782–783.
Hickson, Mary. *Ireland in the Seventeenth Century or the Irish Massacre of 1641–42, Their Causes and Results*. Vol. 1. London: Longman, 1884.
Hickson, Mary. *Ireland in the Seventeenth Century or the Irish Massacre of 1641–42, Their Causes and Results*. Vol. 2. London: Longman, 1884.
Hickson, Mary. "Ireland in the Seventeenth Century." *Academy*, August 9, 1884, 95.
Hickson, Mary. "Ireland in the Seventeenth Century." *Academy*, September 6, 1884, 153–154.
Hickson, Mary. "Ireland in the Seventeenth Century." *Academy*, September 27, 1884, 203.
Hickson, Mary. "The Depositions Relating to the Irish Massacre of 1641." *English Historical Review* 2, no. 5 (1887): 133–137.
Hickson, Mary. "The Depositions of 1641." *English Historical Review* 2, no. 7 (1887): 527.
Hutton, William Holden. *Sir Thomas More*. London: Methuen, 1895.
J.A.F. "Preface." In *Ireland in the Seventeenth Century or the Irish Massacre of 1641–2*, by Mary Hickson, v–xiii. Vol. 1. London: Longman, 1884.
Lang, A. "History as She Ought to Be Wrote." *Blackwood's Edinburgh Magazine*, August 1899, 266–274.
[Laughton, John Knox]. "Frederic II. and Maria Theresa." *Edinburgh Review*, April 1883, 384–422.
Lecky, W. E. H. "Ireland in the Seventeenth Century." *Academy*, August 23, 1884, 121–122.
Lecky, William Edward Hartpole. *Historical and Political Essays*. London: Longman, 1908.

Lee, Sydney L. "Review of Richard Bagwell's *Ireland under the Tudors*." *English Historical Review* 2, no. 6 (1887): 378–379.
Malone, Sylvester. "Ireland in the Seventeenth Century." *Dublin Review*, October 1884, 475–478.
A Memoir of The Right Hon. William Edward Hartpole Lecky, by his wife. London: Longman, 1909.
Orpen, Goddard H. "The Making of Ireland and Its Undoing." *English Historical Review* 24, no. 93 (1909), 129–135.
Robinson, Mary F. "Review of J. Dufresne de Beaucourt's *Histoire de Charles VII*." *English Historical Review* 4, no. 13 (1889): 161–167.
Seeley, J. R. "History and Politics." *Macmillan's Magazine*, August 1879, 289–299.
Seeley, J. R. "History and Politics [II]." *Macmillan's Magazine*, September 1879, 369–378.
Seeley, J. R. "History and Politics [III]." *Macmillan's Magazine*, October 1879, 449–458.
Stopford Green, Alice. *The Making of Ireland and Its Undoing, 1200–1600*. London: Macmillan, 1908.
Stopford Green, Alice. *The Old Irish World*. London: Macmillan, 1912.
Stopford Green, Alice. *The Making of Ireland and Its Undoing, 1200–1600*. Revised ed. London: Macmillan, 1913.
Stubbs, William. *Seventeen Lectures on the Study of Medieval and Modern History and Kindred Subjects*. Oxford: Clarendon Press, 1886.
T.C.D. "*The Irish Rebellion of* 1641." *The British Quarterly Review*, October 1884, 353–364.
Willert, P. F. "Review of W. S. Lilly's *Chapters in European History*." *English Historical Review* 1, no. 3 (1886): 556–561.

Secondary Sources

Adamson, J. S. A. "Eminent Victorians: S. R. Gardiner and the Liberal as Hero." *Historical Journal* 33, no. 3 (1990): 641–657.
Brady, Ciaran. *James Anthony Froude: An Intellectual Biography of a Victorian Prophet*. Oxford: Oxford University Press, 2013.
Collini, Stefan. *Absent Minds: Intellectuals in Britain*. Oxford: Oxford University Press, 2006.
Daston, Lorraine and Peter Galison. *Objectivity*. New York: Zone Books, 2010.
Dunn, Waldo Hilary. *James Anthony Froude: A Biography, 1857–1894*. Oxford: Clarendon Press, 1963.
English, Jeri. "Literary Patronage: Collaboration or Rivalry? Women's Prefaces to Women's Texts in the 20th Century." *Women in French Studies* 3 (2010): 126–139.

Gibney, John. *The Shadow of a Year: The 1641 Rebellion in Irish History & Memory*. Madison: University of Wisconsin Press, 2013.

Garritzen, Elise. "The International Historical Institutes in Rome and Their Scientific and Political Roles c. 1880–1914." *Storia della Storiografia* 64, no. 2 (2013): 37–60.

Garritzen, Elise. "Framing and Reframing Meanings in History Books: The Original and Posthumous Paratexts in J. R. Green's *Short History of the English People*." *History of Humanities* 3, no. 1 (2018): 177–197.

Garritzen, Elise. "Women Historians, Gender, and Fashioning the Authoritative Self in Paratexts in Late-Victorian Britain." *Women's History Review* 30, no. 4 (2021): 651–668.

Genette, Gerard. *Paratexts: Thresholds of Interpretation*. Cambridge: Cambridge University Press, 1997.

Graham, Brian. "Ireland and Irishness: Place, Culture and Identity." In *In Search of Ireland: A Cultural Geography*, edited by Brian Graham, 1–15. London: Routledge, 1997.

Hachey, Thomas E. and Lawrence J. McCaffrey. *The Irish Experience Since 1800: A Concise History*. Armonk, NY: M. E. Sharpe, 2010.

Hesketh, Ian. *The Science of History in Victorian Britain: Making the Past Speak*. London: Pickering & Chatto, 2011.

Hobsbawm, Eric. "Mass-Producing Traditions: Europe, 1870–914." In *The Invention of Tradition*, edited by Eric Hobsbawm and Terence Ranger, 263–307. Cambridge: Cambridge University Press, 1995.

Holton, Sandra. "Gender Difference, National Identity and Professing History: The Case of Alice Stopford Green." *History Workshop Journal* 53 (2002): 119–127.

Jann, Rosemary. *The Art and Science of Victorian History*. Columbus: Ohio State University Press, 1985.

Jones, H. S. "Historical Mindedness and the World at Large: E. A. Freeman as Public Intellectual." In *Making History: Edward Augustus Freeman and Victorian Cultural Politics*, edited by G. A. Bremner and Jonathan Conlin, 293–310. London: British Academy, 2015.

Kingstone Helen, "Feminism, Nationalism, Separatism? The Case of Alice Stopford Green." *Journal of Victorian Culture* 19, no. 4 (2014): 442–456.

Kirkland, Richard. *Irish London: A Cultural History 1850–1916*. London: Bloomsbury Academic, 2022.

Krueger, Christine L. "Why She Lived at PRO: Mary Anne Everett Green and the Profession of History." *The Journal of British Studies* 42, no. 1 (2003): 65–90.

Lane, Leann. "Ireland: Identities and Cultural Traditions." In *Palgrave Advances in Irish History*, edited by Mary Mcauliffe, Katherinee O'donnell, and Leeann Lane, 222–246. Basingstoke: Palgrave Macmillan, 2009.

Leerssen, Joep. *Remembrance and Imagination: Patterns in the Historical and Literary Representation of Ireland in the Nineteenth Century*. Notre Dame: University of Notre Dame Press, 1997.

Malečková, Jitka. "Where Are Women in National Histories?" *The Contested Nation: Ethnicity, Class, Religion and Gender in National Histories*, edited by Stefan Berger and Chris Lorenz, 171–199. New York: Palgrave Macmillan, 2008.

Mandler, Peter. *The English National Character: The History of an Idea from Edmund Burke to Tony Blair*. New Haven: Yale University Press, 2006.

Manniche, Jens Chr. "En umaettelig kundskabstorst: Anna Hude, Danmarks forste kvindelige historiker." In *Clios dotre gennem hundrede år*, edited by Marianne Alenius, Nanna Damsholt, and Bente Rosenbeck, 141–164. Copenhagen: Museum Tusculanums Forlag, 1994.

McCartney, Donald. *W. E. H. Lecky: Historian and Politician 1838–1903*. Dublin: The Lilliput Press, 1994.

McMahon, Timothy G. "'All Creeds and All Classes'? Just Who Made Up the Gaelic League." *Éire–Ireland* 37, no. 3–4 (2002): 118–168.

Mitchell, Angus. "Historical Revisit: Mythistory and the Making of Ireland: Alice Stopford Green's Undoing." *Irish Historical Studies* 44, no. 166 (2020): 349–373.

Mitchell, Angus and Caoilfhionn Ní Bheacháin. "Scholar-Diplomats, Protodiplomacy and the Communication of History: Alice Stopford Green and Jean Jules Jusserand." *Women's History Review* 31, no. 2 (2022): 198–229.

Murphy, Kathryn and Anita Traninger. "Introduction: Instances of Impartiality." In *The Emergence of Impartiality*, edited by Kathryn Murphy and Anita Traninger, 1–29. Leiden: Brill, 2014.

O'Brien, Andrew and Linde Lunney. "O'Connor, Charles." In *Dictionary of Irish Biography*. https://www.dib.ie/index.php/biography/oconor-charles-a6654. Accessed January 4, 2021.

O'Day, Alan. "Perspectives on Irish Identity, Nationalism and Ethnicity." In *History, Nationhood and the Question of Britain*, edited by Helen Brocklehurst and Robert Phillips, 57–72. Basingstoke: Palgrave Macmillan, 2004.

O'Sullivan, William. "O'Connor, Charles." In *Oxford Dictionary of National Biography*. https://doi-org.libproxy.helsinki.fi/10.1093/ref:odnb/20526. Accessed January 4, 2021.

Pašeta, Senia. *Irish Nationalist Women, 1900–1918*. Cambridge: Cambridge University Press, 2013.

Paul, Herman. "Distance and Self-Distanciation: Intellectual Virtue and Historical Method Around 1900." *History and Theory* 50 (2011): 104–116.

Paul, Herman. "The Heroic Study of Records: The Contested Persona of the Archival Historian." *History of the Human Sciences* 26, no. 4 (2013): 67–83.

Paul, Herman. "The Whole Man: A Masculine Persona in German Historical Studies." In *Gender, Embodiment, and the History of the Scholarly Persona: Incarnations and Contestations*, edited by Kirsti Niskanen and Michael J. Barany, 261–286. Cham: Palgrave Macmillan, 2021.
Poncet, Olivier. *Les entreprises éditoriales liées aux Archives du Saint-Siège*. Rome: École Française de Rome, 2003.
Saxer, Daniela. "Monumental Undertakings: Source Publications for the Nation." In *Setting the Standards: Institutions, Networks and Communities of National Historiography*, edited by Ilaria Porciani and Jo Tollebeek, 47–69. Basingstoke: Macmillan, 2012.
Smith, Nadia Clare. *A "Manly Study"? Irish Women Historians, 1868–1949*. Basingstoke: Palgrave Macmillan, 2006.
Spongberg, Mary. *Women and the Nation's Past: Empathetic Histories*. London: Bloomsbury, 2018.
Thirsk, Joan. "Women Local and Family Historians." In *Oxford Companion to Family and Local History*, edited by David Hey. Oxford: Oxford University Press, 2009. https://doi.org/10.1093/acref/9780199532988.001.0001. Accessed December 3, 2021.
Towsey, Mark. *Reading History in Britain and America, c.1750–c.1840*. Cambridge: Cambridge University Press, 2019.
Turner, Frank M. *Contesting Cultural Authority: Essays in Victorian Intellectual Life*. Cambridge: Cambridge University Press, 1993.

PART III

Historians as Entrepreneurs

"I do not profess to understand the art of advertising," Freeman announced to one of Macmillan's press secretaries, and was confident that neither did he. Only a failed marketing campaign could have explained the fiasco of what had become of the sales of *Greater Greece and Greater Britain and George Washington the Expander of England* (1886).[1] Instead of accepting that the awkward title obstructed sales, Freeman insisted that the book failed to stand out on booksellers' shelves and catch the eye in the advertisements because of Macmillan's inadequate marketing.[2] There was nothing exceptional in Freeman's protestation; the omnipresence of advertising spurred historians to blame inefficient marketing and unattractive paratexts for their commercially failed books.[3] The mindset was so widespread among writers that Leopold Wagner's *How to Publish a Book* announced that, against common assumption, publishers had not entered "into a conspiracy" to deny publicity to the books they launched.[4]

A long-standing myth frames publishing as a "noble" business that submits a commercial orientation to the transmission of culture, but historical evidence does not back this claim. The Victorian publishing houses were built to make revenue and publishing evolved into modern business as the century progressed. The prevailing mindset was that writing and publishing could be profitable if managed as commercial enterprises. This market orientation, however, did not necessarily erase the gentility of author–publisher relations which also the scientific historians enjoyed.[5] The university publishers were also forced to modernize their business models and marketing, and to diversify their product range to be able to sustain the production of books with high scholarly but low remunerative value.[6] The learning curve was particularly steep at Cambridge; Seeley admitted that the Delegates were more "pushing and ambitious" and Maitland allowed as late as 1892 that Oxford "manages its books better than does our Press."[7] For the Victorians who lived in a modern age of consumerism, books were objects of conspicuous consumption. The rise of real wages left most of them some disposable income and ingenious marketing lured them to spend their money. Books had been advertised since the invention of printing and many modern marketing techniques originated from the eighteenth century. Nineteenth-century publishers, profiting from technical innovations and the modernizing media space, developed advertising further experimenting with product placement, branding, bandwagon effect, and puffing.[8]

The final part of this book shifts the attention to the entrepreneurial component in the historians' persona as it takes stock of their advertising ventures and promotional paratexts. The following chapters show how the production of such paratexts was teamwork that involved historians, publishers, editors, printers, and family members with strong opinions about styles and designs. Historians accepted that marketing was part of their job description and the domestic financial realities, together with an urge to increase the circulation of their books, inspired them to forge a persona that accommodated the entrepreneurial ethos. Fusing the dimensions of scholar and entrepreneur in the persona came with its own complications. Disputes about promotional paratexts were not only caused by historians' scholarly priorities or inadequate business acumen. Rather, they document the acuteness which historians assigned to their entrepreneurial duties and the difficulty of understanding the publishing trade and the marketplace which were constantly evolving and adapting to the modernizing society. As historians were not merely selling their books, but also their persona and the discipline which the books symbolized, their promotional efforts open up a unique perspective on larger questions that the commodification of knowledge and modern commercial culture posed on Victorian scholarly culture and middle-class morality.

Notes

1. Freeman to Craik, August 15, 1886, Add MSS 55053, BL.
2. Freeman to Macmillan, August 20, 1886, Add MSS 55053, BL.

3. Burrows, *Autobiography*, 243; Swan and Sonnenschein to Browning, December 1891, GBR/0272/OB/1/1597/C, King's Cam.
4. Wagner, *How to Publish a Book*, 80.
5. Miller, *Reluctant Capitalists*, 1–3; Briggs, *History of Longmans*, 251–252, 300–301; Finkelstein, *House of Blackwood*, 130–133; Guy, "Authors, 16–17.
6. Eliot, "Evolution of a Printer," 94–100.
7. Seeley to Browning, April 17, s.a., GBR/0272/OB/1/1455, King's Cam; Maitland to Round, April 10, 1892, in Fifoot, *Letters*, 100.
8. Mason, *Literary Advertising*, 5, 11–13, 24; Raven, *Business of Books*, 269–270.

CHAPTER 8

Commercial yet Scholarly Dignified Historians

According to Powell, there was "no money to be made out of learned books." It was much if they did not leave the writer "poorer." Powell was not too worried about this, for he detested the vulgar "shop" ambition which, according to him, flourished in Oxford. He warned young historians against the great enemy of true learning: "the idea of getting on, getting rich, getting 'biggety'." The pleasure of "having forwarded your subject" should be reward enough for an ambitious historian.[1] If Powell was a classic erudite who detested money and fame, his colleagues had a more intricate relationship with the commodification of historical knowledge. Those with a middle-class background and a limited annuity just could not afford to ignore the business side of history. Even as professors, historians like Seeley were forced to supplement their fee with lecture tours and miscellaneous writing because the endowment of science was meager. The reluctance to use public money for sponsoring scholarship was in tune with the Victorian economic ideology which championed voluntary impulse and the free market.[2] Consequently, historians needed the profits from their books to get on, and Freeman reminded Macmillan that as a father of a family, he could not be indifferent about terms and contracts.[3] Precarious budgets caused considerable stress for historians, as Victorian middle-class men had a duty to keep a family. This was also a matter of reputation and respectability, as personal financial crises

E. Garritzen, *Reimagining the Historian in Victorian England*,
https://doi.org/10.1007/978-3-031-28461-8_8

indicated recklessness and lack of humility, patience, industry, and self-restraint. The proper management of money was a moral responsibility according to the Malthusian doctrine.[4] But this was not all: sales figures also had a direct bearing on historians' scholarly self-esteem, as profit was interpreted as an indicator of success. Scientific worth was equated with economic worth and high sales figures symbolized popularity. It is not surprising at all that the entrepreneurial component in the persona provoked strong emotions in historians.

Byerley Thomson, barrister and Colonial Secretary, contended in *The Choice of a Profession* (1857) that business secured early access to wealth while a profession paved the way to a social position and honor.[5] This captured the essence of the economic, moral, social, and cultural contrasts between the commercial and professional middle classes. The aim of this chapter is to unravel historians' attempts to reconcile their participation in the literary marketplace and the integrity of their persona. Historians and other writers who made a living with the pen were troubled about the close resemblance of their occupation to commerce because the production of manuscripts blurred uncomfortably the distinction between the trades and professionals. The boundary was also undermined because the two groups shared such virtues as industry, perseverance, and thrift. The new professionals promoted their elevated status by presenting the pursuit of money as selfish and immoral, tainted by greed, dishonest competition, deception, fraud, and speculation.[6] These qualities stood in stark contrast with the epistemic goods to which historians aspired, and the fashioning of their entrepreneurial persona was linked to both their disciplinary and social ambitions. Although historians avoided branding themselves as professionals, they were influenced by the discourses of commercial and professional society. In this context, a liberal profession was defined as a possession of such knowledge that was applied for the benefit of society, while trade indicated the manufacturing and selling of objects for personal gain. This idea of professionals using their expert knowledge to the common good was a distinct philosophy of the professional citizenship that helped maintain the hierarchies between commercial and professional society.[7]

The uneasiness about profit making with literary endeavors had its roots in the emergence of modern authorship in the eighteenth century when writing out of love and writing out of a need to make a living were contrasted. The industrialization of publishing intensified such conflicts and Carlyle condemned the "omnipotency of money" which he saw as ruining any literary work. By the mid-century, views had begun to relax, and money and profits were no longer perceived as a threat to the respectability of intellectual work. Nevertheless, the friction between economic and moral convictions did not vanish entirely.[8] Accordingly, Freeman assured the electors of the Chichele Professorship in 1862 that he did not seek the position "as a source of income or as a means of professional advancement," but "as a means of doing some service to the study to which I have devoted my life."[9] In reality, historians were so integrated into the Victorian capitalist system that their persona was poised between the selfless production of useful knowledge and selfish gains. It was critical to accommodate intellectual endeavors and economic aspirations without compromising the integrity of history.

This chapter explores how historians negotiated these seemingly contradictory forces and begins by outlining the contours of the entrepreneurial persona and its adaptation for marking historians as virtuous scholars and owners of a respectable social status. The second part investigates how historians propagated this same message by using paratexts to present history books as dignified, non-essential goods and historians as non-commercial professionals. The selling of manuscripts to publishers appeared ambiguous in terms of the contrast between money and public service, and the professionalizing writers, according to Linda Peterson, drew distance from the trades by highlighting the difference between a text and its material form. The commodification of a text was a publisher's duty, and publishing was presented as a trade, while writing was framed as a profession.[10] On the one hand, historians' attempts to accentuate this boundary suggest its impreciseness and importance. On the other, by engaging with the production of paratexts, historians encroached on that very same boundary. It was pivotal that paratexts communicated high scholarly value that placed history books above ordinary daily goods such as food and clothing. This goal made participation in the production of the book legitimate for historians.

Entrepreneurial Persona

Since historians could not escape interacting with the commercial marketplace, they strove to fashion a persona that promoted skills, knowledge, and traits that were critical for a successful transaction with the publishing industry and adjusted the epistemic and moral virtues for a commercial context. The modernizing publishing industry caused unpleasant surprises for poorly prepared authors, and Percy Russell pronounced in *The Literary Manual* how writers could no longer escape the fact that the modern age was an age of sound technical knowledge. Genius alone did not suffice and even in literary authorship success was denied those who refused to cultivate "the *technique* of the vocation they adopt."[11] The management of the publishing process required from historians an acquisition of special knowledge and rehearsal of skills and a commercial vision. This resonated with the replies which Francis Galton received from the 180 scientists whom he interviewed for *English Men of Science* (1874). He was astonished that so many of them appreciated business habits and considered them useful not just for running daily finances, but as an indication of their epistemic and cognitive competence as well, for both demanded a methodical mind. Scientific inquiry benefited from a business-like practical approach which directed attention to what was important and relevant because that made research itself efficient.[12] This played well together with the boundary work between trade and scholarship as it separated the ambition to maximize profits from the organizational aspects of business and enabled the portrayal of the latter as a noble set of skills. These ideas were familiar to historians as well. As Freeman pronounced in one of his many public lectures, "the pursuit of business and the pursuit of knowledge are not inconsistent." Both required effort and self-discipline.[13] This ethical and epistemic dimension granted additional importance to daily business transactions.

Historians knew that they needed special skills when dealing with the publishing industry. Their approach was mostly pragmatic, and the business side of history was a necessity and a routine that simply had to be taken care of. They had to learn to negotiate, bargain, keep records, and administer personal finances. Such knowledge was mostly gained through trial and error, although some received mentoring from more experienced colleagues. The lessons could be expensive for the self-taught entrepreneurial historians. Creighton once regretted how he had sold the copyright of *The Age of Elizabeth* to Longman without foreseeing the book's potential for sustained popularity. This taught him to retain the copyright of his future books.[14] Oman lamented in similar fashion how he had been naïve and sold the rights of his first book, *Warwick the King-Maker*, to Macmillan for £100. This had been a fine deal for a novice, but after the publisher had produced six editions of the book, he had begun to think differently. He, too, requested royalties for his later titles, as that was "far more profitable" for someone who supplied "sound stuff for the reading world."[15]

At least superficial knowledge of the production process proved to be helpful. Furthermore, historians had to conduct rudimentary market research, as they needed information about competing titles, gaps in publishers' catalogs, circulation, and supply channels. The commercial marketplace was their main concern, but they could not overlook the libraries, either. Freeman and Stubbs were particularly well represented on Mudie's list; Freeman had a total of fifteen of his books on it in 1888.[16] Finally, a grasp of the methods and principles of advertising was a definite asset. Historians excelled in this to varying degrees and their enthusiasm could be greater than their competence. George Longman had to order Browning to stop spreading the news about the new edition of his *Modern England*. The publisher struggled to dispose of the remainder of the first edition because the "trade" had got the wind of a new edition thanks to Browning's hastiness.[17]

Freeman's extensive correspondence provides a unique overview of historians' daily business habits. His letters to Macmillan disclose how he negotiated, bargained, and even pitted publishers against each other to improve his personal finances. He, too, preferred royalties to half profits "as with the royalty one knows where one is while the half profits are influenced by advertisements & other things over which one has no control."[18] His letters tell us about his vigilant monitoring of sales reports and constant speculation about the success and failure of his books. They also show how he canvassed advertisements to see whether his books were prominently announced and made countless suggestions to improve the advertising of his books. But the letters also bear witness to financial worries that drove him to borrow money from Macmillan for unexpected personal expenses and to the apprehension and worry which the economic dependence on book sales caused.

Although historians were concerned about their income, they did not imitate the scientists, who became more vocal about the meager public endowment of science during the latter half of the century. Bury and Powell, who demanded more substantial support for history to prevent the destructiveness of nationalist histories, marked a shift in this. Before them, historians had rarely anchored their persona to public funding but accepted instead the popular belief that state intervention reduced the quality of research by endangering intellectual independence and encouraging laziness.[19] Apart from the universities, historical undertakings were supported through the Public Record Office and British Museum, but these institutions became more involved with conservation and preservation and less with the production of original knowledge as the century progressed. Subsequently, distinct and stringent requirements and personae developed for historians, archivists, and curators. The archivists emerged as facilitators of historical research and experts in auxiliary services. An "archivist" was no longer a historian, and being a historian was no longer sufficient preparation for becoming an archivist.[20] When Gardiner was offered a position as a deputy keeper in 1878, he politely turned it down precisely because he rather saw it going to someone who had been "trained within the office."[21] As a regular patron of the PRO, he had witnessed how archivists and historians had parted ways and how both relied on specific kind of education, career structure, and rewards. Only those who were hired to edit records were an exception to this. Otherwise, the field of history was splintering into subfields, each with marked professional ethos despite the shared interest in the past.

The resistance to public endowment enabled historians to underline the morality of their persona as it foregrounded their independence, dedication, and commitment to the attainment of useful historical knowledge. Public funding could have undermined this. Round, who wrote history with private means and without the commitments of a family man, professed that those who edited documents to the PRO and got paid for their work did not enjoy similar intellectual freedom as those who studied the past privately.[22] This reverberated with the mid-century discussions about intellectual freedom and institutional constraints which the scientists' employment in institutions such as the Geological Survey and the Natural History Museum had instigated.[23] Stubbs profiled himself as an opponent of public funding, fearing that it would only endorse laziness and selfishness. According to him, historians with a steady payment risked surrendering to the temptations of "desultory research" and abandoning their duty to produce useful knowledge. He believed that the kind of research that either amused or benefited historians personally "seldom adds much to the real stock of human knowledge." He accepted provisions for research only when they were paid "for results."[24] His indirect message was that historians should get their patronage from the literary marketplace, not from the public purse, and allow readers to determine their economic worth. The modern notion of academic writing as a form of gift economy that operates with credit, recognition, and honor instead of revenue from selling private property was foreign to Stubbs and his allies.[25] The Victorian historians wanted everything: honor, recognition, and profit, but of course they knew that histories with high scientific value brought at best modest economic returns. Historians believed in the usefulness of scientific history, but as we have already seen, readers found such histories abstruse, uninteresting, and not worth their money. Nonetheless, historians were so steeped in the Victorian value system that demanding higher subsidy contradicted their economic worldview.

It was in fact the economic sacrifice in the name of valuable historical knowledge that made history an ennobling pursuit. One obituarist concluded in the *Western Daily Press* how Freeman had not been "a very wealthy man." The writer explained how Freeman could have employed.

> his literary skill in a way that would have been very remunerative to him. But he not only wrote to live, he lived to write. His primary object was not the acquisition of money, but the dissemination … of knowledge.[26]

This proved his genuine selflessness and "true historical passion." Historians took a cue from the popular Victorian discourse of altruism and from the traditional scholarly virtue of selflessness which had branded science as a calling instead of pursuit for fame and profit. The contrast between altruism and selfishness was an important moral compass for the Victorian educated middle class, although the constant presence of a fear of selfish inclinations and a surrender to egoistic temptations was the flip side of this moral edict. The suppression of these vices demanded self-restraint that merged actions, emotions, and morality in typically Victorian fashion.[27] These conditions were embodied in the ideology of useful knowledge which readers could purchase. This ethos of service for a common good gave historians the necessary intellectual and moral justification for the development of the entrepreneurial side of their persona.

Freeman exemplifies this mindset. He applied the commerce–scholarship dichotomy to erect a boundary between useless, bad history and useful scientific history. He attacked journalists and hacks who disregarded the ideals of scientific history and prioritized quick gains. He admitted that the "hired jesters" ignored the historical method because they wrote "for their dinners," but this did not render their kind of history writing acceptable.[28] It only ruined the reputation of history and historians. In a lecture he delivered at the Birmingham Historical Association, he complained about these vulgar interlopers who unfairly competed against the honest historians and who, to his dismay, won the public ear with their inaccurate and useless histories.[29] He also approached the question about useful knowledge from another angle in the same lecture, as he criticized those who insisted that only knowledge which produced immediate economic gains was worthwhile pursuing. This echoed the general uneasiness about a commercial society where market value was used to measure human action.[30] He ridiculed "Men of some reputation in the world" who.

> have gone about preaching the doctrine that all studies are useless except those which directly tend to fill the pocket. And from this premiss they draw the inference ... that no studies can be less useful than those which deal with the events and the languages of past times.[31]

For Freeman, the idea that the value of knowledge was evaluated according to "the number of guineas" it produced was simply disingenuous, and he went on proving the value of history by maintaining how historical facts had "reference to present affairs of present duties."[32] He revisited the topic in his inaugural lecture six years later, emphasizing how the opportunity to apply historical knowledge to improve modern society compensated the fact that history was not "the path best adapted for the winning of house and land."[33] He could just as well have been citing here Thomas Hughes's popular novel *Tom Brown's Schooldays*, where Tom learned from Dr. Arnold that the sacred goal in life was to do good "whether you make a living or not."[34] This did not mean that Freeman advocated scholarly poverty. Quite the contrary, he yearned for a comfortable living and admitted once how he wished to write a book "that would both sell & do good—make the most of both worlds."[35]

Freeman, just like Powell, discerned a growing tendency toward commercialization in the ancient universities and uncourteously blamed the modern sciences for blurring this critical distinction. He was not a fan of the new scientific disciplines and suspected that the scientists, or the scientific "ninnies" as he called them impolitely, had aligned themselves with the commercial class to infiltrate the trade ethos into his beloved Oxford. Scientists were just like tradesmen: they only "take money from the University to teach their own *trades*." Even worse, he accused the scientists of wanting to abolish Greek from the universities. Greek symbolized for him and the other traditionalists the classical liberal education which scientists now wanted to replace with programs for "tinkers and grocers, with scholarships for skill in weighing sugar."[36] This was as much about the ongoing debate about pedagogical aims as it was about the social standing of men with academic education or the future of disciplines like history. Freeman predicted that the introduction of practical training would have severe ramifications for intellectual life, because eventually such forms of knowledge which did not guarantee immediate financial gains would be discredited.

There was a heavy dose of jealousy in Freeman's anti-science ranting, as he professed how scientists got "all the credit & all the money" and were showered with public "worship."[37] While he defended the status of history, he conveniently ignored how scientists, just like historians, often needed multiple sources of income to earn a living and feared the implications which their commercially oriented writing and lecturing may have for their reputation. Their grievances were aggravated by the fact that the commercial applications of scientific knowledge, technological innovations, and patenting confused the distinction between science as an intellectual occupation and technology as a commercial venture.[38] William Crookes, a chemist who supported his research and family with patents, editorial work, consultation, and business, maintained that scientists had two choices: either fame or money. The former brought an international reputation and immortality, but left one "terribly minus as to the £.s.d." The latter, as Crookes bitterly experienced, secured a steady income but spurred the scientific elite to make accusations of charlatanism which reduced "the dignity of science into a trade."[39]

Freeman's distaste for "shop-ambition" intersected with his paratextual practices as he demanded that all the parts of his books endorsed the distance between historians and the trades. This affected the preparation of the promotional paratexts such as the titles for his books. The renaming of *Old English History for Children* was complicated by his abhorrence for "shoppiness." Macmillan was unhappy with the original title; the word "children" was too restrictive, as he believed that boys were disinclined to read books that purported to be targeted at children. As a corrective, he proposed to replace "children" with "junior students" or "junior classes." Freeman disliked both alternatives but rejected the word "student" altogether because it derived from Latin. His aversion to Latin was well known and his mission was to reinstate "the genuine, oaken, Germanic fibre of the language." Therefore, he proposed either "Readers" or "Learners," since both were in good plain English. Macmillan disapproved of them. Freeman, however, found the word "student" even more revolting because, as he later recalled, it was "shoppy, savouring of a Classical & Commercial Academy."[40] This was not an entirely unfounded argument. Advertisers had grown fond of using Latin and Greek in advertisements and product names and the habit had opened them to ridicule. *Fraser's Magazine* parodied the linguistic fashion of substituting "the honest Saxon" with foreign words and described how advertisers frantically searched dictionaries to find words to describe their wares and how the poor customers consulted Homer and Horace to gain enough competence to purchase a pair of boots.[41] The commercialization of classical languages and motifs vulgarized the dignity of classical culture, its political ideas, and the associations they evoked about education and learning. Freeman did not want to have anything to do with such degradation of ideals that he and his colleagues held in esteem. As a compromise, the later editions were called simply *Old English History* without any designators for the target audience.

If historians embraced the economic opportunities of the literary marketplace, they did it with certain conditions. Their persona had an entrepreneurial dimension which comprised skills, knowledge, personal traits, and a careful balancing of commercial realities with epistemic and moral virtues. As the engagement with publishing evoked associations with trade, the persona was geared to promote both the disciplinary and social status to which historians aspired. In this context, the persona overlapped with the development of the professional society in Victorian England and made claims for a social position for historians by drawing on the argument about the usefulness of historical knowledge that neatly corresponded with the scholarly ideals of the discipline.

History Books as Dignified Commodities

Books were an embodiment of the commodification of historical knowledge. Copyright law granted historians the ownership of their texts and underlined their role as producers of saleable products. Knowledge, which the Victorians, according to Alan Rauch, "fetishized as something valuable for its own sake," became a tradeable good.[42] A tangible paratextual reminder of the union between history and trade was the advertisement apparatus which readers found at the front and back of history books. The books in my paratext sample contain advertisements selling books "from the same author," other volumes of the same series, or other relevant titles. A full catalog of the publisher was often bound in at the back of the book. Unlike popular novels and other cheap prints, these histories were not furnished with advertisements for soap, medicine, or hats. They were obviously considered too dignified for promoting such merchandise. Macmillan granted that publishers perceived prospectuses that were bound to books as useful marketing tools, but he stressed that as a publisher he also had to respect the sanctity of the text. He refused to place advertisements between a half-title page and an index. The breaking of this rule, he explained, would have been a violation against book lovers and a book's integrity.[43] Other publishers were less scrupulous. For example, Creighton's *Life of Simon de Montfort* (1876), which Rivington published in the series "Historical Biographies," had two pages before the half-title page listing "English School Classics" and "Historical Handbooks." Then followed a page-long prospectus between the half-title page and the title page itself. The back of the book, too, was appropriated for commercial purposes, as it promoted "Select Plays of Shakespeare" and "Scenes from the Greek Plays."

This kind of marketing machinery highlighted history books as ordinary saleable goods, and historians used paratexts to counter such impressions by framing books as non-essential goods and articles of erudition. They knew that readers had the competence to decode the messages inscribed in the visual and material paratexts and that smartly designed paratexts could therefore be effective in spotlighting the dignity of history books and the integrity of the persona of their authors. One paratext that served this purpose was non-standardized typography. Paratext scholars have debated the paratextuality of typography, and some have denied it status as a paratext by asserting that it is a quality of a text and cannot perform similar roles in books as do paratexts.[44] The following erases such doubts: the existence and arrangement of words on a page conveyed meanings which readers interpreted and applied to determine the value of the text, its author, and history as a discipline more broadly. The symbolical role of non-standardized typography was at the heart of a conflict that broke out between Freeman and his publishers at Macmillan and Oxford University Press about the preparation of the American edition of Freeman's *The History of the Norman Conquest in England* in 1873. For Freeman, it was a question about scholarly dignity, while for the publishers it was purely a matter of costs, as non-standardized typography was expensive to set in type.

Freeman had originally offered the book to Macmillan, who had quickly realized that it was too erudite for a broad audience. Since

Macmillan had recognized its scholarly value, he had proposed it to Oxford University Press. Freeman had not objected to the arrangement. His only worry had been the marketing, and he inquired from Macmillan whether the "Oxford Press books ever get known or heard of?" because "They used to take every possible means to hinder the world from knowing of their existence."[45] Macmillan, who had entered into a partnership with the Delegates to help them to modernize their processes, handled the North American market for the Press and requested in 1873 from the Delegates a quota for 1500 cheaply and quickly printed copies of *Norman Conquest* to prevent its pirating in America. Bartholomew Price, secretary of the Press, proposed to Freeman that "the book should be an exact reprint." Freeman accepted the offer with this condition. He had made many improvements to his interleaved copy but thought it better not "to give the Americans the advantage of them before ourselves." All he wished was to have the misprints corrected. But problems emerged as soon as Freeman discovered that what Price in fact had suggested was not an exact reprint. Price wanted to make two substantial alterations: to omit the marginal notes and to replace the analytical running heads with general chapter headings.[46] These omissions would have profoundly altered the spirit of the book and Freeman firmly rejected them.

The running heads and marginal notes performed symbolical and interpretive functions in history books. The margins also had a critical duty in documenting chronology and dating the events that were mentioned in the text usually without dates. Canceling them would have left readers without any authorial guidance—the compass and the rudder that John Wilson so desperately missed in the *Opere Inedite di Francesco Guicciardini*.[47] Next to their practical usefulness, the marginal notes in particular conferred symbolic value on books and their writers. A reviewer in the *Exeter* was disappointed with the Oxford Delegates and their economizing with paratexts in books "of real importance." A testimony of this ignorance were the meager marginal annotations in the Clarendon Press Series, which contained such remarkable studies as Stubbs's *The Constitutional History* and Kitchin's *A History of France*. These types of masterpieces, the writer noted with disbelief, the Press had sent to the world forgetting "common decency in the nakedness" of their margins.[48] Nakedness was certainly not a desirable characterization for a scholarly history, as it hinted at deformity and incompleteness, the common traits of unmanliness and indecent nudity that recalled prostitution and lack of gentlemanliness. Neither was it something that Stubbs in general advocated: he was known to be "strongly of [an] opinion" that marginal notes belonged in serious historical works.[49]

Freeman was aware of these connotations and had filled the edges of the pages in *The Norman Conquest* with copious notes (Fig. 8.1). He adopted a similar paratextual strategy later in *The History of Sicily* but did not demand marginal notes for his educational histories. The connection between the margins and high scientific quality was further enhanced by a similar practice in other major histories published at that time, but there were also some exceptions to this. Those books were criticized for their inadequate marginal annotations. Kate Norgate's *England under the Angevin Kings* was issued with plain margins, and when she was preparing her second book, *John Lackland*, she asked Macmillan to allow her to place dates in the margins. Many people, she told, "have remarked to me what a great help marginal dates are in reading any historical book, & what a pity it was that there were none in my 'Angevins'."[50] The practical and symbolic value of marginal notes was captured by a writer in the *Athenaeum* underlining their great importance "to the reader who seeks something more than amusement."[51] Marginal notes, then, defined historians, books, and even readers, as they denoted scholarliness.

For Freeman, the essence of the crisis about the American edition was that the removal of the marginal notes and the analytical running heads would visually alter the scholarly character of his book. He did not take this lightly, because *The Norman Conquest* was, as he reminded his publishers, an "English classic" and he made it clear that he would never allow anyone to treat his masterpiece "like a two-penny-half-penny arithmetic book."[52] It was not just the existence of the marginal notes that fulfilled the symbolic role, but also the unique relationship they had with the narrative. The scientific style in history books was characterized by a minimal use of dates in the narrative. Instead, the chronology was indicated in the margins and Freeman had deliberately minimized the number of dates in the text because their lavish application would have given "the text an inferior look, not that of a standard book, but of some lower kind of that."[53] Thrusting the dates suddenly into the flowing sentences would have been intolerable. In the educational histories, Freeman had worked the dates grammatically into the sentences because that was just fine in books of "less pretension." In *The Norman Conquest*, by contrast, he had crafted his narrative so that the text and the dates in the margin formed a seamless unity, giving the book an air of superior quality and

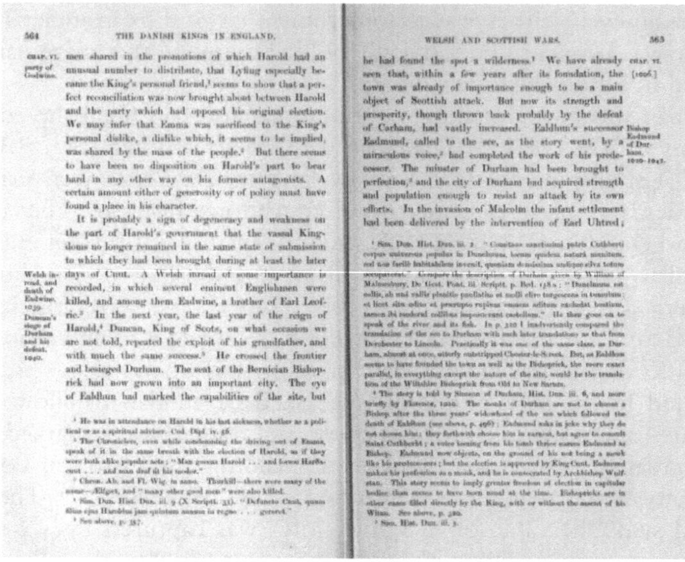

Fig. 8.1 Freeman used marginal notes and chronological annotations in the original version of *The History of the Norman Conquest of England*. This comes from the first volume (1867) (*Source* Internet Archive, original in University of California library)

setting it apart from generic textbook histories. Freeman now announced that he would never give his consent to publish the new edition in such a "slovenly and dis-creditable state" as Price was proposing.[54]

Freeman correctly suspected that the notes in the margins were being eliminated for financial reasons. Craik admitted that he, too, preferred the book to have the dates in the margins, but understood that this was financially not feasible. If Freeman insisted on having marginal notes, there was a risk that the entire project of the American edition would have to be cancelled.[55] Price tried to reason with Freeman: a cheap edition with expensive scholarly paratexts was simply inconceivable. Moreover, the proposed changes greatly exceeded the expenses allowed for a "reprint." Freeman strongly protested this "£.s.d. view of things" and found great irony in the fact that the secretary of Oxford University Press was called Bartholomew *Price*, for so jealously he guarded the Press's purse.[56] He composed a furious letter to Price about the despicable state of a publishing industry which only cared for "expenses, estimates, and calculations" while for him as a historian it was crucial that "a book should not go out in a state disgraceful to all concerned."[57]

Craik and Price realized that a comprehensive historical study without instructive paratexts at the edges of the pages was unlikely to succeed and they partially surrendered to Freeman's wishes. The revised American edition retained the analytical commentary of the running heads, but the publishers were resolute about the removal of the dates in the margins.[58] They knew that this would cause additional work for Freeman because it was impossible to add the dates into the text without carefully considering each sentence. However, a profound reworking of the narrative would have been too time consuming, and Price suggested an alternative solution: to introduce the dates in the text between brackets. This was a common practice, but not what Freeman preferred. Dates in brackets, he explained, destroyed "the *literary* aspect of the book" and appeared "sloppy."[59] Freeman's view was not uncommon. Kegan Paul who issued scholarly titles also claimed that the use of brackets revealed a bad stylist.[60] Yet, Freeman's American edition was released with the dates in brackets. His letters do not reveal what made him change his mind. The tight schedule and pressuring from the publishers must have made him accept the reality. This method also saved him from an extensive rewriting of the text because the dates should have been grammatically worked into the sentences as "A date thrust into a flowing sentence is intolerable." Realism made him compromise his scholarly standards even if inserting the dates in brackets was "barely tolerable."[61] This did not mean that he was pleased with the result. Three years later he was still groaning about the matter and repeating, "A date should either be in the margin or else form part of a grammatical sentence. I don't like the way in which they are done in the American ed. of N.C."[62]

The dispute exemplifies how historians envisioned typographical paratexts as symbols of scholarly value and a means to distinguish scholarly history books from everyday products or mass-produced books. This elevated status helped them to reconcile the moral principles embedded in the persona with the laws of the commercial marketplace. Moreover, for Freeman the clash with Price and Craik had a meaning as a proof of the threats the commercialization of the publishing industry imposed on scholarship and learning. He feared that men such as Price, to whom "a great book is no more than a leg of mutton or a pair of shoes," could not understand that scholarly masterpieces should not be considered in financial terms.[63] While Freeman was desperate to make a living with books, he was also a man of principle and maintained that the value of "big" histories lay somewhere other than in the profits they brought to the publisher and author: they were the true contributions to scholarship and manifestations of the scholarly persona. Their visual image, too, should confirm that.

In the spirit of their age and despite the disappointing sales figures, historians believed in a marketplace that determined the value of their work. To escape the stigma of trade and selfish profit making which participation in the publishing business implied, they forged a persona that accommodated entrepreneurial impulses, skills, and knowledge with their scholarly ideals and intellectual integrity. The discursive hierarchies between the commercial middle class blinded by selfish greed and the professional middle class persistently working to improve society formed the ideological backbone of the scholarly persona. The usefulness of historical knowledge sanctioned the revenue they produced. The cultivation of a persona of an entrepreneurial historian demanded that commercial needs did not undermine moral and scholarly habits and principles, but when we move on to explore the application of these ideals in advertising, endless points of conflict begin to emerge between historians' values and the realities of the literary marketplace, indicating a constant renegotiation between commercial and scholarly needs, values, practices, and hierarchies.

Notes

1. Elton, *Frederick York Powell*, 1:70, 1:107–108, 1:120–122, 1:157.
2. Prothero, *Memoir*, xii; Daunton, "Introduction," 18–20; Heyck, *Transformation of Intellectual*, 88–89.
3. Freeman to Macmillan, July 29, 1870, Add MSS 55049, BL.
4. Perkin, *Rise of Professional Society*, 95; Bellon, *Sincere and Teachable Heart*, 24; Tosh, *Man's Place*, 3, 13; Browne, "Natural Economy of Households," 101–102.
5. Thomson quoted in Searle, *Morality and the Market*, 130–131.
6. Searle, *Morality and the Market*, 34–35, 77–106; Reader, *Professional Men*, 158–159; Cain, "Empire and the Language," 252–255.
7. Egginton and Thomas, "Introduction," 8–12.

8. Peterson, *Becoming a Woman of Letters*, 35–36; Erickson, *Economy of Literary Form*, 105–106; Leary and Nash, "Authorship," 172–173; Searle, *Morality and the Market*, 6–7.
9. Freeman to the electors, April 28, 1862, in Owen, "Chichele Professorship," 219–220.
10. Peterson, *Becoming a Woman of Letters*, 2.
11. Russell, *Literary Manual*, 3.
12. Galton, *English Men of Science*, 104–107.
13. Freeman, "On the Study of History," 319.
14. Covert, *Victorian Marriage*, 155–156.
15. Oman, *Memories*, 155.
16. Colclough, "Press Books," 696–698.
17. Longman to Browning, October 15, 1878, GBR/0272/OB/1/986/C, King's Cam.
18. Freeman to Macmillan, December 14, 1871, Add MSS 55049, BL. Otherwise see the Macmillan Papers Add MS 55049, 55050, 55051, 55052, and 55053, BL.
19. Daunton, "Introduction," 18–20; Heyck, *Transformation of Intellectual*, 88–89.
20. Verschaffel, "Something More than a Storage," 37–38; Levine, *Amateur and the Professional*, 123–134.
21. Gardiner to Acton, June 3, 1878, Add 8009/I/916, CUL.
22. Round, *Studies on the Red Book*, 18–20.
23. Browne, "Natural Economy of Households," 103.
24. Stubbs, *Seventeen Lectures*, 41–42.
25. McSherry, "Uncommon Controversies," 225–231.
26. "Professor Freeman," *Western Daily Press*, March 18, 1892, 5.
27. Collini, *Public Moralists*, 60–66. Of altruism and its reception in Britain see also Dixon, Thomas. *The Invention of Altruism: Making Moral Meanings in Victorian Britain* (Oxford University Press, 2011).
28. Freeman to Thompson, September 4, 1887, U DX9/155, HHC; Freeman to Craik, October 14, 1888, Add MSS 55053, BL; Freeman to Thompson, April 19, 1891, U DX9/185, HHC.
29. Freeman, "On the Study of History," 322–326.
30. Searle, *Morality and the Market*, 3–4.
31. Freeman, "On the Study of History," 319–320.
32. Freeman, "On the Study of History," 319–320.
33. Freeman, *Methods*, 39–40.
34. Hughes, *Tom Brown's Schooldays*, 363.
35. Freeman to Craik, July 9, 1876, Add MSS 55051, BL.
36. Freeman to Margaret Evans, May 10, 1885, in Stephens, *Life and Letters*, 2:334; Freeman to Thompson, February 17, 1884, U DX9/117, HHC.
37. Freeman to Macmillan, February 22, 1874. Add MSS 55050, BL.

38. Heyck, *Transformation of Intellectual*, 102; Fara, *Newton*, 175–178; Lucier, "Court and Controversy," 141.
39. Brock, *William Crookes*, 17, 86.
40. Freeman to Macmillan, April 15, 1870, Add MSS 55049, BL; Macmillan to Freeman, May 13, 1870, Add MS 55390 (2), BL; Freeman to Thompson, July 8, 1870. U DX/9/19, HHC; Freeman to Macmillan, October 4 and 9, 1870, Add MSS 55049, BL; Freeman to Macmillan, July 3, 1879. Add MS 55052, BL; Elton, *Frederick York Powell*, 1:411.
41. "Age of Veneer," *Fraser's Magazine*, January 1852, 90; Loeb, *Consuming Angels*, 34, 174–175.
42. Rauch, *Useful Knowledge*, 3.
43. Macmillan to Price, February 16, 1875, Add MSS (2), BL; Macmillan to William Allington, March 14, 1864, in Macmillan, *Letters*, 168.
44. Birke and Christ, "Paratexts," 69; Ruokkeinen and Liira, "Material Approaches," 116.
45. Freeman to Alexander Macmillan, May 13, 1866. Add MS 55049, BL.
46. Price to Freeman, March 22 and 31, 1873, Price to Craik, March 25, 1873, Letter Books 7, OUP; Freeman to Craik, March 23, 1873, Add. MSS 55050, BL; Craik to Freeman, March 24, 1873, Add. MSS 55393 (2), BL; Garritzen, "Revise, Edit, and Improve," 306–308.
47. [Wilson], "Guicciardini," 439–440.
48. "Mr. Kitchin's History of France," *Exeter*, April 14, 1877, 464.
49. Price to E. J. Payne, June 24, 1878, Letter Books 18, OUP.
50. Norgate to Macmillan, June 8, 1902, Add MSS 55078, BL.
51. "Life and Times of Stein," *Athenaeum*, January 18, 1879, 80.
52. Freeman to Craik and Price, April 4 and 6, 1873, Add. MSS 55050, BL.
53. Freeman to Craik and Price, April 4 and 6, 1873, Add. MSS 55050, BL.
54. Freeman to Craik and Price, April 6, 1873, Add. MSS 55050, BL; Garritzen, "Revise, Edit, and Improve," 306–308.
55. Craik to Freeman, April 5, 7, and 10, 1873, Add. MSS 55393 (2), BL; Craik to Price, April 8, 1873, Add. MSS 55393 (2), BL.
56. Freeman to Craik, April 6, 1873, Add MSS 55050, BL; Price to Freeman, April 5, 1873, Letter Books 7, OUP.
57. Freeman to Price, April 6, 1873, Add. MSS, 55050, BL.
58. Price to Freeman, March 31, 1873, Price to Craik, April 9, 1873, Letter Books 7, OUP.
59. Price to Freeman, March 31, 1873, Letter Books 7, OUP; Freeman to Craik, April 6, 1873, Add MSS 55050, BL.
60. Howsam, *Kegan Paul*, 48.
61. Freeman to Price, April 4, 1873, Freeman to Craik, April 6, 1873, Add MSS 55050, BL; Garritzen, "Revise, Edit, and Improve," 306–308.
62. Freeman to Macmillan, January 30, 1876, Add MSS 55051, BL.
63. Freeman to Craik, April 7, 1873, Add. MSS 55050, BL.

References

Unpublished Primary Sources

British Library: The Macmillan Papers
Cambridge University Library: Acton Papers
Hull History Centre: Letters from Edward Augustus Freeman to Edith Thompson
King's College Archive Centre, Cambridge: Oscar Browning Papers
Oxford University Press Archive: Letter Books

Printed Primary Sources

[Anon.]. "The Age of Veneer. The Science of Puffing." *Fraser's Magazine*, January 1852, 87–93.

[Anon.]. "Mr. Kitchin's History of France." *Exeter*, April 14, 1877, 463–464.

[Anon.]. "The Life and Times of Stein." *Athenaeum*, January 18, 1879, 79–81.

[Anon.]. "Professor Freeman." *Western Daily Press*, March 18, 1892, 5.

Burrows, Stephen Montagu. *Autobiography of Montagu Burrows*, Captain R.N. London: Macmillan, 1908.

Elton, Oliver. *Frederick York Powell. A Life and a Selection from His Letters and Occasional Writings*. Vol. 1. Oxford: Clarendon Press, 1906.

Fifoot, C. H. S. *The Letters of Frederic William Maitland*. London: Selden Society, 1965.

Freeman, Edward A. "On the Study of History." *Fortnightly Review*, spring 1881, 320–339.

Freeman, Edward A. *The Methods of Historical Study*. London: Macmillan, 1886.

Galton, Francis. English Men of Science. *Their Nature and Nurture*. London: Macmillan, 1874.

Hughes, Thomas. *Tom Brown's Schooldays*. Oxford: Oxford University Press, 2008.

Macmillan, George *A. Letters of Alexander Macmillan*. Glasgow: printed for private circulation, 1908.

Oman, Charles. *Memories of Victorian Oxford and Some Early Years*. London: Methuen, 1941

Prothero, G. W. "Memoir." In *The Growth of British Policy*, by J. R. Seeley, vii–xxi. Cambridge: University Press, 1895.

Round, J. H. *Studies on the Red Book of the Exchequer*. Printed for private circulation, [1898].

Russell, Percy. *The Literary Manual; Or, a Complete Guide to Authorship*. London: London Literary Society, 1886.

Stephens, W. R. W. *The Life and Letters of Edward A. Freeman* D.C.L., LL.D. Vol. 2. London: Macmillan, 1895.

Stubbs, William. *Seventeen Lectures on the Study of Medieval and Modern History and Kindred Subjects*. Oxford: Clarendon Press, 1886.

Wagner, Leopold. *How to Publish a Book or Article and How to Produce a Play. Advice to Young Authors*. London: George Redway, 1898.

[Wilson, John]. "Guicciardini's Personal and Political Records." *Quarterly Review*, October 1871, 416–440.

Secondary Sources

Bellon, Richard. *A Sincere and Teachable Heart: Self-Denying Virtue in British Intellectual Life*, 1736–1859. Leiden: Brill, 2015.

Birke, Dorothea and Birthe Christ. "Paratexts and Digitized Narrative: Mapping the Field." *Narrative* 1, no. 21 (2013): 65–87.

Briggs, Asa. *A History of Longmans and Their Books 1724–1990: Longevity in Publishing*. London: The British Library and Oak Knoll Press, 2008.

Brock, William H. *William Crookes (1832–1919) and the Commercialization of Science*. Aldershot: Ashgate, 2008.

Browne, Janet. "The Natural Economy of Households: Charles Darwin's Account Books." In *Aurora Torealis*, edited by Marco Beretta, Karl Grandin, and Svante Lindqvist, 87– 110. Sagamore Beach: Watson Publishing International, 2008.

Cain, Peter J. "Empire and the Language of Character and Virtue in later Victorian and Edwardian Britain." *Modern Intellectual History* 4, no. 2 (2007): 249–273.

Colclough, Stephen "Press Books in the United Kingdom." In *The History of Oxford University Press*, edited by Simon Eliot, 667–703. Vol. 2. Oxford: Oxford University Press, 2013.

Collini, Stefan. *Public Moralists: Political Thought and Intellsectual Life in Britain 1850–1930*. Oxford: Oxford University Press, 1991.

Covert, James. *A Victorian Marriage: Mandell and Louise Creighton*. London: Hambledon and London, 2000.

Daunton, Martin. "Introduction." *The Organisation of Knowledge in Victorian Britain*, edited by Martin Daunton, 1–27. Oxford: Oxford University Press, 2005.

Egginton, Heidi and Zoë Thomas. "Introduction." In *Precarious Professionals: Gender, Identities and Social Change in Modern Britain*, edited by Heidi Egginton and Zoë Thomas, 1–39. London: University of London Press, 2021.

Eliot, Simon. "The Evolution of a Printer and Publisher." In *The History of Oxford University Press*, edited by Simon Eliot, 77–112. Vol. 2. Oxford: Oxford University Press, 2013.

Erickson, Lee. *The Economy of Literary Form: English Literature and the Industrialization of Publishing, 1800–1850*. Baltimore: The Johns Hopkins University Press, 1996.

Fara, Patricia. *Newton: The Making of Genius*. London: Macmillan, 2002.

Finkelstein, David. *The House of Blackwood: Author-Publisher Relations in the Victorian Era*. Pennsylvania: Pennsylvania State University Press, 2002.

Garritzen, Elise. "Revise, Edit, and Improve: Writing and Publishing History as an Unending Process in Victorian Britain." *Clio: A Journal of Literature, History, and the Philosophy of History* 45, no. 3 (2016): 289–314.

Guy, Josephine. "Authors and Authorship." In *The Cambridge Companion to English Literature 1830–1914*, edited by Joanne Shattock, 9–29. Cambridge: Cambridge University Press, 2010.

Heyck, T. W. *The Transformation of Intellectual Life in Victorian England*. London: Groom Helm, 1982.

Howsam, Leslie. *Kegan Paul a Victorian Imprint: Publishers, Books and Cultural History*. London and Toronto: Kegan Paul International and University of Toronto Press, 1998.

Leary, Patrick and Nash Andrew. "Authorship." In *The Cambridge History of the Book in Britain*, edited by David McKitterick, 172–213. Vol. 6. Cambridge: Cambridge University Press, 2009.

Levine, Philippa. *The Amateur and the Professional: Antiquarians, Historians and Archaeologists in Victorian England, 1838–1886*. Cambridge: Cambridge University Press, 1986.

Loeb, Lori Anne. *Consuming Angels: Advertising and Victorian Women*. Oxford: Oxford University Press, 1994.

Lucier, Paul. "Court and Controversy: Patenting Science in the Nineteenth Century." *The British Journal for the History of Science* 29, no. 2 (1996): 139–154.

Mason, Nicholas. *Literary Advertising and the Shaping of British Romanticism*. Baltimore: The Johns Hopkins University Press, 2013.

McSherry, Corynne. "Uncommon Controversies: Legal Mediations of Gift and Market of Authorship." In *Scientific Authorship: Credit and Intellectual Property in Science, edited by Mario Biagioli and Peter Galison*, 225–250. New York: Routledge, 2003.

Miller, Laura J. *Reluctant Capitalists: Bookselling and the Culture of Consumption*. Chicago: University of Chicago Press, 2006.

Owen, Dorothy M. "The Chichele Professorship of Modern History, 1862." *Historical Research* 34, no. 90 (1961): 217–220.

Perkin, Harold. The Rise of Professional Society. New York: Routledge, 1989.
Perronne, Fernanda. "Women Academics in England, 1870–1930." *History of Universities* 12, no. 1 (1993): 339–367.

Peterson, Linda H. *Becoming a Woman of Letters: Myths of Authorship and Facts of the Victorian Market*. Princeton: Princeton University Library, 2009.

Raven, James. *The Business of Books: Booksellers and the English Book Trade 1450–1850*. New Haven: Yale University Press, 2007.

Rauch, Alan. *Useful Knowledge: The Victorians, Morality, and the March of Intellect*. Durham: Duke University Press, 2001.

Reader, W. J. *Professional Men: The Rise of the Professional Classes in Nineteenth-Century England*. London: Weidenfeld and Nicolson, 1966.

Ruokkeinen, Sirkku and Aino Liira. "Material Approaches to Exploring the Borders of Paratext." *Textual Cultures* 11, no. 1–2 (2017): 106–129.

Searle, G. R. *Morality and the Market in Victorian Britain*. Oxford: Clarendon Press, 1998.

Tosh, John. *A Man's Place: Masculinity and the Middle-Class Home in Victorian England*. New Haven: Yale University Press, 1999.

Verschaffel, Tom. "'Something More than a Storage Warehouse': The Creation of National Archives." In *Setting the Standards: Institutions, Networks and Communities of National Historigraphy*, edited by Ilaria Porciani and Jo Tollebeek, 29–46. Basingstoke: Macmillan, 2012.

CHAPTER 9

Sincere and Insincere Advertisers

William Stubbs cautioned his Oxford audience in 1876 about the evils of modern marketing. Even history books, he lamented, were not safe from immoral advertising, and ignorant readers wasted their time on "trashy" histories "which owe[d] their circulation to advertising skill or to pretentious claptrap."[1] He was not entirely mistaken. Rampant advertising was everywhere where Britons cast their eyes—on streets, at train stations, in omnibuses, and even on the stage of their favorite theater. Foreigners were reportedly shocked about the bad taste of the advertisements in London. At the same time, the rising profession of advertisers, the guild of rogues and swindlers, was widely condemned, but without much effect.[2] Science and scholarship were not untouched by the marketing machinery, and some critics were concerned about the ramifications which dishonest advertising could have for the credibility of sciences. *Fraser's Magazine* published in 1869 a satirical essay about a fictional lecture from a "Professor X.Y.Z." about "Mechanical Forces." A disgruntled member of the audience demanded to speak and claimed that the professor had overlooked the greatest of all forces—advertising. A cleverly designed advertisement, the listener maintained, could support.

> thousands of semi-professors, lecturers, and professionals *en masse*, retailing superfluous nothings to ignorant audiences … and all "guaranteed"—just as affirmatively and with as free a conscience as if the whole were the very truth.[3]

E. Garritzen, *Reimagining the Historian in Victorian England*,
https://doi.org/10.1007/978-3-031-28461-8_9

Both Stubbs and William Jerdan, who was the author of the essay in *Fraser's*, were concerned about those among the public who were unqualified to evaluate the reliability of scientific knowledge and who were therefore easy prey to deceitful marketing. Their message was that if the value of scientific discoveries was decided in advertisements, appreciation for properly conducted science could diminish. That was something to take seriously, because it could have severe consequences for the future of British society.

Historians, who oscillated between scholarship and commerce, knew that their books did not sell without some form of marketing as the literary marketplace was flooded with an ever-increasing number of new titles. Powell prosaically observed how scholarly titles kept "popping out like hardy perennials."[4] However, historians were not only competing against other history books, but also against historical novels, periodicals, exhibitions, waxwork spectacles, panoramas, pageants, and other forms of commercialized leisure that drew on Victorians' thirst for nostalgia. In this climate, it was necessary to advertise history books, but in such a fashion that the promotional material would not undermine historians' epistemic and moral virtues or that historians' marketing endeavors were not confused with the deceptive tricks of hacks such as the advertising genius George Robinson in Trollope's *The Struggles of Brown, Jones and Robinson*. Robinson personified the notoriety of the trade as he declared how "the groundwork of advertising is romance" and insisted that advertising relied on deception, lies, exaggeration, and hyperbole.[5] These were qualities that obviously contradicted the honesty, transparency, accuracy, and dignity that historians embraced. While these virtues formed the core of the scholarly persona, they were also considered constitutive for a Victorian gentleman, a man of honor whose word could be trusted.[6] What historians then needed was a marketing vision that enabled them to design promotional paratexts that evaded hyperbole and exaggeration and accentuated their persona. Consequently, the epistemic virtues were converted into normative regulations that gave the necessary moral tenor to historians' promotional paratexts.

At the same time, upholding such high moral standards in advertising proved to be difficult, if not entirely impossible. Historians' advertising ventures highlight both the adaptability of their persona to the varying circumstances and the gap between the persona as an ideal type and the reality which historians experienced. As historians activated different dimensions of their persona, a certain degree of flexibility was needed from the core virtues, as the utmost honesty, for instance, was not the best marketing strategy. This chapter covers the tension between historians' intellectual and moral integrity and the cultural template of dishonesty which was ingrained in Victorian advertising enterprises. Although advertising belonged to historians' toolkit, the subject has hardly attracted scholarly attention, apart from Ian Hesketh's detailed study of Seeley's marketing genius. This chapter, then, enhances our understanding of the advertising of history books by first outlining historians' moral ambivalence. Their silent approval of ethically unsound advertising signals acceptance of marketing ploys as an integral aspect of the Victorian publishing industry. The chapter then moves on to scrutinize the conflicts between honesty and dishonesty in prefaces and titles and shows how moral uprightness was often conditional. As the order of the priorities shifted between the different goals which historians placed on the paratexts, the grade of honesty shifted accordingly. Furthermore, it was impossible to perform the stock of scholarly virtues all at once as paratexts played multiple roles simultaneously. This generated a need for flexibility in interpreting the virtues in specific contexts. One aspect of the persona could make amendments to another as the contexts and goals which historians pursued altered.

Advertising and Historians' Moral Ambivalence

Puffing had plagued the English publishing world since the eighteenth century. By the early nineteenth century, it had come to mean hyperbole and advertising in disguise, and the puffing scandals of the 1820s and 1830s had unmasked the moral corruption of publishers and authors alike. High-minded early-Victorian publishers protested the unscrupulous marketing tricks, but they were not innocent either. If nothing else, orchestrated favorable reviews linked them to puffing.[7] Macaulay belonged to the vocal opponents of puffery as he condemned the practice in one of the classic early-Victorian anti-puff essays. He was scandalized that even men with self-respect condescended to harangue the public with puffery and endangered the reputation of honest authorship with their conduct. They removed respectability from writing by turning books into ordinary commodities like shoes, hosiery, or food. For Macaulay, puffery was proof of the victory of commerce over art; it erased the critical line between men of letters and tradesmen. Because puffery, according to him, shook the Victorian social order, he feared that the genuine literary talents would abandon authorship altogether. That, he predicted, could have devastating consequences for intellectual life in Britain.[8]

Macaulay, importantly, urged that writers ought to take charge of the advertising of their works instead of leaving it to the publishers. In practice, though, he was not rigid about the puffing of his own books. John Lingard, who chose the moral high road in advertising, was appalled that Longman had used pre-release reviews—puffery at its purest—to promote the first installment of Macaulay's *History of England* and that Macaulay had allowed this to happen. Lingard instructed his publisher Charles Dolman several times to abandon any puffing of his histories because that would put his name as a respectable historian into disrepute. He knew that the puffing scandals had taught readers to recognize and condemn insincere and deceptive marketing. Even more alarming, according to Lingard, was that readers assumed that authors supervised the marketing and were hence accountable for the advertisements. Writers widely shared Lingard's conviction that readers held them responsible for the marketing materials of their books.[9] Owing to this assumed connection between marketing and historians' moral character, advertisements were conceived as a critical site for historians to display the persona and emphasize its uprightness and for readers to make judgments about it. It was pivotal to ensure that the advertisements did not harm historians' collective public image.

Although the references to puffery dropped after the mid-century, deceptive advertising or fraudulent marketing tricks did not vanish despite the sustained criticism, satire, and caricatures. Advertising which drew inspiration from early-Victorian puffery simply worked too well for publishers to give it up. Indeed, Leopold Wagner concluded in his manual for authors in 1898 that unpaid advertising was the most efficient method of selling books because the gossip, hype, and news that publishers spread made books and writers more widely known than any commercial could ever do.[10] Scholarly publishing made no exception to this, and historians were therefore entangled in questionable advertising whether they liked it or not. The *English Historical Review* tackled the issue in an anonymously published note in 1890. The publishing house Swan Sonnenschein had issued Jacob Burckhardt's famous *The Civilization of the Renaissance in Italy* without indicating that it was in fact a reprint of the 1878 Kegan Paul version. This was scandalous, and the *Review* charged the publisher with expunging the date of the preface—April 1878—just to "give the work the appearance of a new book."[11] The manipulation of the publishing date was part and parcel of puffery—and so were its public reproofs.[12] While the censoring of the practice in the *Review* might have enhanced the sense of morality among historians, the protestation benefited the publisher as well. The denunciatory note made the readers of the *Review* aware that a fresh stock of *The Civilization of the Renaissance in Italy* was now available.

The *English Historical Review* cast the blame with its collective authority on the publisher, but historians were no less guilty when it came to misleading advertising. Margaret Oliphant, intimately familiar with the literary world, ascribed it as a vice of literary men and of the men who "happen to be 'remembered at the Universities'".[13] While she tried disingenuously to remove the taint of immoral advertising from women, her observation that such a practice was common in learned circles was correct. Advertising was not a popular topic of discussion among historians, and direct references to it are few and far between, but historians, too, appropriated a wide range of marketing strategies that derived from the "puffing pumps." They composed "news" about their projects for publishers to place in the newspapers and, in a spirit of product placement, intentionally praised their friends' books in public lectures. Jessopp reported that he cried about the greatness of Kate Norgate's *Angevins* wherever he was. They, moreover, mastered the art of the bandwagon effect and stimulated demand by creating an illusion of crowds of eager consumers wanting to buy their books. Dean Hook confessed to Freeman how he requested his friends' publications whenever he visited bookstores because the method had, according to him, proven to be effective in generating interest in these books.[14]

However, among historians it was insider reviewing that was the most popular legacy of puffery. Collegial reviewing had been practiced already in the journals of the Republic of Letters, where it had sparked a debate about the objectivity of such reviews.[15] In the nineteenth century, the exchange of favorable criticism was enabled by the culture of anonymous reviewing. Just like Lingard, most historians and publishers drew the line between moral and immoral insider reviews at the publishing date, although newspapers kept requesting advanced sheets from them.[16] Otherwise, the practice of insider reviewing did not provoke ethical uneasiness among historians. Nor did they draw an explicit connection between collegial reviewing and the tradition of puffery. Instead, they talked about the exchange of reviews freely.[17] It is then obvious that some forms of puffery had become, if not acceptable, at least tolerable among scholars and in the respectable publishing houses. Once the

English Historical Review had been established, Gell from Oxford University Press contacted the editors and offered them books for review with instructions about which of their aspects the reviewers should focus on.[18]

When Creighton was editing the *English Historical Review*, he mediated between the authors of the books under review and the reviewers, understanding the intricate relationship between the commercial, scholarly, and social in the book reviews. The curating of the reviews was in line with his mission to establish standards for polite scholarly conduct, but it was also critical because most of the reviews were signed and could make visible personal feuds and other rivalries that eroded the impression of the unity of the historians' community. When Acton offered to review Seeley's *A Short History of Napoleon* the First, Creighton first consulted Seeley about this. He also reminded Acton to be gentle, as Seeley was "the most sensitive (I was about to say *vain*) person."[19] These instructions indicate the role which collegial reviews had for the cohesion of the scholarly community. The same is confirmed by the fact that a failure to comply with the reciprocity of compliments was considered inappropriate. A collegial review was like a gift with an expectation of a return and the violation of this unwritten rule was strictly condemned. As has been recounted many times before, Green was delighted with Freeman's praise of his book but returned the favor with critical reviews. Freeman was infuriated and considered such misplaced honesty as a violation of collegiality, a public humiliation, and injurious to the sale of his books.[20] The social and collegial dimension of the reviewing culture distinguished the friendly appraisals from the other forms of reviews. Historians usually dismissed acerbic and silly reviews in newspapers and magazines, after a moment of self-pity, as commercially useful knowing that controversies stimulated interest in their books.

Although collegial reviews were the most common form of classic puffing among the historians, they could adopt more inventive methods of misleading advertising as well. Seeley was a resourceful advertiser. As Ian Hesketh has shown, before his career as a Cambridge historian he masterminded with Alexander Macmillan an impressive marketing campaign for his anonymously published *Ecce Homo* (1866). They arranged solicited reviews, created hype by quoting the most outrageous reviews in the advertisements, and fed the public curiosity with carefully crafted rumors and news.[21] Five years later, he campaigned for *English Lessons for English People* (1871) which he had co-authored with Edwin Abbott. He composed an announcement for the book but since it was regarded inappropriate for an author to submit so obvious a piece of self-marketing to the papers, he asked his friend Frederick James Furnivall to do this on his behalf. To avoid any accusations of unsound marketing, he begged Furnivall not to add any praise into the text. "Everybody will see," he explained, "that such a paragraph as this must have been furnished either by me or Abbott, and if any praise is mixed with it we shall be suspected of having puffed ourselves."[22] It is hard to miss the hypocrisy in Seeley's letter.

While Seeley's repertoire of marketing tricks was remarkable, his intimate friend Browning also mastered the art of self-promotion. Self-reviewing had been one of the most notorious tricks of the early-Victorian puffing pumps, but its wide condemnation had failed to purge the practice. Browning did not earn high praise as a historian, and he mostly wrote educational histories of the second rank. He also edited several volumes of documents, but they, too, suffered from carelessness and recklessness that became an embarrassment to the societies that had commissioned them.[23] He edited *The Political Memoranda of Francis Fifth Duke of Leeds* (1884) for the Camden Society and without the slightest moral scruple wrote an anonymous and glowing review of it for the *Quarterly Review*. He built his critique essay on the dichotomy between popular and scientific history and applauded the high scholarly character of his own book:

> The form of the book makes no concessions to popularity; it is sternly, even repulsively historical. Lovers of gossip will find nothing to gratify them except the memoranda which refer to the Princess of Wales, which have been partly used by Mr. Fitzgerald. The notes are sufficient to elucidate the text.[24]

Browning did not restrict his self-promotion to praising his own volume but accentuated his accomplishments by contrasting his book with another edited volume which he branded an example of the detrimental impact which the popularizing impulses could have on earnest historical scholarship. This apparently deplorable edition was *Selections from the Letters and Correspondence of Sir James Bland Burges, Bart., Sometime Under-Secretary of State for Foreign Affairs* (1885) which James Hutton had edited, and Murray published. Browning assumed the role of an ardent advocate of scientific history and wrote how Hutton, "a scholar of competent historical knowledge," had lowered his standards and produced a book that was nothing but "a vain attempt to beguile the 'ordinary reader'." Because of this, Hutton had skipped many historical details and this poorly chosen method had made him look ignorant of even the commonest facts to do with the period he was writing about. He had also, Browning wrote, tried to give "liveliness" to his pages with "less legitimate" means. In other words, he professed that Hutton had given priority to picturesqueness. To ensure that readers understood just how weak Hutton's volume truly was, Browning concluded with barely disguised condescension: "Still we are grateful to Mr. Hutton for giving us what he has thought fit to print. Perhaps at some future time the Burges papers will be made to yield more valuable metal."[25] Anonymity protected Browning from accusations of the immorality of self-reviewing. Even if someone suspected something, by cleverly capitalizing on the common stereotypes of popular histories—above literary splendor at the expense of the slightness of facts– and by projecting them onto someone else's document edition, Browning ascribed high value to his own work without direct self-praise.

Seeley and Browning might have been extremes, but most of the scientific historians were involved in one way or another in advertising that resembled traditional puffery. Instead of condemning the mainstream advertising tricks, they added to the entrepreneurial persona a mastery of the protocols that guided the crafting of the scholarly most appropriate forms of advertising such as collegial reviews. Gaining this kind of knowledge was essential, as advertising in all its forms was not just some fringe practice embraced by the seedy side of the late-Victorian publishing world. It was practiced by reputable scholars and publishers alike and its silent approval from the scholarly community enabled it to thrive in the learned book market.

Earnestness as a Marketing Strategy

Edith Thompson, when revising her *History of England* in 1876, advised Macmillan not to hype on the title page the extensions which she had been making to her text because it was "no good calling public attention to an enlarged edition, unless it is up to very last lights." In 1900, when she was once again updating the book, she specified that it should not be labeled as "a *new* or *revised* edition" because she had not given it "such a thorough revision as would justify those expressions."[26] Publishers were eager to promote revisions on a title page, for freshness sold books, but Thompson was anxious about the consequences which exaggerated promises of novelty might have on her reputation. If readers felt misguided by the paratexts, they might lose their confidence in the truthfulness of her book altogether. Macmillan attended to Thompson's wishes and the first one was branded "New Edition. Revised" and the second "Reprinted with Additions." Finding the right label for a revised history was a delicate issue, because overstating the extent of the alterations might irritate readers and reviewers. But balancing between honesty, transparency, and the marketing language was not an issue of revised editions alone. At least some historians expected the same kind of honesty and intellectual integrity from promotional paratexts as they expected from the historical narrative. Mary Bateson complained how Gilbert Child had made an inflated prefatorial claim about "references to published contemporary sources" in his *Church and State under the Tudors*, for the promised sources were "few and far between."[27] Deceptive paratexts indicated lack of self-control, weak intellectual integrity, and a willingness to selfishly surrender scholarly virtues to commercial interest. Such conduct could reduce readers' trust in the reliability of history and impair the dignity of the discipline. The image of a disappointed reader was a powerful subtext for composing promotional paratexts, as historians expected every part of their books to validate their persona. The question about honesty and hype turned out to be the acutest with prefaces.

Before dust jacket blurbs became popular in the early twentieth century, prefaces served the vital function of introducing the book to potential readers.[28] A preface was a real multitasker in a history book. As we have already seen, historians wrote prefaces to introduce the topic, its significance, and historical context; to discuss methods, sources, and epistemic virtues; to indicate the expected audience, to acknowledge collegial assistance and alliances; and to participate in current political and cultural debates. Prefaces bounced between being a text and a paratext, and historians' habit of instilling the prefatorial script with historical facts and resumes bridged them to the ensuing narrative. This in-between status meant that regardless the different modality of a paratext and a text, historians expected prefaces to concur with the disciplinary ideals. The accentuated prefatorial honesty and impartiality obviously contradicted those prefaces that were loaded with subjective observations. What is more, a host of paratext scholars have identified prefaces as ideal zones for authorial self-fashioning, and this can be further extended to their use for forging a collective persona.[29] It is then understandable why some historians were vigilant about the sincerity of their prefatorial utterances.

Historians had a clear understanding of how important an entryway paratext their preface was for selling the book. As Maitland observed, even the "learned Englishman" needed some prefatorial assurance "that important conclusions are to be gained thereby."[30] He was outspoken about the impact prefaces had on readers and especially on reviewers and admitted that "As a matter of prudence therefore I put into an Introduction a passage about the book which I mean critics to copy." He was quite amused by reviewers who caught "the bait."[31] Reviewers indeed relied heavily on prefaces and filled their submissions with long extracts from them. While this indicated intellectual laziness to some, it created indirect opportunities for historians to access a far larger audience than their prefaces would ever command. Because of this, the prefatorial words truly mattered and it is quite striking how little attention publishers showed to historians' prefaces. Ambiguity or excessive honesty about the book's shortcomings were among the reasons for publishers' rare interventions with historians' prefaces.

Freeman's prefatorial honesty illustrates the tensions which the conflict between historians' scholarly and publishers' commercial aspirations could cause. Freeman's early prefaces had been self-laudatory in terms of marketing the originality, novelty, methodological rigor, and epistemic virtuousness of his books. Patrick Marot's term *function monumentale* describes them well.[32] In the 1880s, however, Freeman revised his prefatorial habits, and adopted honesty as an additional guideline for preface writing. Perhaps he calculated that his name had become so established that his books did not need any other advertising and he could therefore afford prefatorial modesty and sincerity. In this new spirit, he began to minutely report the defects that spoiled his books. In *The Historical Geography of Europe* (1881), he confessed, "It has been impossible to make it, what a book should, if possible, be."[33] Virtually every paragraph in the preface began with an admission of deficiencies and Freeman alarmed readers, for instance, about slips in dates and errors in the spelling of foreign names. And this came from the man who prided himself on unyielding accuracy and who scanned other historians' books for their imperfections. It is unclear what the publisher C. J. Longman thought about such an unattractive preface, but another, equally frank preface from Freeman drove him to despair and he begged the eminent historian to make it more commercially appealing.

Freeman wrote the controversial preface for a short history of Exeter in a series called *Historic Towns* which Longman had invited him to edit with William Hunt. The draft preface has not survived, but Longman's response gives enough clues to presume that it was the same that was eventually published. This preface began with a stunningly honest confession: "The present volume, I wish it distinctly to be understood, does not represent any independent research into the Exeter archives." Then Freeman went on to recognize the high importance of municipal and ecclesiastical archives in Exeter yet admitting that "to study them as they must be studied in manuscript would call for the offering of no small part of life, and such an offering is clear that I cannot make." As a justification for his chosen method, he assured that even the consultation of printed records had provided him with some gains because he had written the history of Exeter "as part of the history of England, and the history of England mainly as part of the history of Europe." After once again establishing his view of the interdependence between local, national, and international history which the antiquaries failed to recognize, he closed the preface with an appeal to his readers in Exeter: "I have no doubt that local knowledge will be able to suggest and correct many things, and all suggestions and corrections will be received with real thankfulness."[34] Longman's apprehensions are understandable: originality and scholarliness sold histories, but the preface contained only the feeble promises of either of them.

The dismissal of archival research in the preface must have come as a surprise to many readers who did not know Freeman personally. After all, the representatives of scientific history had already been touting for several decades how documents, manuscripts, and the consultation of original authorities were indispensable for historical research. Freeman had been at the forefront of this campaign. For those who knew him, the prefatorial confession was less surprising. They were aware of his aversion to libraries and archives. If the preface was an honest reflection of Freeman's views of original authorities, as a marketing device it was hopeless.

Longman was terrified of the preface and suggested revisions because first, it was too sincere, and second, the educational character of the book did not entail exhaustive original research. Longman did not think that Freeman's candor fitted within the limits of conventional prefatorial apologies.[35] An apology was a traditional prefatorial gesture, but the habitual humility and apologies both amused and annoyed Victorian writers. Thackeray parodied the fashion in "Prefaces and Dedications" and demanded that a text's significance should remain "transparent through all its [prefatorial] conventional humility." Percy Russell warned aspiring authors against writing absurdly apologetic prefaces just to forestall criticism.[36] Historians mostly disregarded the warnings, but Creighton questioned the meaningfulness of apologies. According to him, they were useless for historians because it should have gone "without saying that we have done our best, and it equally goes without saying that that is not much to begin with."[37] Freeman was fond of defending himself in his prefaces but in *Exeter*, according to Longman, he overstated the prefatorial repentance.

Another issue was that Freeman seemed to ignore the distinction between histories "which are written from original research and books which are not," as Longman put it. *Exeter* fell into the latter category, and no one expected Freeman to have conducted extensive original research "for so small a book." The first paragraph in the preface was therefore "over scrupulous."[38] But as Chapter 5 shows, the matter was not this straightforward, for readers had different expectations about the level of scholarliness in the educational histories. Freeman was aware of this and just when he and Longman were debating about the preface, the *Athenaeum* issued a review of W. J. Loftie's volume about London in the same series. The reviewer was displeased precisely about Loftie's slight use of unpublished sources in the book. When the *Athenaeum* later reviewed Burrows's contribution to the series, *Cinque Ports*, the writer congratulated him for conducting original research to an extent that was extraordinary—but desirable—for histories in this category.[39] Against this light, it is understandable why Freeman wanted to caution readers against expecting original research, as he did not wish to appear as a historian who made unfounded promises.

Longman saw the matter differently and warned Freeman that the preface "may have a considerable effect on the future of the book." Yet, he promised to accept "without a murmur" whatever Freeman chose to do with it. As we may guess, Freeman was unmoved and held tight to honesty and transparency. Longman then went ahead and published *Exeter* with a preface which he feared was "likely to injure the book."[40] The preface, however, did not cause irreparable damage to sales and a second edition was issued in 1890 and a third in 1892. Yet, it did not go fully unnoticed, either. While the reviews were in general quite generous, at least the *Athenaeum, Saturday Review, Newcastle Courant*, *Exeter and Plymouth Gazette*, *Western Times*, and *Graphic* reported Freeman's methodological confessions in the preface. The *Newcastle Courant* went as far as saying, "Mr Freeman wishes it to be distinctly understood that the volume does not represent any independent research into the Exeter archives" and the *Athenaeum* found Freeman's "almost nervous anxiety to assure that he has not worked from original records, in his case, quite needless."[41] Freeman was annoyed about such smirking at "the 'limitations' of my mind," but refused to connect the dots between his preface and the reviews.[42]

Longman was alarmed at the attention which the preface attracted, and to offset the criticism, he addressed the issue in a prospectus of the series with strategically chosen quotes from reviews. Macaulay had condemned review quotes as the worst sort of puffery, but with no effect.[43] The quotes were like the testimonials from scientists and other experts that were planted in advertisements to sell anything from soap to baby food and medicine. The advertising industry considered expert opinions as an antidote to consumers' suspicions about puffery, for they evoked authority with their scientifically proven "facts" about the product. Such facts tended to be mere fabrications.[44] Publishers relied on the same mechanism, and their catalogs, newspaper advertisements, flyleaves, prospectuses, and other marketing materials were decorated with short extracts from the reviews. The advertisements were put together with the scissor-and-paste method and appropriate sentences were removed from their original context, which might have been anything between hyperbolic adulation and ferocious criticism.

Longman inserted the prospectus into the front matter of the 1890 edition of *Exeter*. Freeman's volume was introduced by quoting the *Saturday Review*: "The limitations of Mr. Freeman's method do not affect the goodness of his work within the lines which he has chosen; and the oddity of his recondite knowledge only adds a pleasant flavour to his pages."[45] The message was more pronounced in the 1892 edition, where Longman quoted only the first part of the sentence, repeating nothing else other than "The limitations of Mr. Freeman's method do not affect the goodness of his work within the lines which he has chosen."[46] The passage stood out from the rest of the quotes in the prospectus which promoted the other volumes, referring in conventional fashion to their originality, readability, and accuracy. For example, an excerpt from the *Scotsman* was prefixed to William Hunt's *Bristol* to prove how the book was.

> written with the accuracy and from research of a scholar; at the same time he is careful never to lose sight of the popular character aimed at by this

> series of books, and what might in less skilful hands have been a jejune chronicle becomes a fresh and engaging narrative.[47]

The *Oxford Review* was quoted to complement C. W. Boase's *Oxford* as "A most careful, interesting, and trustworthy book, which every Oxford man and every Commemoration visitor would do well to purchase." Amidst these careful, accurate, scholarly, yet popular books, Longman's decision to make a reference to Freeman's method is a clear indication that he wanted to use the advertisement to remove any doubts the preface might arouse in readers.

The reviews indicate how prefatorial confessions could capture reviewers' attention and Freeman was not the only historian whose exaggerated prefatorial earnestness was discussed in reviews. Gardiner's first volume of *History of England* (1863) contained a preface that competed in candor with Freeman's *Exeter* as he opened it with the dramatic confession, "I cannot pretend that the period of history which I have made the object of my especial study is possessed of that striking dramatic interest which is to be found elsewhere."[48] The *Athenaeum* called into question Gardiner's marketing strategy:

> Mr. Gardiner opens his undertaking with a complaint that his theme is wanting in the dramatic interest which may be found elsewhere. This is unfortunate, and, we think, it is also wrong. If it had been true, the writer should have concealed the fact with the utmost art. A reader warned at the outset that the poet himself has found his legend poor, is not likely to pursue the story with ardent zest.[49]

Three years later, the *Athenaeum* again complained about poorly chosen prefatorial words. James E. Thorold Rogers had furnished his preface to *A History of Agriculture and Prices in England* with a warning about its dryness and dullness as he anticipated how "the facts and comments" in his book "will attract but few readers." He even proposed that the book would be "repulsive" for an average reader.[50] The writer in the *Athenaeum* was appalled by Rogers's arrogance and pointed out how the preface revealed the author's gross underestimation of the general audience's interest in carefully composed histories.[51] He should not have openly discouraged readers in the preface because that, the writer implied, was plain condescending. While such rhetoric might have helped Rogers to elevate his scholarly status, denigrating readers was questionable as a marketing ploy.

Excessive honesty and transparency in prefaces distanced historians from the unethical advertising business and highlighted the morality of their persona. Publishers, however, were alarmed by prefatorial earnestness and reviewers found it absurd. A preface's unique position as a mediator between historians, texts, historical facts, and readers prompted at least some historians to resort to extreme earnestness while they might be less scrupulous about the other marketing paratexts in their books. Titular disagreements between historians and publishers reveal dissonance between their pledges of honesty and the actual paratextual practices.

Fluidity of Honesty

Titles were marketing devices *par excellence* and, unsurprisingly, the promotional paratext that repeatedly caused friction between historians and publishers. Readers had begun to regard titles as authorial statements in the eighteenth century when the introduction of copyright legislation and the modern culture of authorship enhanced the link between a text and its author.[52] This authorial ownership of a title made the scientific historians cautious, as they realized that readers conceived titles as the reflections of their persona. Consequently, they expected that their titles awarded respectability to their texts, themselves, and the discipline they represented. Yet, they were willing to make concessions about the truthfulness of the titles when they prioritized their economic, cultural, or ideological aims over their scholarly ones. Both Genette and John Fisher have dismissed the commercial component in titles as evident and hence uninteresting. Fisher even proposed ignoring altogether the commercial aspect as conceptually unhelpful because publishers reject titles that jeopardize sales.[53] Such a comment is hasty, for the negotiations about titles demonstrate how late-Victorian publishers made allowances for historians who insisted on giving commercially harmful titles to their books.

Tittles engrossed the public. Reviewers, essayist, readers, publishers, and writers of manuals for aspiring authors all had opinions about titles and were happy to share them. For historians, a good title was catchy and a precise description of the content and the genre. The long narrative titles common in early modern books had been abandoned in histories and were used occasionally only by amateur historians. An ideal title was adjustable to its many locations in a book: a spine, cover, half-title page, title page, and running head.[54] A good title also adapted to the discursive spaces beyond the book as titles circulated in advertisements, reviews, dinner conversations, and correspondence between friends, family, and colleagues, reaching a far larger audience than the books themselves ever did. As Macmillan put it, a title should "run easily in the mouth" and be easily abbreviated in daily talk.[55] Yet, historians regularly suggested titles that were too ambiguous, generic, or pedantic. According to Gell, from the trade's point of view there was not a "more unhopeful title than 'A History of England'." Doble concurred. It was the kind of title that only caused "bewilderment" in readers.[56] Getting titles in order, then, could be challenging and when Freeman once again struggled to find the right title for one of his many books, he inquired from Macmillan with desperation how it was possible that titles could "effect sales" so much. That, he allowed, was a great "mystery" to him.[57]

Since titles were for historians a representation of their persona, they adopted a strategy of resisting titular trends. Trendy titles might have diminished scholarly respectability and eroded the boundary between history as science and history as literature. As we already saw, Freeman detested the fashion for adopting words with foreign roots for product names. Stubbs, for his share, was profoundly shocked by James Robertson's title *Becket, Archbishop of Canterbury. A Biography* because it carried a "catchpenny" genre-indicating subtitle. The flashiness of the title nearly prevented him from buying the book which, he realized, was more scholarly and valuable than its terrible title suggested. He was so upset about the title that even several years later he alluded to it when he parodied the titular trends that had crept into history books. "Shall we have in time," he jeered, "Beard a Biography, Pepys a Prolusion, Eden an Essay, Langley a Life."[58] Such genre-denoting titles were, indeed, in vogue. Thackeray commented on titles with genre indicators by explaining how the earlier metaphorical ingenuity had been substituted with such clever titles. The "Life of Smith" was now called "Smith" with a subtitle "'Life History' or a 'Soul's Struggle,' or something of that."[59] Historians admitted that a genre indicator was an integral part of a good title, but Freeman, Stubbs, Gardiner, Creighton, and many others avoided trendy subtitles and instead integrated "history," "sketch," "essay," or some other genre-denoting description into the title itself.

The persona project demanded that titles echoed historians' virtues, but the different functions which they assigned to the titles made it difficult to find labels that served all the purposes equally well. An excellent illustration of the competing interests which a title embodied is Freeman's *General Sketch of European History*. Freeman and Macmillan strongly disagreed about the word "European" in the title. Macmillan preferred either "General History" or "Origins of the Nations," because he anticipated a large demand in the colonies and suspected that the overseas audience—especially the Americans—would not get excited about *European* history. Freeman replied by appealing to accuracy: he wrote about European history, not about non-European people, which a broader title would have implied.[60] He warned Macmillan that a misleading title could upset readers, bring bad publicity, incite unsympathetic reviews, and harm the success of the book.

As such, Freeman was right: reviewers took pleasure in humiliating historians whose titles were ambiguous, deceptive, or promised more than the book delivered. One commentator found Gardiner's title *Prince Charles and the Spanish Marriage* historically false as "everybody knows that the wretched intrigue and humiliating journey of Prince Charles ended in matrimonial failure."[61] Duke de Broglie's title *The King's Secret: Being the Secret Correspondence of Louis X* was rated as a misnomer as the book did not contain any secrets. Ulick Burke, lawyer and scholar, was even more merciless when he reviewed W. J. Fitzpatrick's *Secret Service under Pitt*. Burke conceived Pitt's name in the title as a cheap marketing trick because in reality "the great Minister was not more concerned with the secret history of the spies employed by the authorities at Dublin Castle than with that of gentleman who blackened the ministerial shoes in Downing Street, or drank the ministerial port wine at Putney."[62] Freeman wanted to make sure that his title would not give any reason for this kind of sneering or accusations of historical inaccuracy.

However, Freeman did not only oppose Macmillan's alternatives because they described the topic inaccurately. His titular determination was also colored by his personal biases. He explained to Macmillan how the book was about "*General European History*, that is a history of *Dutch and Welsh. General History* would take in *something nasty*, with which I have nothing to do." By "nasty" he meant people of non-European denominations, and he listed several racialized groups whose history he had excluded from his narrative just to ensure that Macmillan grasped how important it was to retain the word "European" in the title.[63] The debate lingered on, and time and again Freeman pointed out how Macmillan's alternative titles were "simply misleading and dishonest" and how the deletion of the word "Europe" would give "an utter false view of the book, its object and contents."[64] While Freeman justified his title with the readers' right to know what they were buying, he objected to Macmillan's suggestions largely for ideological reasons. He distanced himself from physiological race theory and materialist ethnology but nonetheless embraced a hierarchical organization of races based on culture, language, and institutions.[65] He drew an explicit parallel between civilization and Europe in the book's preface, explaining how it offered a "sketch of the history of the civilized world, that is, of Europe and of the lands which have drawn their civilization from Europe."[66] The question about racialized hierarchies between the Aryans and non-Aryans gained acuteness in the 1870s when English progress seemed to be faltering. Freeman's anxiety about the title was one manifestation of this growing concern.

Freeman also pleaded *historical* accuracy when he defended his choice for the title, but in this case, too, he fused his historical views and biases. He was confident that the American audience would comprehend and find the title attractive precisely because of its historical accuracy. Inspired by Amero-Teutonism, Freeman believed in the shared history of the Americans and Europeans and claimed that because of this historical reality, an American "cannot expect less than 100 years to be put on a level with 2640 [years]."[67] Drawing inspiration from the comparative method, he argued that "the forms of government of the Aryan nations ... all spring from a common source, an *Urbrunnen*."[68] Within this theoretical framework, he maintained that the Americans and English derived from the same division of the Teutonic race and that the shared language, history, and institutions tied them closely together. The Teutons were the foremost race in the world and the English its dominant branch; they were the torchbearers of Western civilization, with language and institutions that were the highest refinement of this process.[69] It was this common past that made "General Sketch of European History" both a historically appropriate and commercially attractive title. Claiming anything else in a title would have been simply dishonest toward the readers, Freeman was convinced that the word "European" was not a threat, but an advantage, for the book in America.

Macmillan, perhaps too tired to argue any longer about the title and eager to launch the book, gave in and allowed it to appear as *General Sketch of European History*. Freeman confidently announced that this was the "best" possible title for it.[70] But Macmillan's suspicions were not unfounded. The North American publishers knew their audience, and when Holt & Williams released the book in 1872 in New York, it was called *Outlines of History*. Henry Holt issued a new edition several years later as *General Sketch of History*, while in Canada, the book was published as *Outlines of History*. Freeman was furious about these alterations and begged the publisher to do something about the new "ill" sounding title that failed to "express the object of the book."[71] He also complained about the "evil" act of changing the title to Thompson, raging about foreign publishers who had the freedom to alter books without consulting the author. He was particularly disillusioned by the Canadians for their inability to "see that they are Europeans carrying Europe with them."[72]

The dispute about an appropriate title for what was to become *General Sketch of European History* betrays the multiple motives that informed the making of titles. A good title appealed to different audiences, described the content accurately, and resonated with the historians' persona. But titles could carry additional meanings and, for instance, bolster historians' ideological agendas. Titles were expected to give a clear explanation of the content, style, and genre of the book and tell something about its author. Yet, historians could be less vigilant about their moral standards when they designed titles than when they wrote prefaces. The different relationships that a title and a preface had with the text was the main reason for this. Freeman tweaked titles out of ideological concerns but was willing to do the same for commercial reasons as well. When he persuaded Macmillan to publish in book form the three lectures about the history of Wells Cathedral which he had delivered for the locals, the publisher doubted the success of such a venture as he correctly anticipated only limited, local attraction to it. Freeman, knowing very well that this was true, proposed to adjust the title to create an impression that the book's interest was not "purely local."[73] Macmillan acceded and issued the book with a title *History of the Cathedral Church of Wells: As Illustrating the History of the Cathedral Churches of the Old Foundation*. Readers were able to see through the title and the book failed to gain popularity outside the Wells region.

*

Historians were entangled in Victorian advertising culture, for marketing was part and parcel of the publishing process. Although they were aware of the afterglow of the early-Victorian puffery and of the harm which misleading advertising could cause to the reputation of their discipline, they nonetheless silently accepted some forms of puffing just as they accepted that their input was needed in the development of the advertising for their books. They strove to create promotional paratexts that did not contradict their persona. If they loudly condemned Froude and the other belles-lettres historians for sacrificing truth to exaggeration and inaccuracies, they could not take the risk of compromising their moral stature with misleading or hyperbolic paratexts. Powell, critical of the commodification of historical knowledge, compared historians who descended "from the pursuit of truth to the lower methods" with "the quack advertiser."[74] For the publishers, historians' insistence on honesty appeared as ignorance about the realities of the literary marketplace. When Alexander Macmillan once again brought up the topic of paratexts in a conversation with Bartholomew Price, he could have been talking about historians as he complained about stubborn writers who were "hard to persuade" when they had set their heart on certain kinds of paratexts.[75] The reality, though, was different, and historians were often flexible about the virtues that constituted their persona when it came to advertising. As their overlapping roles—and dimensions of their persona—were embedded in their paratexts, the tensions between different dispositions were unavoidable. The difficulties with the promotional paratexts derived from the struggle to find the right kind of equilibrium between the different goals book parts served in books and the performance of the persona in them. Engagement with the commercial marketplace was unavoidable but could not be accepted at any cost to the persona project either.

Notes

1. Stubbs, *Seventeen Lectures*, 52.
2. White, "Chapter on Puffs," 42–46; "Age of Veneer," *Fraser's Magazine*, January 1852, 87–93; "Thoughts on Puffing," *All the Year Round*, March 4, 1871, 329–332.
3. W. J., "The Grand Force," 380–381.
4. Elton, *Frederick York Powell*, 1:381.
5. Trollope, *The Struggles of Brown, Jones, and Robinson*, 68.
6. Secord, *Victorian Sensation*, 373, 405–406.
7. Mason, *Literary Advertising*, 11–12, 31–34, 135–139.
8. Macaulay, "Mr. Robert Montgomery's *Poems*, and the *Modern Practice of Puffing*," 196–200.
9. Lingard to Coulston, undated, and Lingard to Walker, [1849–1850], in Haile and Bonney, *Life and Letters*, 342, 350; Howsam, *Kegan Paul*, 37–38.

10. G. W. Y. "Confessions of Puff," 24; Campbell. "Reviewing Oneself," 164; Wagner, *How to Publish a Book*, 81–82.
11. Notice in *EHR*, 1890, 808.
12. Clifford, "Warning to Authors," 187.
13. Oliphant to Blackwood, February 6, 1873, in Coghill, *Autobiography and Letters*, 240–241.
14. Maine to Freeman, December 22, 1873, FA1/7/51–533, JRL; Jessopp to Stopford Green, May 3, 1887, MS 15,081/4/5, NIL; Dean Hook to Freeman, January 24, 1865, FA 1/7/297–320, JRL.
15. Goldgar, *Impolite Learning*, 76–77, 107–111.
16. George Macmillan to Ward, September 15, 1875, Add MSS 55397, BL; Stopford Green to Macmillan, March 30, [1904], Add MSS 55095, BL.
17. Macmillan to Freeman, March 3, 11, and 23, 1863, FA 1/7/493–510, JRL; Macmillan to Freeman April 1, 1863 and May 12, 1863, Add MSS 555381 (A), BL; Freeman to Macmillan, November 22, 1864, Add MSS 55049, BL; Benjamin Thorpe to Freeman, March 21, 1866, FA 1/7/758–777, JRL; Freeman to Green, December 18, 1871, FA 1/8/1–30, JRL; Freeman to Craik, August 16, 1874, MSS 55050, BL; Max Müller to Freeman, December 21, 1875, FA 1/7/584–622, JRL; Ihne to Freeman, May 9, 1882, FA 1/7/389–423, JRL; Freeman to Macmillan, November 17, 1882, Add MSS 55052, BL; Max Müller to Freeman, November 12, s.a. FA 1/7/584–622, JRL.
18. Gell to Lane Poole, September 21, 1885, Letter Books 38, OUP.
19. Acton to Creighton, December 30, 1885, Add MSS 6871/29, CUL; Creighton to Acton, January 26, 1886, Add MSS 8119/1/C253, CUL.
20. Brundage, *People's Historian*, 58, 101–102; Hesketh, *Science of History*, 121–124.
21. Hesketh. *Victorian Jesus*, 68–101.
22. Seeley to Furnivall, [1871], FU 806, Furnivall Papers, Huntington Library.
23. Laughton to Browning, December 1 and 5, 1897, GBR/0272/OB//1/939/C, King's Cam.
24. Browning, "Pitt's Foreign Policy," 112. The review is attributed to Browning in *Wellesley Index to Victorian Periodicals*.
25. Browning, "Pitt's Foreign Policy," 112–113.
26. Thompson to Macmillan, August 30, 1876, December 16, 1900, Add MSS 55078, BL.
27. Bateson, "Review of Gilbert Church," 382; Child, *Church and State*, vi.
28. Tanselle, *Book-Jackets*, 13–19, 64.
29. Marot, "Pour une poétique," 11; Masson, "Marginalité de la préface," 14; Pender, "Framing the Reformation Woman Writer," 29–30; Murray, "The Politics of the Preface," 50.
30. Maitland to Vinogradoff, September 1888, in Fifoot, *Letters*, 48.

31. Maitland to Vinogradoff, March 12, 1889, in Fifoot, *Letters*, 60.
32. Marot. "Pour une poétique," 24–25.
33. Freeman, *Historical Geography of Europe*, v.
34. Freeman. *Exeter*, v–vii.
35. Longman to Freeman, January 10, 1887, FA1/7/476–492, JRL.
36. W.M.T., "Prefaces and Dedications," 25–28; Russell, *Literary Manual*, 175.
37. Creighton to Lane-Poole, December 23, 1885, in Creighton, *Life*, 338–339.
38. Longman to Freeman, January 10, 1887, FA1/7/476–492, JRL.
39. "Our Library Table," *Athenaeum*, January 22, 1887, 127; "Historic Towns—Cinque Ports," *Athenaeum*, November 3, 1888, 587.
40. Longman to Freeman, January 10, 1887, FA1/7/476–492, JRL.
41. "New Books and Editions," *Newcastle Courant*, February 11, 1887, 6; "Historic Towns," *Athenaeum*, March 19, 1887, 378; "New History of Exeter," *Exeter and Plymouth Gazette*, February 19, 1887; "Reader," *Graphic*, March 26, 1887, 331.
42. Freeman to Thompson, April 12, 1887, U DX9/152, HHC.
43. Macaulay, "Mr. Robert Montgomery's *Poems* and the *Modern Practice of Puffing*," 197–198.
44. Loeb, *Consuming Angels*, 75–79.
45. Freeman, *Exeter*, 2nd ed., front matter.
46. Freeman, *Exeter*, 3rd ed., front matter.
47. Freeman, *Exeter* 2nd ed., front matter.
48. Gardiner, *History of England from the Accession of James I.*, v.
49. "History of England," *Athenaeum*, March 21, 1863, 392.
50. Rogers, *History of Agriculture and Prices*, vi.
51. "History of Agriculture and Prices," *Athenaeum*, June 23, 1866, 825.
52. Shevlin, "To Reconcile *Book*," 57–61.
53. Fisher, "Entitling," 287; Genette, *Paratexts*, 92.
54. Doble to C. Plummer, August 15, 1885, Letter Books 38, OUP.
55. Macmillan to Sir Roundell Palmer, November 25, 1861, in Macmillan, *Letters*, 98; Macmillan to Price, January 20, 1874, Add MSS 55394 (2), BL.
56. Gell to James Ramsay, November 16, 1891, Letter Books 54, OUP; Doble to Powell, November 4, 1891, Letter Books 54, OUP; Edward Bell to Florence Freeman, June 7, 1898, FA1/9/4, JRL.
57. Freeman to Macmillan, July 2, 1878, Add MSS 55051, BL.
58. Stubbs to Freeman, January 3, 1860, in Hutton, *Letters*, 78; Stubbs to Freeman, August 16, [1863], MS. Eng. Misc. e. 148, Bodleian.
59. W.M.T. "Prefaces and Dedications," 25.
60. Freeman to Macmillan, October 4, 1870, April 16, 1871, BL, Add MSS 55049.

61. "Prince Charles," *Athenaeum*, May 8, 1869, 629.
62. Hayward, "*Secret Correspondence of Louis XV*," 468; [Burke], "Irish Spies," 82; "Secret Service under Pitt," *Athenaeum*, May 21, 1892, 658.
63. Freeman to Macmillan, October 4, 1870, and April 16, 1871, Add MS 55049, BL.
64. Freeman to Macmillan, May 21, 1871, Add MS 55049, BL; Freeman to Macmillan, April 25, 1872, Add MSS 55050, BL.
65. Morrisroe, "'Sanguinary Amusement'," 31–37; Koditschek, "A Liberal Descent?" 215–216.
66. Freeman, *General Sketch*, v.
67. Freeman to Macmillan, April 25, 1872, Add MS 55050, BL.
68. Freeman to E. B. Tylor, July 20, 1872, in Stephens, *Life and Letters*, 2:57.
69. Conlin, "Consolation of Amero-Teutonism," 108–111; Bell, "Alter Orbis," 222.
70. Freeman to Macmillan, April 28, 1872, Add MSS 55050, BL.
71. Freeman to Craik, November 26, 1876, Add MSS 55051, BL.
72. Freeman to Thompson, August 31, 1879, U DX9/82, HHC.
73. Macmillan to Freeman, January 17, 1870, February 17, 1870, and February 19, 1870, Add MSS 55390 (2), BL; Freeman to Macmillan, February 20, 1870, Add MSS 55049, BL.
74. Elton, *Frederick York Powell*, 2:40.
75. Macmillan to Price, January 19, 1876, Add MSS 55398, BL.

References

Unpublished Primary Sources

Bodleian Library, Oxford: Stubbs Papers.
British Library: The Macmillan Papers.
Cambridge University Library: Acton Papers.
Hull History Centre: Letters from Edward Augustus Freeman to Edith Thompson.
Huntington Library, San Marino, California: Furnivall (F.J.) Papers.
John Rylands Library, Manchester: E. A. Freeman Archive.
King's College Archive Centre, Cambridge: Oscar Browning Papers.
National Library of Ireland, Dublin: Alice Stopford Green Additional Papers.
Oxford University Press Archive: Letter Books.

Printed Primary Sources

[Anon.]. "The Age of Veneer. The Science of Puffing." *Fraser's Magazine*, January 1852, 87–93.
[Anon.]. "History of England." *Athenaeum*, March 21, 1863, 392–393.
[Anon.]. "A History of Agriculture and Prices in England." *Athenaeum*, June 23, 1866, 825–827.
[Anon.]. "Prince Charles and the Spanish Marriage." *Athenaeum*, May 8, 1869, 629–630.
[Anon.]. "Thoughts on Puffing." *All the Year Round*, March 4, 1871, 329–332.
[Anon.]. "Our Library Table." *Athenaeum*, January 22, 1887, 127.
[Anon.]. "New Books and Editions." *Newcastle Courant*, February 11, 1887, 6.
[Anon.]. "A New History of Exeter." *Exeter and Plymouth Gazette*, February 19, 1887.
[Anon.]. "Historic Towns." *Athenaeum*, March 19, 1887, 377–378.
[Anon.]. "The Reader." *Graphic*, March 26, 1887, 331.
[Anon.]. "Historic Towns—Cinque Ports." *Athenaeum*, November 3, 1888, 586–587.
[Anon.]. "Notice." *The English Historical Review* 5, no. 20 (1890): 808.
[Anon.]. "Secret Service under Pitt." *Athenaeum*, May 21, 1892, 658.
Bateson, Mary. "Review of Gilbert W. Child's *Church and State under the Tudors*." *English Historical Review* 6, no. 22 (1891): 381–383.
[Browning, Oscar]. "Pitt's Foreign Policy: The Leeds and Bland Burges Papers." *Quarterly Review*, July 1885, 110–141.
[Burke, Ulick Ralph]. "Irish Spies and Informers." *Edinburgh Review*, July 1892, 81–113.
Campbell, J. Dykes. "Reviewing oneself." *The Athenaeum*, August 3, 1889, 164.
Child, Gilbert W. *Church and State under the Tudors*. London: Longman, 1890.
Clifford, Lucy. "Warning to Authors." *Athenaeum*, February 11, 1893, 187.
Coghill, Harry Mrs. *The Autobiography and Letters of Mrs M. O. W. Oliphant*. Edinburgh: William Blackwood, 1899.
Creighton, Louise. *Life and Letters of Mandell Creighton, Sometime Bishop of London*. Vol. 1. London: Longman, 1904.
Elton, Oliver. *Frederick York Powell. A Life and a Selection from His Letters and Occasional Writings*. Vol. 1. Oxford: Clarendon Press, 1906a.
Elton, Oliver. *Frederick York Powell. A Life and a Selection from His Letters and Occasional Writings*. Vol. 2. Oxford: Clarendon Press, 1906.
Fifoot, C. H. S. *The Letters of Frederic William Maitland*. London: Selden Society, 1965.
Freeman, Edward. A. *The General Sketch of European History*. London: Macmillan, 1872.
Freeman, Edward A. *The Historical Geography of Europe*. Vol. 1. London: Longman, 1881.

Freeman, Edward A. *Exeter*. London: Longman, 1887.
Freeman, Edward A. *Exeter*. 2nd ed. London: Longman, 1890.
Freeman, Edward A. *Exeter*. 3rd ed. London: Longman, 1892.
Gardiner, Samuel Rawson. *History of England from the Accession of James I. to the Disgrace of Chief-Justice Coke*. Vol. 1. London: Hurst and Blackett, 1863.
G. W. Y. "The Confessions of Puff." *Athenaeum*, January 6, 1866, 24.
Haile, Martin and Edwin Bonney. *Life and Letters of John Lingard 1771–1851*. London: Herbert & Daniel, 1911.
[Hayward, Abraham]. "*The Secret Correspondence of Louis XV*." *Quarterly Review*, April 1879, 468–511.
Hutton, William Holden. *Letters of William Stubbs Bishop of Oxford 1825–1901*. London: Archibald Constable, 1904.
[Macaulay, Thomas Babington]. "Mr. Robert Montgomery's *Poems*, and the *Modern Practice of Puffing*." *Edinburgh Review*, July 1830, 193–220.
Macmillan, George A. *Letters of Alexander Macmillan*. Glasgow: Printed for Private Circulation, 1908.
Rogers, James E. Thorold. *A History of Agriculture and Prices in England*. Vol. 1. Oxford: Clarendon Press, 1866.
Russell, Percy. *The Literary Manual; or, a Complete Guide to Authorship*. London: London Literary Society, 1886.
Stephens, W. R. W. *The Life and Letters of Edward A. Freeman D.C.L., LL.D.* Vol. 2. London: Macmillan, 1895.
Stubbs, William. *Seventeen Lectures on the Study of Medieval and Modern History and Kindred Subjects*. Oxford: Clarendon Press, 1886.
Trollope, Anthony. *The Struggles of Brown, Jones, and Robinson by One of the Firm*. London: The Trollope Society, 1999.
Wagner, Leopold. *How to Publish a Book or Article and How to Produce a Play. Advice to Young Authors*. London: George Redway, 1898.
White, Mrs. "A Chapter on Puffs and Advertisements." *Ainsworth's Magazine*, 1849, 42–46.
W. J. "The Grand Force!" *Fraser's Magazine*, March 1869, 380–383.
W. M. T. "Prefaces and Dedications." *Living Age*, April 1859, 25–28.

Secondary Sources

Bell, Duncan. "Alter Orbis: E. A. Freeman on Empire and Racial Destiny." In *Making History: Edward Augustus Freeman and Victorian Cultural Politics*, edited by G. A. Bremner and Jonathan Conlin, 217–235. London: Proceedings of the British Academy, 2015.
Brundage, Anthony. *The People's Historian: John Richard Green and the Writing of History in Victorian England*. Westport: Greenwood Press, 1994.

Conlin, Jonathan. "The Consolation of Amero-Teutonism: E. A. Freeman's Tour of the United States, 1881–2." In *Making History: Edward Augustus Freeman and Victorian Cultural Politics*, edited by G. A. Bremner and Jonathan Conlin, 101–118. London: The British Academy, 2015.

Fisher, John. "Entitling." *Critical Inquiry* 11, no. 2 (1984): 286–298.

Genette, Gerard. *Paratexts: Thresholds of Interpretation*. Cambridge: Cambridge University Press, 1997.

Goldgar, Anne. *Impolite Learning: Conduct and Community in the Republic of Letters 1680–1750*. New Haven: Yale University Press, 1995.

Hesketh, Ian. *The Science of History in Victorian Britain: Making the Past Speak*. London: Pickering & Chatto, 2011.

Hesketh, Ian. *Victorian Jesus: J.R. Seeley, Religion, and the Cultural Significance of Anonymity*. Toronto: University of Toronto Press, 2017.

Koditschek, Theodore. "A Liberal Descent? E. A. Freeman's Invention of Racial Traditions." In *Making History: Edward Augustus Freeman and Victorian Cultural Politics*, edited by G. A. Bremner and Jonathan Conlin, 199–216. London: Proceedings of the British Academy, 2015.

Loeb, Lori Anne. *Consuming Angels: Advertising and Victorian Women*. Oxford: Oxford University Press, 1994.

Marot, Patrick. "Pour une poétique historique des textes luminaires." In *Les textes liminaires*, edited by Patrick Marot, 7–27. Toulouse: Presses Universitaires du Mirail, 2010.

Mason, Nicholas. *Literary Advertising and the Shaping of British Romanticism*. Baltimore: The Johns Hopkins University Press, 2013.

Masson, Pierre. "Marginalité de la préface autoriale." In *L'Art de la préface*, edited by Philippe Forest, 11–23. Nantes: Éditions Cécile Defaut, 2006.

Morrisroe, Vicky L. "'Sanguinary Amusement': E. A. Freeman, the Comparative Method and Victorian Theories of Race." *Modern Intellectual History* 10, no. 1 (2013): 27–56.

Murray, Jessica. "The Politics of the Preface: Lady Anne Barnard's Gendered Negotiations in a Liminal Textual Space." *English Studies in Africa* 56, no. 2 (2013): 49–59.

Pender, Patricia. "Framing the Reformation Woman Writer: John Bale's Preface to Anne Askew's *Examinations*." *Parergon* 29, no. 2 (2012): 29–45.

Secord, James A. *Victorian Sensation: The Extraordinary Publication, Reception, and Secret Authorship of* Vestiges of the Natural History of Creation. Chicago: University of Chicago Press, 2000.

Shevlin, Eleanor F. "'To Reconcile *Book* and *Title*, And Make 'em Kin to One Another': The Evolution of the Title's Contractual Functions." *Book History* 2 (1999): 42–77.

Tanselle, G. Thomas. *Book-Jackets: Their History, Forms, and Use*. Charlottesville: Bibliographical Society of the University of Virginia, 2011.

CHAPTER 10

The Air of a Dignified Historian

Freeman once noted pungently how Froude's latest book sold "because of its cover" even if Froude as a historian was known to be a liar.[1] Freeman's comment uncovers the allure of aesthetically pleasing books among Victorian readers. Even the Queen was known to fancy books in "bright colours," one of the secretaries of Oxford University Press revealed.[2] Indeed, no matter how Carlyle had insisted that books were nothing but "bits of paper" and "black ink" or that the "outward form of the thing" was of no consequence, it became more and more evident that books' materiality mattered as the century progressed.[3] Readers engaged with the physical and aesthetic traits of books and read them as external signs of books' genre and value. But this was not all. Visually arresting books were desirable status objects that conveyed their owners' class and gender identities and were used as accessories that reflected taste, cultivation, and personality. Historians were not immune to this, and their opinions about the appropriate visual image of their books were shaped not just by the needs of the persona project or advertising but also by personal vanity and fascination for handsome-looking histories.[4] Lecky wanted all his books to be as equally sized as possible and allowed Longman to tinker with their margins to ensure uniformity, an important characteristic in a gentleman's library.[5]

The materiality of history books is yet another under-studied historiographical topic that deserves more attention. Binding, size, type,

E. Garritzen, *Reimagining the Historian in Victorian England*,
https://doi.org/10.1007/978-3-031-28461-8_10

paper quality, layout, and margins, all of which George Bornstein and Jerome McGann call a book's "bibliographical code," contextualize a text and generate meanings and value for it. They also make claims and assumptions about the preferences, tastes, and values of the anticipated audiences.[6] Historians were aware of all this and realized that it would be disingenuous to separate the form and content of history books as readers were so engrossed with their material and visual conditions. Since scholarly authority was gained at least partially through a history book's material and aesthetic qualities, it was critical to find a style that accurately promoted the persona. Historians were aware of a potential disharmony in their and their readers' tastes. Maitland dreaded that what was for him the admirably grim look of the *Brockton's Note-Book* "will frighten readers."[7] The first part of this chapter then focuses on the fashioning of a dignified look for scholarly histories and suggests that this was achieved mostly by ruling out styles that performed personae of which the scientific historians disapproved. After this, the chapter discusses the practical challenges historians and publishers encountered when they created a look for those histories that fell into several categories in terms of style, audience, and persona. The last part considers the quality of different paratexts as an endorsement of the dignity of the persona.

The craze for pretty, handsome, or ornamental books was facilitated by nineteenth-century advances in technology. As David McKitterick has explained, the look of a book altered profoundly in the nineteenth century for the first time since the Middle Ages. The introduction of case-binding allowed the rapid preparation of identical covers. Improved methods of printing letters and decorations on cloth together with the invention of the embossing machine broadened the repertoire of cover designs. The development of chemical dyes expanded the available color palette for cloth covers.[8] If customers had earlier arranged binding by themselves, books were now sold prebound. This had a dramatic impact on the marketing of books as aesthetically pleasing objects that could serve as adornments in middle-class homes. Publishers such as Macmillan, Longman, and Murray understood that scholarly books should not be excluded from the visual makeover of the literary marketplace, and the university presses followed their example from the 1860s onwards. The Oxford Delegates created for their scholarly prestigious titles a restrained yet elegant look that pleased the eye while accentuating their "learned character."[9]

The question of the aesthetics of books was complex and comprised a number of factors spanning from technological parameters to trends. The technology defined the repertoire of designs and hues for history books. The use of color is a good example of this and a tangible illustration of the visual transformation of the book market during the nineteenth century. Thackeray captured this change in 1859, writing how, four decades earlier, books had been "clad in homely, drab-colored boards, with white paper labels in the backs," but were now dressed in "crimson, green, gold, azure, scarlet, orange, purple." The colorful volumes were like "a tulip bed lying in the sun."[10] To fully embrace the possibilities of the color palette, historians and publishers needed technical knowledge and an understanding of the limitations which existed despite the innovations. Longman advised Browning in 1880 against choosing blue covers because "the booksellers object to it on account of its fading so much."[11] Ten years later, Doble had to admit that "There are some dyes which cloth refuses to take, and we sometimes have a good deal of difficulty in securing an attractive binding."[12] Then, color was also a powerful vehicle for creating associations about a theme, value, and genre. Joseph Zaehnsdorf went as far as proposing in *The Art of Bookbinding* (1880) that the publishing industry should introduce a standardized color scheme for books. He suggested dark red for history, "dull red" for archeology, and "some fancy colour" for poetry.[13] This did not take off, but books nonetheless were often identified by their color. Stubbs, for instance, casually referred to the "stout green volumes" of the Public Record Office in his inaugural lecture and Oxford University Press reserved green for history in its Classic Series at the beginning of the twentieth century.[14] Finally, color was about fashion and decorativeness, and thus intimately connected with the ongoing debates about readers' unhealthy fixation with pretty or elegant books. There was a good reason why Mr. Pooter, the archetype of middle-class suburbia in *The Diary of Nobody* (1892), painted the backs of the family's Shakespeare edition with red to give them a facelift that would better match his position in life.[15]

These debates about the content and form were familiar to historians as well. While they acknowledged that materiality was an unavoidable act in the paratextual performance of their persona, Freeman's comment about the look of Froude's book indicates that they were not entirely comfortable with the prominence that appearances had gained in the literary marketplace. This apprehension connected them to the essayists, novelists, artists, and other intellectuals who believed that the obsession about books as aesthetically desirable objects was a sign of the intellectual degeneration of the nation. The discussions were triggered by a fear that readers were becoming incapable of separating books' material and textual worth and that publishers could take advantage of readers' ignorance by serving them "bad" literature as long as it was wrapped in a pretty package. Obviously, defining what was good or bad—high or low—literature was problematic and highly subjective, and as Mary Hammond notes, it was hard to define a book's category by merely looking at it.[16] This, however, did not stop the critics who, armed with a fair amount of cultural snobbery, were confident that ordinary readers were unable to resist the dishonest publishers' attempts to disguise low quality texts with an elegant outfit. The mania for gaudy books was seen as proof of the superficiality of middle-class readers, who were allured by the "glories of typography" and "elaborate bindings" instead of serious study.[17] Leah Price has insightfully observed how publishers were required to pay attention to all the details of a book's appearance, but the aesthetic attention of consumers was interpreted as self-indulgence.[18] The controversy was waged simultaneously with historians' attempts to establish history as a scientific discipline and gave additional weight to the paratextual performance of the persona. Since the public discourse was informed by the cultural elite's urge to protect the class hierarchies, the question of the look of a history book was interlinked with historians' attempts to forge themselves a respectable social status. Historians championed aesthetics that bolstered their scholarly and social position.

Dressing up the Persona

The making of material and visual paratexts was a collaborative venture. Publishers provided historians with information about the implications of their choices on a book's size, legibility, and price. They also supplied historians with models of different types, papers, dyes, and layouts. Historians, for their part, made decisions about books' visual aspects in several roles: as authors of their own books and as Delegates and Syndicates. According to Doble, the Oxford Delegates, for example, were much exercised about paper quality in the mid-1880s.[19] As members of the committees of learned societies which produced document editions, historians also considered books' appearance. Gardiner indicated that in this role they were the guardians of regularity in the look from one volume to another.[20] Historians knew that the selling and buying of history books was a transaction of ideas and images of a historian whose persona was engraved in the book's aesthetics.

Yet, the investigation of history books' dress code is complicated by the incomplete survival of sources. Many history books come down to us in nondescript library bindings and a surprisingly large number of the digitized books in the platforms like the Internet Archive do not contain images of the covers, let alone the bindings. Furthermore, despite historians' constant preoccupation with paratexts, the discussions of the visual dimensions of history books almost entirely excluded scholarly history—apart from the constant nagging about their unwieldy size. Hence, the views on an appropriate design for a scientific history must be unearthed from historians' sporadic remarks about the outward form of their books and from the more frequent comments on the looks of the other types of histories. These discussions were premised on an imagined ideal style of a scientific history which served as an evaluative category for the other kinds of histories. Even from these scattered sources emerges a dress code that prioritized somber colors and plain or restrained design with minimal ornamentation. This style resonated with historians' striving to frame their books as a particular kind of dignified object.

When Creighton negotiated with Longman about the visual image for the *English Historical Review*, he appeared to be speaking on behalf of the entire scholarly community when he stressed, "We do not wish to look ornamental or gaudy." He was content with the chosen "type" and "get-up" because they reflected the agreed standards and expectations.[21] The unadorned look historians preferred emphasized their earnestness and dedication as scholars. As with titles that eschewed fashionable genre indicators, the somber look denoted similar resistance to trends. But the look also mediated class and gender identities and I would like to argue that there was a striking resemblance between the scientific histories and the three-piece suit, the unofficial uniform of a Victorian middle-class man.

Victorians read garments as expressions of morality, inner worth, and class status, and sartorial choices were considered as the maintenance of public identities.[22] While the plain suit epitomized the middle-class virtues of hard work, sobriety, duty, and self-control, the somber and restrained bindings of history books conveyed industry, modesty, humility, duty, gravity, and self-control that rejected the temptations of the latest fashions. Both the suit and the bindings were a performance and, as Brent Shannon has argued, the sartorial choices of middle-class men made visible the categories of gender and class as the three-piece suit was expected to erect boundaries between the hard-working middle-class men and the idle aristocracy, coquettish women, and effeminate dandies.[23] By drawing a parallel between the middle-class sartorial matrix and the visual and material paratexts in history books, the link between the scholarly persona and social status becomes evident once again.

The desired style was validated with marked quietness which reverberated through the middle-class sartorial ideal. While men were instructed not to fuss about their dress, they were reminded that their clothing legitimized their social and professional belonging, and as such, had to be carefully monitored and maintained. Fashion, in other words, was not something in which only women were absorbed.[24] It was crucial that men picked the correct vestments but did not publicly show interest in or awareness of sartorial questions. When Freeman appeared in front of an American audience, the *New Haven Evening Register* noted how his clothing affirmed his profession, for "he had apparently not time to give to the conventionalities of dress, for there was a bagginess about the trousers which indicate the work-a-day of a man of might."[25] The baggy pants dramatized the scholarly persona for the journalist. This impression of dedication to hard work and indifference to worldly matters was precisely what Freeman wanted to communicate with his dress. He claimed that the only thing that he cared about in an attire was that the clothing were "big enough."[26] Of course with this statement he already showed comprehension of the symbolism embedded in his vestments. Although historians abstained from elaborating the aesthetics of their books, moderation was approved silently in many quarters as an ideal look for serious historical works. This is confirmed by the reproof of paratexts that deviated too much from this style.

Publishers issued histories in a wide spectrum of colors, designs, and materials, ranging from primers in inexpensive trade bindings to richly illustrated histories of material culture and majestically decorated folios that tapped into the pictorial turn in Victorian book culture. Generously decorated histories such as Mrs. Bury Palliser's *A History of Lace* catered to readers' fascination with visual spectacle, and the elaborate histories of notable families and individuals appealed to the tastes of the affluent. The latter in particular provoked both adoration and hesitation in reviews. Reviewers did not spare superlatives when they described their flamboyance and elegance—the carefully executed illustrations and rich use of color, the superior quality of paper and typography, large margins, and decorative bindings. The priciness added to their air of exclusiveness and aristocratic flair, and it was not uncommon for the production of these stout folios to be sponsored by wealthy patrons. At the same time, the exorbitant price and intentionally low print run propelled some reviewers to condemn such an undemocratic way of disseminating valuable historical knowledge as the Second and Third Reform Acts which enlarged the electorate made access to knowledge a critical question. They demanded the production of cheaper versions so that the latest historical discoveries were available for a larger audience. The suitability of visual richness for serious historical works was also questioned because the lavish paratextual apparatus overwhelmed the narrative and eclipsed the books' real merits and demerits. The critics worried that readers would be carried away by the dashing appearance and forget what truly mattered in history books.[27] Altogether, flamboyance was too expensive and had too strong a connotation of aristocracy to make an acceptable outfit for histories that took seriously the duty of circulating reliable historical knowledge to the public.

However, the many compliments which commentators paid to decorative elements indicate that they were not able to resist altogether the very same sumptuousness that they considered harmful for an ordinary reader. Although the literary and scholarly elite condemned the vanities of bindings, among the public there was a persistent fascination for deluxe editions and extravagant bindings such as elephant or rhinoceros skin.[28] The reviewers of history books often placed themselves above such extravaganza, but even they, while demanding democratization of historical knowledge and fearing that sumptuousness distracted readers, had a weak spot for elegance. They showed understanding toward the affluent layer of society who adorned their libraries with handsome folios that were not just aesthetically pleasing but also contained knowledge that was unavailable to less-endowed readers. The reviewers silently accepted that often these histories were purchased to satisfy an appetite for luxury rather than out of a thirst for learning.

"Prettiness" was another visual qualification that cropped up in reviews. Prettiness denoted reduced size and "all the lovely hues" which created an impression that "slices of rainbow occupied the shelves instead of books," as one writer described the modern family bookcase.[29] Prettiness was a quality that was mostly associated with popular and feminine history writing. Freeman had already ridiculed in 1855 the obsession with small, delightful history books, contrasting the old "stately quartos in stout leather bindings," with the trendy "pretty-looking, small octavos, in garments alike so frail and so gaudy that one is well-nigh afraid to handle them."[30] He targeted his criticism at short biographies which women wrote about princesses, queens, and other famous women, and whose scientific worth he seriously questioned. This gave an immediate gendered undertone to his remarks and created an illusion that these books were feminine objects, fragile and shallow maidens, pleasing to the eye, but intellectually empty. This corresponded with the cultural stereotype of clothing and beautification of the body being associated with feminine vanity from which the middle-class man should distance himself.[31] This was a style, both in terms of content and format, that scientific historians most certainly should not have emulated.

Through negation, scientific historians' books and persona were distanced from both apparent aristocratic exclusiveness and flamboyance and feminine decorativeness and shallowness. The challenge was that, just as the cultural elite professed, publishers could replicate respectable designs to make weak histories appear first-rank contributions to history. This was a serious threat to the public image of the historians who publicly distanced themselves from the deceptive marketing ploys of the professional advertisers. The predicament was addressed in the *Athenaeum* in 1874. Mr. J. A. Fox, a concerned reader from Kilburn, was upset about Rev., Dr. J. A. Wylie's *The History of Protestantism*. Wylie had written a three-volume history which Cassell had first published in 1869. The later versions were sold with a title page that promised "five hundred and fifty illustrations *by the best artists*." Fox was unimpressed and judged the book to be a "pure romance" and feared that readers might be fooled by the careful look of the volumes: they were well printed and the illustrations made them attractive. The book also contained footnotes, the emblem of science, but Fox accused Wylie of such a deliberate misuse of his references that was too gross to be considered "ignorance or carelessness." Fox had to adopt "a less charitable conclusion" about the author's honesty, and as a conclusion he announced how it was his "Christian duty" and his concern for "the interest of historical truth" that encouraged him to unmask the scam and warn readers about the book's deceptive appearances.[32]

As with all the paratexts with a marketing role, the aesthetics of history books demanded balancing between historian, publisher, and readers' preferences. One form of compromise was to customize books for different groups of readers at different points during a book's life-cycle. Publishers had much experience of this. Although most histories appeared just with one look, there were some notable exceptions such as the Library Editions. Publishers developed these to accelerate the sales of scholarly acclaimed histories among those who appreciated both handsome wrappings and serious content. These histories had already earned praise for their high scholarly standards, so there was no risk that more opulent covers or thicker paper could have undermined their scholarly stature. The publishers' eagerness to invest in the bindings, typography, and paper of established scholarly histories attest to the importance of the "non-reading" aspect in the marketing and owning of respectable scientific works.

This was understood at Oxford University Press. Stubbs's *Select Charters* and *The Constitutional History* were published originally with a modest look that emphasized their learned character. They were even used in the Press as a model for later scientific histories in terms of binding, typography, and paper.[33] Macmillan, however, advised Price to choose a more attractive type for the envisioned library edition of *The Constitutional History*. He was certain that a higher price would not obstruct its anticipated buyers whom he called "the class of '*every gentleman's library*' purchasers."[34] Reviewers welcomed the new design with enthusiasm. *The Times* reported, "Many will be glad to possess it in the larger and more sumptuous form in which it now appears." The *Leeds Mercury* agreed but suspected that the price may restrict the audience. According to the writer, the more luxurious edition would find "a cordial welcome from the very considerable though limited class to whom it appeals." This class, the writer specified, was comprised of "arm-chair politicians."[35]

The look of a history book was a performative act of the persona and historians expected that the material and visual paratexts endorsed their inner qualities. The somber and restrained look conveyed dedication to serious intellectual work and the ability to resist fashions and temptations. The dark colors and discreet decorativeness created a compelling image of the severity of the contents and their creators. Such design conveyed a message that it was the content that mattered, but historians knew that the look of their books—just like their own dress—was treated as a reflection of their virtuousness. The contrast between the air of scientific gravity and the opulence and gaudiness that adorned other types of histories also allowed the scientific historians to make claims about their social position and its distance from aristocratic ostentatiousness. The preferred design, then, had both scholarly and social importance for historians.

Paratextual Design, the Persona, and Multiple Audiences

Visually attractive bindings helped to make books desirable objects that stood out in bookstores where they were exhibited in windows, piled up on tables, and placed on bookshelves. Alexander Macmillan was convinced that booksellers' displays were essential to the business. Unlike advertising and reviewing—"the one costly, the other fitful"—stores offered readers a chance to *see* and *feel* the available selection.[36] But what about books that did not have stunning covers or had a look that was so ambiguous that readers struggled to interpret their genre, style, or intended readership? How did one win acceptance from multiple audiences with one design? What did such aesthetics say about the persona their writers cultivated? Successful paratexts should assign a book its own identity and contain enough familiar features to make it recognizable for customers; effective promotional paratexts were unique and familiar at the same time. This posed a challenge for historians and publishers alike when history books and their writers did not neatly fit into one prescribed category of style and type or one segment of readers. How historians and publishers confronted these challenges is illuminated by the production, publishing, and marketing of Green's *A Short History of the English People* (1874). For a brief moment, it seemed that the book's success could have been eclipsed by its misleading look.

Green fostered a business-like approach to history books. He grasped the importance of investing his books with appropriate material and visual paratexts and wished to take charge of every detail of his publications and their accompanying marketing materials. The relationship between Green and the Macmillan family exemplifies the amicable relations authors and publishers could foster in late-Victorian England. Alexander Macmillan viewed Green with a mix of collegial respect, friendship, and paternal sentimentality. He appreciated Green's insights into the state of historical research and shared his friend's conviction that what readers needed was well-researched and well-written histories that summarized the latest scholarship. Thus, he commissioned Green to edit several textbook series and sought his advice about book proposals, manuscripts, paratexts, and overall strategies for publishing histories.[37] The process of making the *Short History* was, however, somewhat different, because the two men disagreed about the book's type and, consequently, about its intended audience. Macmillan treated the book from the beginning as a textbook for schools, whereas Green fostered more ambitious authorial goals and had a more heterogeneous audience in mind. On one point they agreed: their expectations were low. Macmillan offered Green only modest terms, £350 for the copyright, but when the book became an immediate phenomenon, selling a whopping 32,000 copies in the first year, he was gentleman enough to alter the contract, offering Green royalties.[38]

Green's *Short History* was not a traditional textbook or handbook, and it was its unique position between a schoolbook and a more ambitious attempt to sketch the social history of the English people that caused the paratextual predicament. When Green presented in 1869 the first detailed plan for his book to Macmillan, he envisioned it functioning as "a school-manual for the higher forms, and as a handbook for the universities."[39] However, a year later, he informed Macmillan that he and some of his colleagues considered the book more suitable for the general reader than to pupils. When Freeman read the draft in 1873, he anticipated the largest success to be among those who already knew "a good deal" about history.[40] When Green eventually submitted the manuscript to Macmillan in 1874, the book had drifted away from its original idea of a handbook for schools. This, as is suggested in Chapter 5, came to be the widespread impression of its character.

Macmillan, on the contrary, still fostered the idea of a textbook and so it was important to be able to offer the book at a competitive price and with inexpensive paratexts that placed it in that category. Green had already realized in 1870 when the first specimen pages were printed that the look Macmillan advocated identified the book too narrowly as a manual. Green assured him that he had nothing against his book being a handbook, but he did not "wish it to be a manual and nothing else." He did not protest the book being cheap as that made it accessible to a wider audience and was in tune with his persona of a historian with a democratic and social consciousness. It was the too strong emphasis on paratexts that branded it as a textbook that irritated him. To emphasize the educational nature of the book, genealogical tables, chronological annals, and marginal notes were added to aid students in managing and organizing historical facts and knowledge. But Green found the words in heavy black in the text particularly indicative of the schoolbook look and therefore did not care for them. Macmillan disagreed, believing that the blackened words were "aidful" to a general reader and "his sluggish mind."[41] Since Green was adamant about the black letters and insisted on their removal, Macmillan complied with his wishes. Green was not the only historian who protested the black letters that were a standard navigational paratext in primers. Price had to coax Freeman to accept them for *A Short History of the Norman Conquest of England* by arguing how such typographical design was used in the educational books of all the major publishers.[42] It is possible that historians feared that the use of black letters would make the textbook character too pronounced, needlessly restricting the audience. Nonetheless, in a spirit of true compromise, as Macmillan gave in on the black lettering, Green, for his part, gave his consent to the marginal notes which he disliked for reasons that he unfortunately did not elaborate in his correspondence.[43]

Once Green had completed his manuscript, he suggested to Macmillan that some of the paratexts should be reconsidered. Despite Green's protestations, Macmillan adamantly believed that schools were the book's main market, although he admitted that it had qualities that made it "well suited for that somewhat vague entity—the general reader." He also acknowledged the risks that were involved when the form of Green's book would not quite suit a general reader: "It *looks* a school book, and oh; our somewhat vague friend is rather stupid and goes by appearance."[44] He only agreed to make some minor alterations to the title page which both his editors and Green "seemed to think needful to meet the somewhat changed character" of the book. Certainly, Macmillan knew that format indicated a book's style and audience. Two years earlier he had disapproved of Price's orders for a schoolbook-like bindings for a book that was not exactly a schoolbook and had warned that the look would mislead consumers.[45] Now he was about to make a similar mistake with Green's book by holding on to his conviction that the best way to promote it was to frame it as a schoolbook. Indeed, the discrepancy between the form and content seemed to confuse readers and undermine the success of the *Short History*.

Margaret Oliphant drew attention to the disparity between the modesty of the appearance and the richness of the narrative, thus confirming Green's fears about the false impression the paratexts created. She observed how the book lacked "a vestige of the *luxe* of printing," for which she listed "glories of typography, smooth velvety leaves, and elaborate binding." Instead, the print was small, the paper flimsy, and the margins reduced. The small crown octavo was modest at best and nothing like its pirated copy in America which, to Green's astonishment, was "gorgeous in form, and margin, and type."[46] Furthermore, Green had omitted footnotes and Oliphant congratulated him for not tormenting readers with "authorities, footnotes, quotations, and a full sight of all that scaffolding" that broke the narrative and ruined "the reader's enjoyment"[47] (Fig. 10.1). For Green's expected schoolroom audience, the unpretentious look was a pulling factor: without the bells and whistles of luxury editions, the book was affordable and expanded the potential market beyond the well-to-do classes. As such, this went well with the persona which Green fostered, as it had a strong educational and democratic ethos. The problem was that the book did not have just one audience to please with the paratexts.

Oliphant was not the only reviewer who was concerned about the humble covers that the narrative outshone. The comments about the book's misleading appearance must have made Macmillan ask whether the chosen look was obstructing even bigger sales. To offset such a risk, he availed himself of the same strategy which Longman later employed

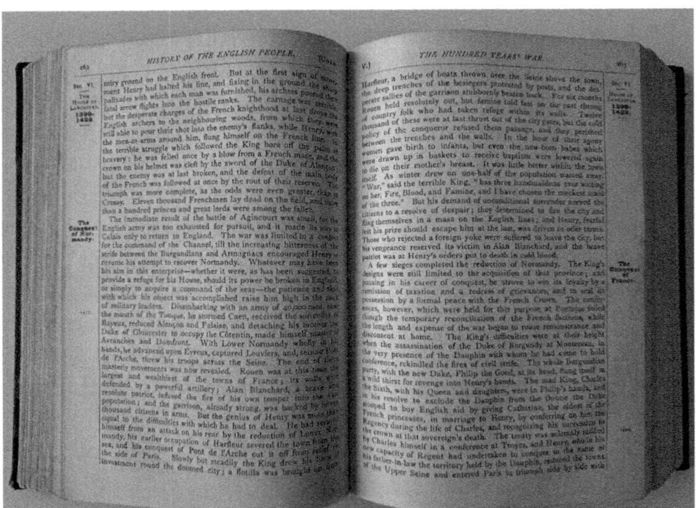

Fig. 10.1 *A Short History of the English People* had minuscule type and narrow margins (*Source* Author's personal collection)

to redress the undesirable attention which Freeman's prefatorial sincerity attracted: giving prominence to it in the advertisements. It was not uncommon to praise books' materiality in advertisements when there was something extraordinary in the paratexts. For example, Macmillan's Globe edition of Shakespeare was promoted for its "elegant and readable type" and "strong and handsome" binding.[48] However, educational histories were rarely advertised for their looks; the size and price were the only indications of their materiality in Macmillan's catalogs and notices. The advertisements that Macmillan prepared for the *Short History* were exceptional as they called attention not just to the content, but to the form as well. This is a clear indication of the deliberateness of the selected extracts.

Macmillan tackled the issue for the first time only weeks after the release of the *Short History* in an advertisement that was published at least in the *Athenaeum*. He chose for it an extract from the *Academy* that mentioned the problematic look.[49] In the following month, he dealt with the matter with more force in an advertisement that had six testimonies attesting to the book's value. The broad message that they imparted was that both young and old readers should adopt the *Short History* as their guide to English history. The deceptive appearance was explicitly confronted twice in the advertisement. First, the *Academy* was quoted again and now the review was attributed to Gardiner, who had written that "The size and general look of the book [...] would place it among school books," but "its fresh and original views, and its general historical power" distinguished it from this class of books. Second, Freeman's anonymous review in the *Saturday Review* was cited to clarify that, "This is a single volume, of modest dimensions and unpretending appearance of a school book," nevertheless containing a "wealth of material, of learning, of thought, and fancy," making it a much more significant contribution to English history than its appearance suggests.[50] The announcement was so loud in underlining the mismatched appearance that it seems safe to say that Macmillan was anxious about the matter and wished to stress that *A Short History of the English People* was not only aimed at school use.

Oliphant was pleased to state in her review that despite the fact that the book had been supplied with "all the poverty of a school edition," it was a success. She was clearly not a fan of aristocratic flamboyance, because she announced triumphantly that the internal qualities of Green's book were enough to make "sufficient noise in the world" while hundreds of trifling histories with all the paratextual glories passed into oblivion the day after their production. By challenging the "vanities of appearance," the *Short History* also proved that readers still recognized the "real excellence" of a text no matter how it had been dressed up.[51] Indeed, in all its unpretentiousness, it mocked the opulent paratexts in the luxury editions and demonstrated that ordinary readers were capable of recognizing textual value, unlike the cultural elite insisted. The hype that the book created eventually undid the harm its moderate paratexts could have caused. Macmillan nonetheless continued to cite Gardiner and Freeman in the advertisements even years later to reassure readers that the modest covers hid a gem of history writing.[52]

Green's book retained its humble dress in later impressions as well. Green, ignoring the duties of a scientific historian, never prepared a properly revised edition despite Macmillan's repeated pleas. Instead, he only corrected typos and factual errors and made a four-volume long version of the book. It was politely received but reviewers were disappointed because it contained very little changes or new knowledge. The *Athenaeum* admitted that it was "more dignified and luxurious" than the "closely-printed" original version, but the writer was clearly disappointed that although it took more space on a bookshelf, the content hardly exceeded the shorter version.[53] The profound revising of the original version was left to Green's widow, Alice Stopford Green. The new editions which she prepared were given covers in different colors, but otherwise the style and design remained unaltered.[54] The book's unostentatiousness was hinted at as late as 1899 in the *English Historical Review*, where the French translation was complimented for "appearance and arrangement … superior to those of the English edition."[55]

There was one exception to this: the four-volume illustrated edition which Stopford Green composed with Kate Norgate between 1892 and 1894. The visually sumptuous edition was aimed at an affluent audience and bore no resemblance to the standard version which embraced in its cheapness Green's democratic worldview. The opulent burgundy covers had gilt stamped lettering, an ornamental scepter, and what Stopford Green called a "curly scroll." Thick paper, large print, and broad margins completed the elegant look. Hundreds of illustrations from small engravings to photographs and lavishly colored plates added to the air of luxury. Green had deemed illustrations to be useless "in a volume of this sort" and only for didactic reasons had the book been issued with five maps recycled from Freeman's *Old English History for Children*.[56] In the preface to the illustrated edition, Stopford Green professed how it had been "a favourite wish" of her husband to have English history illustrated with pictures "which should tell us how men and things appeared to the lookers-on of their own day" and how their contemporaries had "aimed at representing them."[57] Whether Green had meant that his text would provide the narrative backbone for such a venture is another matter. Nevertheless, the illustrated edition certainly fulfilled this goal, but it was anything but a "people's edition." While it provided a new dress for the *Short History*, it also offered a new interpretation of the persona which Green had fashioned; but, as so often happened, yet another edition followed. This time Macmillan took a turn toward the original ethos of the *Short History* and issued a cheaper illustrated version in 1898 in 40 sixpence parts and with half Morocco bindings instead of cloth to cut the production costs.[58] Eventually, Green and the paratextual interpretations of his persona circulated in parallel illustrated editions in the literary marketplace.

Quality Marks the Persona

The *English Historical Review* issued a short notice to announce the English translation of Fustel de Coulanges's *The Origin of Property in Land* in 1891 and, after complimenting its scholarly merits and the carefully executed translation, the writer added, "it is a pity that so good a book should not be better printed."[59] Quality was an important, but often overlooked, feature in paratexts. According to Mary Hammond, Victorian publishers recognized quality as a marketing device during the last quarter of the century. They strove to disassociate especially their popular classics series from earlier cheap ones that had been characterized by shoddiness and an idea of disposability. New series were announced as inexpensive but handsome, printed on good paper in readable type, durable, and bound in cloth.[60] The quality of paratexts was important for history as well. E. R. Roscoe was furious about the cheap paper covers and bad print which the Manuscript Commission used in its editions. He considered it "disgraceful that an important national publication" was issued in such a regrettable state and that documents from historical collections of great consequence "have not been published in a more desirable and dignified form."[61] This was not an insignificant issue: glorious national history deserved equally stunning wrappers and it was not a secret that European nations also competed in the art of bookbinding.[62] For historians, the quality of paratexts was primarily a matter of the dignity of their discipline and persona. They demanded the same attention to detail in their books as from their methodological accuracy or from what was expected in the attire of professional men. Plainness also meant neatness, high-quality fabrics, and a careful finish in tailoring.[63]

The bond between the quality of the paratexts, the book's value, and the scholarly persona was captured by Stopford Green. She was determined to ensure that all the paratexts in her own histories and the posthumous editions of her husband's conformed to her high standards. As her schemes for customizing *The Making of Ireland and Its Undoing* for different audiences indicated, she was an astute businesswoman who had learned how paratexts can shape a book's commercial success. She was not ignorant about production costs and pricing but did not accept expense as an excuse for sloppiness. Several times she wrote to Macmillan how "careful attention to details" translated into good sales.[64]

Nevertheless, Stopford Green was not concerned about the standards of the paratexts for commercial reasons alone. Her zeal was also inspired by her emotional investment in the protection of Green's scholarly legacy. The editorial work consoled her, as it allowed her to maintain an intellectual connection to him. Victorians had rediscovered the ancient idea of writers continuing to live posthumously through their works, and Stopford Green was clearly moved by the metaphysical idea of eternal life.[65] This makes understandable her vigilance about the paratexts in Green's books; it was important that they mirrored the prestige the books and their author enjoyed. Then again, her paratextual strategies were also informed by her personal aesthetic taste and fondness for handsome books. She expected that her own books should also look welcoming and that the pages, for instance, should have "an engaging air." She evaluated the typeface, font size, and margins accordingly and requested to see samples of typography, layout, and bindings before she gave her permission for printing.[66] If her friend Beatrice Webb once described her dress as "natty" with "the negligence of a woman with intellectual pretentions," she did not allow any such carelessness with her paratexts.[67] Yet, and despite all her attentiveness, the printers made mistakes. She had given instructions to wrap *Woman's Place in the World of Letters* (1913) in brown paper boards with a white spine. The sample copy which she had received had been correct, but the book was accidentally issued with covers in "exceedingly unpleasant blue" and with unsatisfactory lettering. This mishandling of her literary work made her "very much annoyed."[68]

When Macmillan invited Stopford Green to prepare the first revised edition of the *Short History*, she went through the text with extreme care, and Gell, writing to Stubbs, hoped that she would receive appreciative reviews as compensation for her conscientious effort.[69] The new edition was released in 1888 and she was disappointed in the quality of its paratexts. Since it continued to be moderately priced, it had been printed on very thin paper, and the pages were marred by show-through. She was vexed about this defect and reported to Macmillan how she had overheard customers complain about the paper and how the booksellers felt "very strong about this." She wanted to know whether paper could be at least a "degree better" in the future impressions.[70] She was even more distressed about the poor quality of paper in the new edition of Green's *The Making of England* (1897), which she also had composed. *The Making of England* had been the first "big" book which Green had written to gain respect as a historian of serious standing and the book's paratexts should have supported the image of an author who cultivated the persona of a scientific historian. On the contrary, the book did not speak for meticulousness or attention to detail. The frontispiece and the maps were printed on so soft a paper that the images were blurred. The binding, too, was poorly executed. The pages were stitched so weakly that the uncut leaves "cracked at the first opening" of the book "in a most disagreeable way."[71] The paratexts screamed of carelessness and cheapness and did not do justice to Green or to her as a devoted widow who protected his scholarly legacy.

Genette has suggested that paper quality is mostly an aesthetic or economic issue or a matter of durability, and hence has only minor relevance to the text.[72]However, Stopford Green's complaints and the discussions which the physical state of the *Short History* spurred indicate the opposite: paper was of course an aesthetic and economic matter, but it shaped the textual reception as well. It assigned value to the text and affected the actual reading process and the use of history books. The experiences of the readers show how a handbook such as *A Short History of the English People* lost something essential of its function if its materiality rendered its consultation inconvenient. Freeman was exceedingly delighted when he learned that Green was making a four-volume version of the book, as he expected the new shape to mean larger type, wider margins, and thicker paper. He had read the book with pleasure in its present condition, but rarely consulted it because the condensed type and thin paper were "grievous to the eye."[73] Freeman was not the only reader who struggled with the reduced margins, minuscule print, and flimsy paper. The *Examiner*, when announcing the new four-volume edition, congratulated Green for relieving readers of the agonies that the type, paper, and absence of margins produced. The writer was amazed that so many had read the book despite its paratextual obstacles.[74] The quality of paper had a direct bearing on a book's attractiveness. It was an integral element of a page, a book's most central cognitive space. Stopford Green recognized this and was anxious to ensure that the material and physical conditions of the *Short History* did not make readers abandon it. In such a case, the public would have forgotten Green, his achievements, and the persona of a pioneering social historian of the English people which he had emulated.

Stopford Green returned once more to the book's weak paratextual conditions when another revised edition was issued in 1898. This time she disliked the extremely narrow margins which prevented those who wished to have their copy "nicely bound" at their own expense. She reported how booksellers were complaining "a good deal" about the book's material shortcomings.[75] As in previous times, she appealed to the authority of the anonymous "booksellers," calculating that the trade's view would give additional force to her arguments. She was highly gifted in using this kind of anonymized authority to achieve her goals. Another example of this same strategy was the "Irish scholar" in the preface to her revised edition of *The Making of Ireland*. This was not all. In negotiations about the textbook on English and Irish history for the American market with Macmillan's New York agent George Brett, she tried to convince him of its value by telling how "a History Professor, whom Mr. York Powell thinks the best History teacher he has ever met" held a "most encouraging" view of her manuscript.[76] The unnamed figures were for her types that represented collective expert knowledge, skills, and stature in their respective fields. She drew on these bywords to sanction her claims and to promote her and her husband's histories.

The look, the feel, and an air of dignity were important qualities in paratexts for Stopford Green. Book parts had economic, emotional, and aesthetic currency for her, but she also understood how paratexts influenced the legibility and usability of a book. They were for her, moreover, expressions of the personae which she and her late husband embodied. She captured the essence of paratexts when she wrote to Macmillan about the paratexts in *Letters of John Richard Green* which Leslie Stephen edited.

It might be appropriate to allow her one more time to explain her paratextual vision, as it shows so well how important paratexts could be to late-Victorian historians and readers alike:

> I like the page of print you sent me—and am very glad I feel you have put the letters in a good and easily-read type. I don't know how many pages there will be, and therefore whether there is any fear of the thick paper you send me making too bulky and heavy a volume. I do so greatly dislike a heavy book to hold ... Also people have grown tired of the big biographies ... I think it would be an advantage to decrease its bulk in appearance ... I think [it] would make more people read it. So I very much hope the volume may be kept as thin as convenient. Every one says the same thing now about the *size* of biographies.[77]

*

Thomas Carlyle maintained that books were nothing but shells which should be dismissed as soon as the "kernel is extracted"; that is, the text, "read with intensity of attention."[78] This was an unpopular idea among the Victorians, who engaged with the materiality of books in multiple ways. They interpreted books' physical and aesthetic paratexts as indicators of genre, style, and value, and approached books as desirable objects that transmitted messages about their owners' cultural and social standing. Such fetishization of books was condemned by the cultural and scholarly elite, and George Gissing criticized this craze for beautiful covers in *New Grub Street* when he made Mrs. Reardon enthusiastically exclaim how Mr. Reardon's latest book did not "look like a book that fails."[79] Regardless of the parody that the fascination with elegant and handsome books invited, historians could not afford to ignore the materiality of their books. They knew that the look was critical for both selling their books and promoting their persona. It was crucial that their books emitted the right sort of scholarly charisma. Indeed, Maitland confirmed this by hoping that the appropriately scholarly look of the *Borough Book* "will do us credit."[80] Since material and visual paratexts served a key role as regulators of cultural, social, and intellectual demarcations, the commercial use of a book's bibliographical code was nothing but straightforward. Yet, historians and publishers agreed that materiality was pivotal to making books desirable and were most of the time eager to find compromises that left everyone content. Whether this produced paratexts that readers found pleasing or accurate depictions of the historians' persona, was of course another matter.

Notes

1. Freeman to George Macmillan, September 19, 1886, Add MSS 55,053, BL.
2. Cannan to Powell, December 7, 1897, Letter Books 70, OUP.
3. Carlyle, *On Heroes*, 164–165.
4. Hammond, *Reading*, 113; Phillips, "How Books are Positioned," 20–21; Hosgood, "Mrs Pooter's Purchase," 148; Dettmar, "Bookcases," 5.
5. Lecky to Booth, autumn 1876, *Memoir*, 111.
6. Bornstein, *Material Modernism*, 1–2; McGann, *Textual Condition*, 12–14.
7. Maitland to Melville M. Bigelow, October 7, 1886, in Fifoot, *Letters*, 23.
8. McKitterick, "Changes in the Look," 96–102.
9. Gell to Plummer, October 7, 1885, Letter Books 38, OUP; Gell to Macmillan, February 23, 1887, Add MSS 54,886, BL; Green, "Look of the Books," 273–274; Eliot, "Evolution of a Printer," 94–100; McKitterick, *History of Cambridge University Press*, 3:126.
10. W.M.T., "Prefaces and Dedications," 25.
11. Longman to Browning, May 28, 1880, GBR/0272/OB/1/986/C, King's Cam.
12. Doble to Miss Freeman, December 16, 1890, Letter Books 51, OUP.
13. Zaehnsdorf, *Art of Bookbinding*, 96.
14. Stubbs, *Seventeen Lectures*, 11; Hammond, *Reading*, 110.
15. Grossmith and Grossmith, *Diary of a Nobody*, 19.
16. Hammond, *Reading*, 4–5; Collin, "Bookmaking," 66–68.
17. [Oliphant], "New Books," 89.
18. Price, *How to Do Things with Books*, 29–30.
19. Doble to Powell, September 3, 1886, Letter Books 40, OUP.
20. Gardiner to Browning, February 3, 1884, GBR/0272/OB/619/A, Kings' Cam.
21. Creighton to Longman, September 24, 1885, and to Lane Poole, January 18, 1885, in Creighton, *Life and Letters*, 338, 340.
22. Breward, *Hidden Consumer*, 41, 76–88.
23. Shannon, *Cut of His Coat*, 25–26.
24. Shannon, *Cut of His Coat*, 4–5, 22–34; Breward, *Hidden Consumer*, 2–3.
25. Cited in Conlin, "Consolation of Amero-Teutonism," 111–112.
26. Freeman to Thompson, May 10, 1874, U DX9/47, HHC.
27. [Reeve], "*Chief Victories of Charles V*," 68; [Reeve], "Annals of the House of Percy," 396; [Elton], "Old Dorset," 35; [Stanhope], "Countess of Nithsdale," 78; [Russell], "Chiefs of Grant," 63.

28. T.H.L. "Chat with a Bibliopegist," 58.
29. St.Swithin, "Few Mems. On Books," 270.
30. [Freeman], "Art of History-Making," 52.
31. Shannon, *Cut of His Coat*, 4.
32. Fox, "Strange History," 829.
33. Doble to Gardiner, August 23, 1886, Letter Books 48, OUP; Doble to Ramsay, April 29, 1890, Letter Books 49, OUP.
34. Macmillan to Price, January 7, 1879, Add MSS 55,407 (3), BL; Price to Stubbs, February 25, 1880, Letter Books 22, OUP.
35. "Canon Stubbs's Constitutional History," *Times*, March 31, 1880, 4; "Literature," *Leeds Mercury*, April 21, 1880.
36. Macmillan to Herbert Spencer, June 13, 1882, in Macmillan, *Letters*, 316; McKitterick, "Changes in the Look," 114–116; Matthews, "Introduction," xi–xii.
37. See Green's correspondence with Macmillan, Add MSS 55,058, BL; Macmillan to Price, January 27, 1875, Add MSS 55,396 (2), BL; Macmillan to Green, January 26, 1876, Add MSS 55,398, BL; George Macmillan to Green, November 8, 1879, Add MSS 55,409 (2), BL.
38. Brundage, *People's Historian*, 1.
39. Green to Macmillan, [December 1869], Add MSS 55,058, BL.
40. Green to Macmillan, undated [1870], Add MSS 55,058, BL; Green to Freeman, September 16, 1873, in Stephen, *Letters*, 357; Freeman to Green, September 21, 1873, FA 1/8/1–30, JRL.
41. Green to Macmillan, undated [1870], Add MSS 55,058, BL; Macmillan to Green, May 18, 1870, Add MSS 55,390 (2); Macmillan to Green, June 1, 1870, Add MSS 55,391 (1), BL.
42. Price to Freeman, May 6 and 11, 1880, Letter Books 23, OUP.
43. Macmillan to Green, June 1, 1870, Add MSS 55,391 (1), BL.
44. Macmillan to Green, undated [1874], PP. Green 17, Jesus College Archive.
45. Macmillan to Price, November 20, 1872, Add MSS 55,393 (1), BL; Garritzen, "Framing and Reframing," 183–185.
46. [Oliphant], "New Books," 89; Green to Olga von Glehn, June 21, 1875, in Stephen, *Letters*, 419.
47. [Oliphant], "New Books," 90–91.
48. Quoted in McKitterick, *History of Cambridge University Press*, 2:393; Garritzen, "Framing and Reframing," 184–185.
49. *Athenaeum*, December 12, 1874, 779.
50. *Athenaeum*, 16 January 1875, 75.
51. [Oliphant], "New Books," 89.
52. "By the same author," in Green, *History of the English People*, back matter.
53. Green to Stopford Green, March 30, 1877, in Stephen, *Letters*, 453–454; Macmillan to Green, February 10, 1880, Add MSS 55,410 (1), BL; "History of the English People," *Athenaeum*, August 7, 1880, 167.

54. Stopford Green to Frederick Macmillan, November 4, 1915, Add MSS 55,060, BL.
55. "Review of the French Translation," *EHR*, 1889, 398.
56. Green to Macmillan, [December 1869], Add MSS 55,058, BL.
57. Stopford Green, "Preface to Illustrated Edition," v.
58. Frederick Macmillan to Stopford Green, April 28, 1898, and May 20, 1898, PP. Green 17, Jesus College Archive..
59. "Review of Fustel de Coulanges," *EHR*, 1891, 804.
60. Hammond, *Reading*, 91–93.
61. [Roscoe], "*Harley Papers*," 152.
62. T.H.L. "Chat with a Bibliopegist," 58.
63. Breward, *Hidden Consumer*, 30.
64. Stopford Green to Macmillan, April 1, [1889], Add MSS 55,059, BL.
65. O'Gorman, "Dead," 259.
66. Stopford Green to Macmillan, May 22, 1908. Add MSS 55,059, BL; Stopford Green to Macmillan, August 19, 1924, Add MSS 55,061, BL.
67. Diary of Webb, August 4, 1889, in MacKenzie, *Diary*, 288.
68. Stopford Green to Macmillan, May 30, 1913, and undated [1913], Add MSS 55,060, BL.
69. Gell to Stubbs, January 23, 1888, Letter Books 44, OUP.
70. Stopford Green to Macmillan, March 24, [1888], Add MSS 55,059, BL.
71. Stopford Green to Macmillan, November 21, 1898, and March 21, 1899, Add MSS 55,059, BL.
72. Genette, *Paratexts*, 34–35.
73. Freeman to Green, January 23, 1876, and February 13, 1876, FA 1/8/31–60, JRL.
74. "Mr. Green's English History," *Examiner*, December 15, 1877, 1584.
75. Stopford Green to Macmillan, November 21, 1898, Add MSS 55,059, BL.
76. Stopford Green to Brett, June 1898, MS 15,124/3, NIL.
77. Stopford Green to Macmillan, April 24, [1902], Add MSS 55,059, BL.
78. Yates, *Celebrities at Home*, 182.
79. Gissing, *New Grub Street*, 233.
80. Maitland to Fossett Lock, November 23, 1904, in Fifoot, *Letters*, 320.

References

Unpublished Primary Sources

British Library: The Macmillan Papers.
Hull History Centre: Letters from Edward Augustus Freeman to Edith Thompson

Jesus College Archive, Oxford: J. R. Green Papers
John Rylands Library, Manchester: E. A. Freeman Archive
King's College Archive Centre, Cambridge: Oscar Browning Papers
National Library of Ireland, Dublin: Alice Stopford Green Additional Papers
Oxford University Press Archive: Letter Books

Printed Primary Sources

[Anon.]. "Mr. Green's English History." *Examiner*, December 15, 1877, 1584–1585.

[Anon.]. "Literature." *Leeds Mercury*, April 21, 1880.

[Anon.]. "Canon Stubbs's Constitutional History." *The Times*, March 31, 1880, 4.

[Anon.]. "History of the English People." *Athenaeum*, August 7, 1880, 167–170.

[Anon.]. "Review of the French Translation of J. R. Green's *Short History of the English People*." *English Historical Review* 4, no. 14 (1889): 397–398.

[Anon.]. "Review of Foustel de Coulanges's *The Origin of Property in Land*." *English Historical Review* 6, no. 24 (1891): 803–804.

Bury Palliser Mrs. *A History of Lace*. London: Sampson, Low, Son, and Marston, 1865.

Carlyle, Thomas. *On Heroes, Hero-Worship and the Heroic in History*. Lincoln: University of Nebraska Press, 1966.

Creighton, Louise. *Life and Letters of Mandell Creighton, Sometime Bishop of London*. Vol. 1. London: Longman, 1904.

[Elton, C. I.]. "Old Dorset." *Edinburgh Review*, July 1894, 35–60.

Fifoot, C. H. S. *The Letters of Frederic William Maitland*. London: Selden Society, 1965.

Fox, J. A. "A Strange History." *Athenaeum*, 19 December 1874, 829.

[Freeman, Edward A.]. "The Art of History-Making." *Saturday Review*, November 17, 1855, 52–54.

Gissing, George. *New Grub Street*. London: Penguin Books, 1985.

Green, John Richard. *History of the English People*. Vol. 2. London: Macmillan, 1878.

Grossmith, George, and Weedon Grossmith. *The Diary of a Nobody*. Oxford: Oxford University Press, 2008.

MacKenzie, Norman and Jeanne (eds.). *The Diary of Beatrice Webb*. Vol. 1. London: Virago, 1982.

Macmillan, George A. *Letters of Alexander Macmillan*. Glasgow: printed for private circulation, 1908.

A Memoir of The Right Hon. William Edward Hartpole Lecky, by his wife. London: Longman, 1909.

[Oliphant, Margaret]. "New Books." *Blackwood's Magazine*, July 1875, 82–99.

[Reeve, Henry]. "*The Chief Victories of Charles V*." *Edinburgh Review*, July 1870. 67–99.

[Reeve, Henry]. "Annals of the House of Percy." *Edinburgh Review*, October 1888, 373–396.

[Roscoe, E. S.]. "*The Harley Papers*." *Edinburgh Review*, January 1898, 151–178.

[Russell, John]. "The Chiefs of Grant." *Edinburgh Review*, July 1884, 63–89.

[Stanhope, Philip Henry]. "The Countess of Nithsdale." *Quarterly Review*, July 1874, 77–104.

Stephen, Leslie. *Letters of John Richard Green*. London: Macmillan, 1902.

Stopford Green, Alice. "Preface to the Illustrated Edition." In *A Short History of the English People*, by J. R. Green, v–x. Vol. 1. London: Macmillan, 1892.

St. Swithin. "A Few Mems. On Books and Reading." *The Englishwoman's Domestic Magazine*, March 1861, 270–272.

Stubbs, William. *Seventeen Lectures on the Study of Medieval and Modern History and Kindred Subjects*. Oxford: Clarendon Press, 1886.

T.H.L. "A chat with A Bibliopegist—Mr. J. W. Zaehnsdorf at Home." *The Album*, October 28, 1891, 57–58.

W. M. T. "Prefaces and Dedications." *Living Age*, April 1859, 25–28.

Yates, Edmund. *Celebrities at Home*. Vol. 1. London: Office of 'The World', 1877.

Zaehnsdorf, Joseph Z. *The Art of Bookbinding. A Practical Treatise*. London: George Bell and Sons, 1880.

Secondary Sources

Bornstein, George. *Material Modernism: The Politics of the Page*. Cambridge: Cambridge University Press, 2001.

Breward, Christopher. *The Hidden Consumer: Masculinities, Fashion and City Life 1860–1914*. Manchester: Manchester University Press, 1999.

Brundage, Anthony. *The People's Historian: John Richard Green and the Writing of History in Victorian England*. Westport: Greenwood Press, 1994.

Collin, Dorothy W. "Bookmaking: Publishers' Readers and the Physical Book." *Publishing History* 44 (1998): 59–76.

Conlin, Jonathan. "The Consolation of Amero-Teutonism: E. A. Freeman's Tour of the United States, 1881–2." In *Making History: Edward Augustus Freeman and Victorian Cultural Politics*, edited by G. A. Bremner and Jonathan Conlin, 101–118. London: The British Academy, 2015.

Dettmar, Kevin J. H. "Bookcases, Slipcases, Uncut Leaves: The Anxiety of the Gentleman's Library." *Novel* 39, no. 1 (2005): 5–24.

Eliot, Simon. "The Evolution of a Printer and Publisher." In *The History of Oxford University Press*, edited by Simon Eliot, 77–112. Vol. 2. Oxford: Oxford University Press, 2013.

Garritzen, Elise. "Framing and Reframing Meanings in History Books: The Original and Posthumous Paratexts in J. R. Green's *Short History of the English People*." *History of Humanities* 3, no. 1 (2018): 177–197.

Genette, Gerard. *Paratexts: Thresholds of Interpretation*. Cambridge: Cambridge University Press, 1997.

Green, Maureen. "The Look of the Books." In *The History of Oxford University Press*, edited by Simon Eliot, 227–274. Vol. 2. Oxford: Oxford University Press, 2013.

Hammond, Mary. *Reading, Publishing, and the Formation of Literary Taste in England, 1880–1914*. London: Routledge, 2019.

Hosgood, Christopher P. "Mrs Pooter's Purchase: Lower-Middle-Class Consumerism and the Sales, 1870–1914." In *Gender, Civic Culture and Consumerism: Middle-Class Identity in Britain 1800–1940*, edited by Alan Kidd and David Nichols, 146–163. Manchester: Manchester University Press, 1999.

Matthews, Nicole. "Introduction." In *Judging a Book by Its Covers: Fans, Publishers, Designers, and Marketing Fiction*, edited by Nicole Matthews and Nickianne Moody, xi–xxi. Aldershot: Ashgate, 2007.

McGann, Jerome J. *The Textual Condition*. Princeton: Princeton University Press, 1991.

McKitterick, David. *A History of Cambridge University Press*. Vol. 2. Cambridge: Cambridge University Press, 1998.

McKitterick, David. *A History of Cambridge University Press*. Vol. 3. Cambridge: Cambridge University Press, 2004.

McKitterick, David. "Changes in the Look of the Book." In *The Cambridge History of the Book in Britain*, edited by David McKitterick, 75–116. Vol. 6. Cambridge: Cambridge University Press, 2009.

O'Gorman, Francis. "The Dead." In *The Cambridge Companion to Victorian Culture*, edited by Francis O'Gorman, 255–272. Cambridge: Cambridge University Press, 2010.

Phillips, Angus. "How Books Are Positioned in the Market: Reading the Cover." *Judging a Book by Its Covers: Fans, Publishers, Designers, and Marketing Fiction*, edited by Nicole Matthews and Nickianne Moody, 19–30. Aldershot: Ashgate, 2007.

Price, Leah. *How to Do Things with Books in Victorian Britain*. Princeton: Princeton University Press, 2012.

Shannon, Brent. *The Cut of His Coat: Men, Dress, and Consumer Culture in Britain, 1860–1914*. Athens, Ohio: Ohio University Press, 2006.

CHAPTER 11

Conclusion: Heavenly Historians and Their Personae

There was a "historic heaven" where Aristotle, Machiavelli, and Sarpi rubbed shoulders with a selected band of late-Victorian historians. Gardiner and Firth had their "quiet harbour" on the Delectable Mountains and Freeman and Stubbs resided in a place of their own, away from the self-deceiving philosophers, journalists, and Froude. The latter bunch belonged to the Eternal hell, where they drank, quarreled, and sneered at each other. The "plain female historian," too, was in this heaven, though "well concealed." All in all, this was a place full of good company. On Sundays, there were even "a few 'weel favour'd hizzies'" who supplied solace "in the intervals of high talk." This historic heaven was a highly exclusive resting place for retired historians; "for many are called, few are chosen."[1] This was a fantasy of Frederick York Powell. If we put aside his masculinist daydreaming and gendered prejudices, his historic heaven from 1900 is an evocative testament of the late-Victorian historians' creativity in consolidating their persona. Taxonomies, hierarchical structures, and boundaries were all aimed at shoring up the position of history as a discipline and the scientific historian as the only eligible and authoritative producer of reliable historical knowledge. Out of the historians' communal act of creation emerged a scholarly persona that was not uncontested, but nonetheless uniform enough to ascribe a recognizable physiognomy to the new kind of scientific historian. Powell's heavenly rendition is also a testimony of the symbolic importance of the

E. Garritzen, *Reimagining the Historian in Victorian England*,
https://doi.org/10.1007/978-3-031-28461-8_11

persona in forming and regulating communities and creating a sense of belonging between like-minded scholars. Only those who had cultivated the accepted virtues, skills, character traits, emotions, and manners, and who also possessed the required innate qualities, were received to this celestial community of self-consciously scientific historians.

Exploring how the historians' scholarly persona took shape in England during a critical historiographical period when history was developing into an academic discipline and the professional standards were evolving has been less about testing whether historians were truly as virtuous as they presented themselves to be than about decomposing the idealized image of a scientific historian. Accordingly, the chapters above have recounted how that persona was drawn from a range of possibilities, the meanings that were ascribed to its components, who participated in its construction, and how it gained currency among expert and non-expert audiences. As the formation of the persona predated any major impetus toward professionalization and the founding of the markedly professional institutions, the persona itself became a means to institutionalize the novel disciplinary status of history. The persona embodied features which were considered essential for acquiring truthful historical knowledge and, as historians used it for corroborating their authority, it was vital that their public image corresponded with the idealized persona. This makes understandable their sensitivity about their reputation and why the management of the public image of the discipline turned out to be an integral part of the persona project. The shifting of the focus from the products of historical inquiry to their producers has also made visible how the late-Victorian historians were multitaskers who fused overlapping scholarly, pedagogical, and entrepreneurial duties into the persona and also applied it to claim a respectable social status. Historians were much more than archive scholars and producers of original knowledge, and they therefore needed an ability to adjust the persona to diverse positions and to acquire special skills and knowledge instead of narrowly limiting the compass to methods and research practices.

Since English historians had slender institutional space for promoting their persona, they enlisted their books for this purpose and developed a repertoire of cultural and scholarly practices for the bookish performance of their expertise. They used in the paratexts textual, visual, and material symbols and ritualized activities like footnoting to consolidate their persona and edge out undesirable rival models of being a historian. This collective disciplinary self-fashioning makes visible the important non-reading roles which history books can play. Turning books into a meaningful sphere for producing and managing the persona meant a constant balancing between historian, publisher, and readers' interests, aims, and preferences. The realities of the book trade and the literary marketplace set the conditions for the use of paratexts in history books, but the collaborative nature of the author–publisher relationship ensured ample opportunities for historians to design paratexts which they conceived as appropriate expressions of their persona.

The promotion of the persona in paratexts had significant consequences for the circulation of the image of the scientific historian in the specialist and non-specialist realms. The persona was replicated and multiplied in different editions, versions, and formats. Since the fragmentation of the readership was a fundamental feature of the Victorian culture of reading, historians began to write many types of histories to appeal to various groups of readers. Because of this, their persona was available in multi-volume studies, handbooks, schoolbooks, and children's histories. It was important that historians reached the widest possible audience, for they needed public approval to legitimize their persona. Therefore, they were willing to accommodate the needs of readers who preferred histories in bite-size portions even though they were otherwise concerned about the impact the superficial consumption of texts could have on the mental habits of the ordinary reader. Moreover, the paratextual projection of the persona was shaped by revising, which was a core scholarly practice during the nineteenth century. While historians revised their texts, they also had a chance to edit the presentation of their persona by adding new paratexts or removing and altering earlier ones. Publishers contributed to this refashioning as well when they proposed new colors, designs, and materials for the bindings in the later editions. Altogether, the scholarly persona circulated simultaneously in many versions in the literary marketplace. Readers encountered and re-encountered the persona in multiple formats and in different temporal, cultural, and spatial contexts. Victorian publishing culture encouraged the multiplication of the enactment of the persona, and while its essence remained unaltered from one title to another, the type of book determined how extensively and thoroughly the historians could display their persona.

One feature that cut through every aspect of the persona and the talk about it was the reliance on the discourse of virtue. This vocabulary was so pervasive and intimately bound to historians' personal qualities that it undermined all appeals to disinterested and disembodied science.

The link between the reliability of scientific knowledge and the skills, qualities, personal traits, and conduct of the scientific practitioners was self-evident for Victorian historians and their audiences. They were so familiar with this vocabulary that they applied it fluently publicly and privately. Historians' use of the virtue discourse was characterized by flexibility as they reinterpreted the core virtues according to the different goods they pursued. The virtue discourse gave the reviewers and readers tools to evaluate histories and historians' performances. Publishers, for their part, instilled virtues in their advertisements. Since they were more occupied with profit than disciplinary boundary work, they used the virtues to promote scholarly and non-scholarly histories and authors indiscriminately. Strikingly, the parroting of the virtues did not seem to trivialize them. Otherwise, it would be hard to explain why historians drew so confidently on the virtue discourse when they related themselves as scientific historians.

Historians' scholarly personae fascinated the public, who tried to make sense of the new scientific history. As the Victorians had a voracious appetite for stories of exemplary lives, they were curious about its protagonists. The heroization of Stubbs together with the endless attempts to portray the new kind of historian in newspapers and magazines catered to this interest and weaved the persona into the public imagination. The scholarly community had mixed feelings about such attention but understood that the visibility was necessary to make scientific history and historians known outside the scholarly community. This public interest in history did not spring only from the attraction of exceptional biographies. The attention also derived from the popularity history enjoyed. The Victorians were engrossed with the past and looked back when they sought to understand who they were. The olden times provided respite from the hustle of modern life and the narrative of an onward marching nation bolstered the sense of national pride, as it offered convincing evidence for Britain's leading role in the world. It was indeed significant how the past was remembered, what kind of lessons were drawn from history, and who was entitled to deliver those lessons. History was not an innocent storehouse of entertaining stories, and the deluge of nationalist histories alarmed the Victorians, who realized that the reimagining of national histories in Europe could profoundly shape the future of nations and continents. In this climate, it truly mattered who were permitted to call themselves historians. Since historians used the persona to reinforce their authority, the wide attention their collective self-fashioning attracted attests to the high social and political relevance of both history and the historians' persona.

The persona that the late-Victorian historians crafted reflected historians' disciplinary needs at a specific historical moment. But personae do not resist change. When the very last edition of Edith Thompson's *History of England* was published in 1924, more than 50 years after the release of the original, she realized that she and her book both appeared hopelessly outdated. With a sense of melancholy, she sighed how her book was not "pacifist or socialist enough" for the post World War I climate.[2] While I have traced the manifestations of historians' persona in a particular temporal, geographical, and disciplinary context, the persona as an analytical concept is not bound by time, space, or specialization. The potential of the persona is precisely in its ability to subvert such limitations and create opportunities for research that could, for example, go beyond the current emphasis on Western science in persona studies or cross the mental boundary that now separates the history of science and the history of humanities.

What is more, pausing to consider personae can be a helpful exercise in terms of self-reflection. Personae continue to shape our perceptions of different disciplines and their representatives at a moment when academia is absorbed with unforeseen zeal in branding, personal profiles, and marketing of superstar scientists, the modern-day versions of the nineteenth-century heroic historians. In this climate where the public and private, persona and personal often collapse, the persona exerts influence on how we perceive and present ourselves as historians and how we envision our careers. Institutional vetting systems, application forms for funding, and peer reviewing define and redefine the boundaries between acceptable and unacceptable personae and make tangible the impact which personae have on the intellectual goods which we chose to pursue. Yet, personae also manifest themselves in subtler ways, as they are present when we teach, explain our work to non-expert audiences, or draw mental maps of the disciplinary subfields. We might not fall back on the nineteenth-century virtue vocabulary when we narrate what it takes to be a certain kind of historian in the 2020s, but nonetheless assume that historians with shared methodological preferences, topical interests, or professional ethos may develop collectively approved traits and habits. Because of this, the pedagogical implications of personae should not be overlooked; it does matter how we relate the historian's craft to students, which models we introduce them to, and which ones we discreetly discourage them from emulating. Admitting the existence of personae in lecture halls and other academic spaces would be a step forward in recognizing how personae make their presence felt, as they are not necessarily endorsed today with similar boldness and panache of virtuosity to how our Victorian predecessors championed them.

Notes

1. Powell to a young historian, 1900, in Elton, *Frederick York Powell*, 1:300.
2. Thompson to Macmillan, March 12, 1924, Add MS 55078, BL.

References

Unpublished Primary Sources

British Library: The Macmillan Papers

Printed Primary Sources

Elton, Oliver. *Frederick York Powell. A Life and a Selection from His Letters and Occasional Writings*. Vol. 1. Oxford: Clarendon Press, 1906.

Appendix: The Titles in the Paratext Database

Aikin, Lucy. *Memoirs of the Court of King Charles the First*. Vol. 1. 2nd ed. London: Longman, 1833.
Aikin, Lucy Aikin. *The Life of Joseph Addison*. Philadelphia: Carey and Hart, 1846.
Allardyce, Alexander. *Scotland and Scotsmen in the Eighteenth Century*. Vol. 1. Edinburgh and London: William Blackwood, 1888.
Allardyce, Alexander. *Scotland and Scotsmen in the Eighteenth Century*. Vol. 2. Edinburgh and London: William Blackwood, 1888.
Amherst, Alicia. *A History of Gardening in England*. 2nd ed. London: Bernard Quaritch, 1896.
Armitage, Ella S. *The Childhood of the English Nation, or the Beginnings of English History*. London: Longman, 1877.
Ballantyne, Archibald. *Lord Carteret a Political Biography 1690–1763*. London: Richard Bentley, 1887.
Ballantyne, Archibald. *Voltaire's Visit to England 1726–1729*. London: Smith, Elder, & Co. 1893.
Bateson, Mary. *A Narrative of the Changes in the Ministry 1765–1767. Told by the Duke of Newcastle in a Series of Letters to John White, M.P.* London: Longman, 1898.
Bateson, Mary. *Records of the Borough of Leicester*. London: C. J. Clay, 1899.
Beesly, Edward Spencer. *Queen Elizabeth*. London: Macmillan, 1892.
Boase, Charles W. *Register of the University of Oxford*. Vol. 1. Oxford: Oxford Historical Society, 1884.
Boase, Charles W. *Oxford*. London: Longman, 1890.
Brewer, J. S. *Giraldi Cambrensis Opera*. London: Longman, 1861.

E. Garritzen, *Reimagining the Historian in Victorian England*,
https://doi.org/10.1007/978-3-031-28461-8

Brewer, J. S. *Letters and Papers, Foreign and Domestic, of the Reign of Henry VIII*. London: Longman, 1864.
Brewer, J. S. *Registrum Malmesburiense*. London: Longman, 1879.
Bright, J. Franck. *A History of England. Period I: Mediæval Monarchy*. London: Rivingtons, 1877.
Bright, J. Franck. *A History of England. Period II: Personal Monarchy Henry VII to James II*. London: Rivingtons, 1880.
Bright, J. Franck. *A History of England. Period III: Constitutional Monarchy*. London: Rivingtons, 1880.
Bright, J. Franck. *A History of England. Period III: Constitutional Monarchy*. 4th ed. London: Rivingtons, 1884.
Bright, J. Franck. *A History of England. Period I: Mediæval Monarchy*. 4th ed. London: Rivingtons, 1887.
Bright, J. Franck. *Joseph II*. London: Macmillan, 1897.
Bright, J. Franck. *Maria Theresa*. London: Macmillan, 1897.
Bright, J. Franck. *A History of England. Period II: Personal Monarchy*. 8th impression. London: Longman, 1901.
Bright, J. Franck. *A History of England Period III: Constitutional Monarchy*. 9th impression. London: Longman, 1902.
Bright, William. *Chapters of Early English Church History*. London: Macmillan, 1878.
Browning, Oscar. *Historical Reader*. London: Griffith and Farran, [1884].
Browning, Oscar. *The New Illustrated History of England*. Vol. 1. London: J. S. Virtue, [1888].
Browning, Oscar. *The New Illustrated History of England*. Vol. 2. London: J. S. Virtue, [1888].
Browning, Oscar. *The Life of Bartolomeo Colleoni, of Anjou and Burgundy*. London: Arundel Society, 1891.
Browning, Oscar. *Guelphs & Ghibellines. A Short History of Mediaeval Italy from 1250–1409*. London: Methuen, 1893.
Browning, Oscar. *Guelphs & Ghibellines. A Short History of Mediaeval Italy from 1250–1409*. 2nd ed. London: Methuen, 1894.
Browning, Oscar. *The Age of Condottieri: A Short History of Mediaeval History from 1409–1530*. London: Methuen, 1895.
Browning, Oscar. *Peter the Great*. London: Hutchinson, 1898.
Browning, Oscar. *Charles XII of Sweden*. London: Hurst and Blackett, 1899.
Bryce, James. *The Holy Roman Empire*. London: Macmillan, 1864.
Bryce, James. *The Holy Roman Empire*. New ed. London: Macmillan, 1866.
Bryce, James. *The Holy Roman Empire*. 3rd ed. London: Macmillan, 1871.
Bryce, James. *The Holy Roman Empire*. 4th ed. London: Macmillan, 1873.
Bryce, James. *The Holy Roman Empire*. 5th ed. London: 1875.
Bryce, James. *The Holy Roman Empire*. 6th ed. London: 1876.

Bryce, James. *The Holy Roman Empire*. Reprint. London: Macmillan, 1889.
Bryce, James. *The Holy Roman Empire*. New ed. London: Macmillan, 1904.
Burrows, Montagu. *The Relations of Church and State Historically Considered*. Oxford: Parker, 1866.
Burrows, Montagu. *Constitutional Progress*. London: John Murray, 1869.
Burrows, Montagu. *Worthies of All Souls. Four Centuries of English History*. London: Macmillan, 1874.
Burrows, Montagu. *Imperial England*. London: Cassell, 1880.
Burrows, Montagu. *The Register of the Visitors of the University of Oxford from A.D. 1647 to A.D. 1658*. London: Camden Society, 1881.
Burrows, Montagu. *Wiclif's Place in History*. London: Wm. Ibister, 1882.
Burrows, Montagu. *History of the Families of Larcom, Hollis, and McKinley*. Oxford: for private circulation, 1883.
Burrows, Montagu. *The Life of Edward Lord Hawke. Admiral of the Fleet, Vice-Admiral of Great Britain, and First Lord of the Admiralty from 1766 to 1771*. London: W. H. Allen, 1883.
Burrows, Montagu. *The Family of Brocas of Beaurepaire and Roche Court. Hereditary Masters of the Royal Buckhounds*. London: Longman, 1886.
Burrows, Montagu. *Cinque Ports*. London: Longman, 1892.
Burrows, Montagu. *Commentaries on the History of England from the Earliest Times to 1865*. London and Edinburgh: Blackwood, 1893.
Burrows, Montagu. *The History of the Foreign Policy of Great Britain*. London and Edinburgh: Blackwood, 1895.
Burrows, Montagu. *The Life of Edward Lord Hawke. Admiral of the Fleet, Vice-Admiral of Great Britain, and First Lord of the Admiralty from 1766 to 1771*. New ed. London: W. H. Allen, 1896.
Burrows, Montagu. *The History of the Foreign Policy of Great Britain*. New ed. London and Edinburgh: Blackwood, 1897.
Burton, John Hill. *A History of the Reign of Queen Anne*. Vol. 1. Edinburgh: William Blackwood, 1880.
Burton, John Hill. *A History of the Reign of Queen Anne*. Vol. 2. Edinburgh: William Blackwood, 1880.
Burton, John Hill. *A History of the Reign of Queen Anne*. Vol. 3. Edinburgh: William Blackwood, 1880.
Camperdown, Earl of. *Admiral Duncan*. London: Longman, 1898.
Cartwright, James J. (ed.) *The Memoirs of Sir John Reresby of Thrybergh, Bart., M.P. for York &c*. London: Longman, 1875.
Cleveland, Duchess of. *The Battle Abbey Roll*. Vol. 1. London: John Murray, 1889.
Cleveland, Duchess of. *The Battle Abbey Roll*. Vol. 2. London: John Murray, 1889.

Cleveland, Duchess of. *The Battle Abbey Roll*. Vol. 3. London: John Murray, 1889.
Cox, George W. *The Mythology of the Aryan Nations*. Vol. 1. London: Longman, 1870.
Cox, George W. *The Mythology of the Aryan Nations*. Vol. 2. London: Longman, 1870.
Cox, George W. *A History of Greece*. Vol. 1. London: Longman. 1874.
Cox, George W. *A History of Greece*. Vol. 2. London: Longman, 1874.
Cox, George W. *The Mythology of the Aryan Nations*. London: Kegan Paul, 1882.
Cox, George W. *The Mythology of the Aryan Nations*. London: Kegan Paul, 1887.
Creighton, Mandell. *The Age of Elizabeth*. London: Longman, 1876.
Creighton, Mandell. *Life of Simon de Montfort Earl of Leicester*. London: Rivingtons, 1876.
Creighton, Mandell. *Life of Simon de Montfort Earl of Leicester*. 2nd ed. London: Rivingtons, 1877.
Creighton, Mandell. *The Shilling History of England*. London: Longman, 1879.
Creighton, Mandell. *A History of the Papacy during the Period of the Reformation*. Vol. 1. London: Longman, 1882.
Creighton, Mandell. *A History of the Papacy during the Period of the Reformation*. Vol. 2. London: Longman, 1882.
Creighton, Mandell. *A History of the Papacy during the Period of the Reformation*. Vol. 3. London: Longman, 1887.
Creighton, Mandell. *A History of the Papacy during the Period of the Reformation*. Vol. 4. London: Longman, 1887.
Creighton, Mandell. *The Age of Elizabeth*. 7th ed. London: Longman 1888.
Creighton, Mandell. *Cardinal Wolsey*. London: Macmillan, 1888.
Creighton, Mandell. *Carlisle*. London: Longman, 1889.
Creighton, Mandell. *Cardinal Wolsey*. Reprint. London: Macmillan, 1891.
Creighton, Mandell. *The Shilling History of England*. New ed. London: Longman, 1891.
Creighton, Mandell. *A History of the Papacy during the Period of the Reformation*. Vol. 5. London: Longman, 1894.
Creighton, Mandell. *The Early English Renaissance*. Cambridge: Cambridge University Press, 1895.
Creighton, Mandell. *Life of Simon de Montfort Earl of Leicester*. New ed. London: Longman, 1895.
Creighton, Mandell. *Queen Elizabeth*. Paris: Boussod, Valadon, 1896.
Creighton, Mandell. *The Age of Elizabeth*. 13th ed. London: Longman, 1897.
Creighton, Mandell. *A History of the Papacy from the Great Schism to the Sack of Rome*. Vol. 4. London: Longman, 1897.
Creighton, Mandell. *A History of the Papacy from the Great Schism to the Sack of Rome*. Vol. 5. London: Longman, 1897.

Creighton, Mandell. *A History of the Papacy from the Great Schism to the Sack of Rome*. Vol. 6. London: Longman, 1897.
Creighton, Mandell. *Queen Elizabeth*. New ed. London: Longman, 1899.
Creighton, Mandell. *A History of the Papacy from the Great Schism to the Sack of Rome*. Vol. 3. New impression. London: Longman, 1903.
Creighton, Mandell. *A History of the Papacy from the Great Schism to the Sack of Rome*. Vol. 2. New impression. London: Longman, 1904.
Creighton, Mandell. *A History of the Papacy from the Great Schism to the Sack of Rome*. Vol. 2. New impression. London: Longman, 1909.
Dasent, George Webbe. *The Story of Burnt Njal or Life in Iceland at the End of the Tenth Century*. Vol. 1. Edinburgh: Edomonston and Douglas, 1861.
Dasent, George Webbe. *The Story of Burnt Njal or Life in Iceland at the End of the Tenth Century*. Vol. 2. Edinburgh: Edomonston and Douglas, 1861.
Donne, W. Bodham. *The Correspondence of King George the Third with Lord North from 1768 to 1783*. Vol. 1. London: John Murray, 1867.
Donne, W. Bodham. *The Correspondence of King George the Third with Lord North from 1768 to 1783*. Vol. 2. London: John Murray, 1867.
Doran, Dr. *Lives of the Queens of England of the House of Hanover*. Vol. 1. London: Richard Bentley, 1855.
Dormer Harris, Mary. *The Coventry Leet Book: or Mayor's Register*. Vol. 1. London: Early English Text Society, 1907.
Elliot, Hugh. *The Life of Sidney, Earl of Godolphin, K.G. Lord High Treasurer of England 1702 to 1710*. London: Longman, 1888.
Everett Green, Mary Anne. *Lives of the Princesses of England, from the Norman Conquest*. Vol. 1. London: Henry Colburn, 1850.
Everett Green, Mary Anne. *Lives of the Princesses of England, from the Norman Conquest*. Vol. 3. London: Henry Colburn, 1851.
Everett Green, Mary Anne. *Lives of the Princesses of England, from the Norman Conquest*. Vol. 4. London: Henry Colburn, 1852.
Everett Green, Mary Anne. *Lives of the Princesses of England, from the Norman Conquest*. Vol. 5. London: Henry Colburn, 1854.
Everett Green, Mary Anne. *Lives of the Princesses of England, from the Norman Conquest*. Vol. 6. London: Henry Colburn, 1855.
Everett Wood, Mary Anne. *Letters of Royal and Illustrious Ladies of Great Britain, from the Commencement of the Twelfth Century to the Close of the Reign of Queen Mary*. Vol. 1. London: Henry Colburn, 1846.
Everett Wood, Mary Anne. *Letters of Royal and Illustrious Ladies of Great Britain, from the Commencement of the Twelfth Century to the Close of the Reign of Queen Mary*. Vol. 2. London: Henry Colburn, 1846.
Everett Wood, Mary Anne. *Letters of Royal and Illustrious Ladies of Great Britain, from the Commencement of the Twelfth Century to the Close of the Reign of Queen Mary*. Vol. 3. London: Henry Colburn, 1846.

Ferguson, Alex. *The Honourable Henry Erskine Lord Advocate for Scotland*. Edinburgh and London: William Blackwood and Sons, 1882.
Firth, C. H. *The Journal of Joachim Hane*. Oxford: Blackwell, 1886.
Firth, C. H. *The Clarke Papers*. Vol. 1. London: Camden Society, 1891.
Firth, C. H. *The Clarke Papers*. Vol. 2. London: Camden Society, 1894.
Firth, C. H. *The Memoirs of Edmund Ludlow, Lieutenant-General of the Horse in the Army of the Commonwealth England 1625–1672*. Vol. 1. Oxford: Clarendon Press, 1894.
Firth, C. H. *The Memoirs of Edmund Ludlow, Lieutenant-General of the Horse in the Army of the Commonwealth England 1625-1672*. Vol. 2. Oxford: Clarendon Press, 1894.
Firth, C. H. *Scotland and the Commonwealth*. Edinburgh: Scottish History Society, 1895.
Firth, C. H. *The Clarke Papers*. Vol. 3. London: Longman, 1899.
Firth, C. H. *Scotland and the Protectorate*. Edinburgh: Scottish History Society, 1899.
Fitzgerald, Percy. *The Life of David Garrick*. Vol. 1. London: Tinsley Brothers, 1868.
Fitzgerald, Percy. *The Life of David Garrick*. Vol. 2. London: Tinsley Brothers, 1868.
Fitzgerald, Percy. *The Kembles. An Account of the Kemble Family, Including the Lives of Mrs. Siddons, And Her Brother John Philip Kemble*. Vol. 1. London: Tinsley Brothers, [1871].
Fitzgerald, Percy. *The Kembles. An Account of the Kemble Family, Including the Lives of Mrs. Siddons, and Her Brother John Philip Kemble*. Vol. 2. London: Tinsley Brothers, [1871].
Fitzgerald, Percy. *The Life of David Garrick*. New ed. London: Simpkin, Marshall, Hamilton, Kent & Co., 1899.
Fitzmaurice, Lord Edmond. Life of William, Earl of Shelburne, Afterwards First Marquess of Lansdowne. Vol. 1. London: Macmillan, 1875.
Fitzmaurice, Lord Edmond. Life of William, Earl of Shelburne, Afterwards First Marquess of Lansdowne. Vol. 2. London: Macmillan, 1876.
Fitzmaurice, Lord Edmond. Life of William, Earl of Shelburne, Afterwards First Marquess of Lansdowne. Vol. 3. London: Macmillan, 1876.
Forster, John. *Arrest of the Five Members by Charles the First. A Chapter of English History Rewritten*. London: John Murray, 1860.
Forster, John. *The Debates on the Grand Remonstrance, November and December 1641*. London: John Murray, 1860.
Forster, John. *Sir John Eliot: A Biography*. Vol. 1. London: Longman, 1864.
Forster, John. *Sir John Eliot: A Biography*. Vol. 2. London: Longman, 1864.
Forster, John. *Walter Savage Landor: A Biography*. Vol. 1. London: Chapman and Hall, 1869.

Forster, John. *Sir John Eliot: A Biography*. Vol. 1. 2nd ed. London: Chapman and Hall, 1872.
Foxcroft, H. C. *The Life and Letters of Sir George Saville, Bart*. Vol. 1. London: Longman, 1898.
Foxcroft, H. C. *The Life and Letters of Sir George Saville, Bart*. Vol. 2. London: Longman, 1898.
Fraser, William. *The Book of Carlaverock. Memoirs of the Maxwells, Earls of Nithsdale Lords Maxwell & Herries*. Vol. 1. Edinburgh: Privately Published, 1873.
Fraser, William. *The Book of Carlaverock. Memoirs of the Maxwells, Earls of Nithsdale Lords Maxwell & Herries*. Vol. 2. Edinburgh: privately published, 1873.
Freeman, Edward A. *History of Federal Government. From the Foundation of the Achaian League to the Disruption of the United States*. Volume 1. London: Macmillan, 1863.
Freeman, Edward A. *History of the Norman Conquest, Its Causes and Results*. Vol. 1. Oxford: Clarendon Press, 1867.
Freeman, Edward A. *The History of the Norman Conquest of England, Its Causes and Its Results*. Vol. 2. Oxford: Clarendon Press, 1868.
Freeman, Edward A. *The History of the Norman Conquest of England, Its Causes and Its Results*. Vol. 3. Oxford: Clarendon Press, 1869.
Freeman, Edward A. *Old English History for Children*. London: Macmillan, 1869.
Freeman, Edward A. *History of the Cathedral Church of Wells as Illustrating the History of the Cathedral Churches of the Old Foundation*. London: Macmillan, 1870.
Freeman, Edward A. *The History of the Norman Conquest of England, Its Causes and Its Results*. Vol. 1. 2nd ed. Oxford: Clarendon Press, 1870.
Freeman, Edward A. *The History of the Norman Conquest of England, Its Causes and Its Results*. Vol. 2. 2nd ed. Oxford: Clarendon Press, 1870.
Freeman. Edward A. *The History of the Norman Conquest of England, Its Causes and Its Results*. 4. Oxford: Clarendon Press, 1871.
Freeman, Edward A. *General Sketch of European History*. London: Macmillan, 1872.
Freeman, Edward A. *The Growth of the English Constitution from the Earliest Times*. London: Macmillan, 1872.
Freeman, Edward A. *General Sketch of European History*. 2nd ed. London: Macmillan, 1873.
Freeman, Edward A. *The Growth of the English Constitution from the Earliest Times*. 2nd ed. London: Macmillan, 1873.
Freeman, Edward A. *The History of the Norman Conquest of England, Its Causes and Its Results*. Vol. 3. 2nd ed. Oxford: Clarendon Press, 1875.

Freeman, Edward A. *The History of the Norman Conquest of England, Its Causes and Its Results*. Vol. 4. 2nd ed. Oxford: Clarendon Press, 1876.
Freeman, Edward A. *The History of the Norman Conquest, Its Causes and Its Results*. Vol. 5. Oxford: Clarendon Press, 1876.
Freeman, Edward A. *The Growth of the English Constitution from the Earliest Times*. 3rd ed. London: Macmillan, 1876.
Freeman, Edward A. *Historical and Architectural Sketches: Chiefly Italian*. London: Macmillan, 1876.
Freeman, Edward A. *The History of the Norman Conquest of England, Its Causes and Its Results*. Vol. 1. 3rd ed. Cambridge: Clarendon Press, 1877.
Freeman, Edward A. *The History of the Norman Conquest of England, Its Causes and Its Results*. Vol. 2. 3rd ed. Oxford: Clarendon Press, 1877.
Freeman, Edward A. *The History of the Norman Conquest of England, Its Causes and Its Results*. Vol. 6. Oxford: Clarendon Press, 1879.
Freeman, Edward A. *A Short History of the Norman Conquest of England*. Oxford: Clarendon Press, 1880.
Freeman, Edward A. *A Short History of the Norman Conquest of England*. 2nd ed. Oxford: Clarendon Press, 1880.
Freeman, Edward A. *The Historical Geography of Europe*. Vol. 1. London: Longman, 1881.
Freeman, Edward A. *The Historical Geography of Europe*. Vol. 2. London: Longman, 1881.
Freeman, Edward A. *Old-English History*. 2nd ed., reprint. London: Macmillan, 1881.
Freeman, Edward A. *General Sketch of European History*. 5th ed., reprint. London: Macmillan, 1882.
Freeman, Edward A. *The Historical Geography of Europe*. Vol. 1. 2nd ed. London: Longman, 1882.
Freeman, Edward A. *The Historical Geography of Europe*. Vol. 2. 2nd ed. London: Longman, 1882.
Freeman, Edward A. *The Reign of William Rufus and the Accession of Henry the First*. Vol. 1. Oxford: Clarendon Press, 1882.
Freeman, Edward A. *The Reign of William Rufus and the Accession of Henry the First*. Vol. 2. Oxford: Clarendon Press, 1882.
Freeman, Edward A. *Exeter*. London: Longman, 1887.
Freeman, Edward A. *A Short History of the Norman Conquest of England*. 3rd ed. Oxford: Clarendon Press, 1887.
Freeman, Edward A. *Exeter*. 2nd ed. London: Longman, 1890.
Freeman, Edward A. *The History of Sicily from the Earliest Times*. Vol. 1. Oxford: Clarendon Press, 1891.
Freeman, Edward A. *The History of Sicily from the Earliest Times*. Vol. 2. Oxford: Clarendon Press, 1891.

Freeman, Edward A. *Exeter*. 3rd ed. London: Longman, 1892.
Freeman, Edward A. *The History of Sicily from the Earliest Times*. Vol. 3. Oxford: Clarendon Press, 1892.
Edward A. Freeman, *The History of Sicily from the Earliest Times*. Vol. 4. Ed. Arthur J. Evans. Oxford: Clarendon Press, 1894.
Freeman, Edward A. *Atlas to the Historical Geography of Europe*. 3rd ed. London: Longman, 1903.
Freeman, Edward A. *The Historical Geography of Europe*, 3rd ed. London: Longman, 1903
Friedmann, Paul. *Anne Boleyn. A Chapter of English History 1527–1536*. Vol. 1. London: Macmillan, 1884.
Friedmann, Paul. *Anne Boleyn. A Chapter of English History 1527–1536*. Vol. 2. London: Macmillan, 1884.
Froude, James Anthony. *History of England from the fall of Wolsey to the death of Elizabeth*. Vol. 1. London: John W. Parker and Son, 1856.
Froude, James Anthony. *History of England from the fall of Wolsey to the death of Elizabeth*. Vol. 3. London: John W. Parker and Son, 1858.
Froude, James Anthony. *History of England from the fall of Wolsey to the death of Elizabeth*. Vol. 5. London: John W. Parker and Son, 1860.
Froude, James Anthony. *The English in Ireland in the Eighteenth Century*. Vol. 1. London: Longman, 1872.
Froude, James Anthony. *The English in Ireland in the Eighteenth Century*. Vol. 2. London: Longman, 1874.
Froude, James Anthony. *The English in Ireland in the Eighteenth Century*. Vol. 3. London: Longman, 1874.
Froude, James Anthony. *History of England from the Fall of Wolsey to the Defeat of the Spanish Armada*. Vol. 1. New ed. London: Longman, 1875.
Froude, James Anthony. *The Divorce of Catharine of Aragon. The Story Told by the Imperial Ambassadors Resident at the Court of Henry VIII*. London: Longman, 1891.
Fulton, Forrest. *A Manual of Constitutional History*. London: Butterworths, 1875.
Gairdner, James. *The Houses of Lancaster and York with the Conquest and Loss of France*. London: Longman, 1875.
Gairdner, James. *History of the Life and Reign of Richard the Third*. 2nd ed. London: Longman, 1879.
Gairdner, James. *Henry the Seventh*. London: Macmillan, 1889.
Gairdner, James. *History of the Life and Reign of Richard the Third*. New ed. Cambridge: Cambridge University Press, 1898.
Gardiner, Samuel Rawson. *History of England from the Accession of James I. to the Disgrace of Chief-Justice Coke*. Vol. 1. London: Hurst and Blackett, 1863.

Gardiner, Samuel Rawson. *History of England from the Accession of James I. to the Disgrace of Chief-Justice Coke*. Vol. 2. London: Hurst and Blackett, 1863.
Gardiner, Samuel Rawson. *Prince Charles and The Spanish Marriage: 1617–1623*. Vol. 1. London: Hurst and Blackett, 1869.
Gardiner, Samuel Rawson. *The Thirty Years' War 1618–1648*. London: Longman, 1874.
Gardiner, Samuel Rawson. *A History of England under the Duke of Buckingham and Charles I*. Vol. 1. London: Longman, 1875.
Gardiner, Samuel Rawson. *The Personal Government of Charles I. A History of England from the Assassination of the Duke of Buckingham to the Declaration of the Judges of Ship-Money 1628–1637*. Vol. 1. London: Longman, 1877.
Gardiner, Samuel Rawson. *The First Two Stuarts and the Puritan Revolution 1603–1660*. 3rd ed. London: Longman, 1878.
Gardiner, Samuel Rawson. *The First Two Stuarts and the Puritan Revolution 1603–1660*. 4th ed. London: Longman, 1880.
Gardiner, Samuel Rawson. *Outline of English History, First Period B.C. 55–A.D. 1603*. London: Longman, 1881.
Gardiner, Samuel Rawson. *Outline of English History, Second period A.D. 1603–A.D. 1880*. London: Longman, 1881.
Gardiner, Samuel Rawson. *The Thirty Years' War*. 5th ed. London: Longman, 1881.
Gardiner, Samuel Rawson. *The Fall of the Monarchy of Charles I. 1637–1649*. Vol. 1. London: Longman, 1882.
Gardiner, Samuel Rawson. *History of England from the Accession of James I. to the Outbreak of the Civil War 1603–1642*. Vol. 1. London: Longman, 1883.
Gardiner, Samuel Rawson. *History of England from the Accession of James I. to the Outbreak of the Civil War 1603–1642*. Vol. 2. London: Longman, 1883.
Gardiner, Samuel Rawson. *History of England from the Accession of James I. to the Outbreak of the Civil War 1603–1642*. Vol. 3. London: Longman, 1883.
Gardiner, Samuel Rawson. *History of England from the Accession of James I. to the Outbreak of the Civil War 1603–1642*. Vol. 4. London: Longman, 1883.
Gardiner, Samuel Rawson. *History of England from the Accession of James I. to the Outbreak of the Civil War 1603–1642*. Vol. 5. London: Longman, 1883.
Gardiner, Samuel Rawson. *History of England from the Accession of James I. to the Outbreak of the Civil War 1603–1642*. Vol. 6. London: Longman, 1884.
Gardiner, Samuel Rawson. *History of England from the Accession of James I. to the Outbreak of the Civil War 1603–1642*. Vol. 7. London: Longman, 1884.
Gardiner, Samuel Rawson. *History of England from the Accession of James I. to the Outbreak of the Civil War 1603–1642*. Vol. 8 London: Longman, 1884.
Gardiner, Samuel Rawson. *History of England from the Accession of James I. to the Outbreak of the Civil War 1603–1642*. Vol. 9. London: Longman, 1884.

Gardiner, Samuel Rawson. *History of England from the Accession of James I. to the Outbreak of the Civil War 1603–1642*. Vol. 10. London: Longman, 1884.
Gardiner, Samuel Rawson. *History of the Great Civil War 1642–1649*. Vol. 1. London: Longman, 1886.
Gardiner, Samuel Rawson. *Outline of English History B.C. 55–A.D. 1886*. New ed. London: Longman, 1888.
Gardiner, Samuel Rawson. *History of the Great Civil War 1642–1649*. Vol. 2. London: Longman, 1889.
Gardiner, Samuel Rawson. *The First Two Stuarts and the Puritan Revolution 1603–1660*. 9th ed. London: Longman, 1890.
Gardiner, Samuel Rawson. *History of the Great Civil War 1642–1649*. Vol. 3. London: Longman, 1891.
Gardiner, Samuel Rawson. *Student's History of England from the Earliest Times to 1885*. Vol. 2. London: Longman, 1891.
Gardiner, Samuel Rawson. *Student's History of England from the Earliest Times to 1885*. Vol. 3. London: Longman, 1891.
Gardiner, Samuel Rawson. *Student's History of England from the Earliest Times to 1885*. Vol. 1. 3rd ed. London: Longman, 1892.
Gardiner, Samuel Rawson. *History of the Commonwealth and Protectorate 1649–1660*. Vol. 1. London: Longman, 1894.
Gardiner, Samuel Rawson. *Cromwell's Place in History*. London: Longman, 1897.
Gardiner, Samuel Rawson. *History of the Commonwealth and Protectorate 1649–1660*. Vol. 2. London: Longman, 1897.
Gardiner, Samuel Rawson. *What Gunpowder Plot Was*. London: Longman, 1897.
Gardiner, Samuel Rawson. *History of the Commonwealth and Protectorate 1649–1660*. Vol. 3. London: Longman, 1901.
Gardiner, Samuel Rawson. *Oliver Cromwell*. London: Longman, 1901.
Gardner, Alice. *Synesius of Cyrene, Philosopher and Bishop*. London: Society for Promoting Christian Knowledge, 1886.
Gardner, Alice. *Julian Philosopher and Emperor and the Last Struggle of Paganism against Christianity*. London: G. P. Putnam's Sons, 1895.
Gardner, Alice. *Rome. The Middle of the World*. London: Edward Arnold, 1897.
Gardner, Alice. *Studies in John the Scot (Erigena). A Philosopher of the Dark Ages*. London: Henry Frowde, 1900.
George, Hereford B. *Genealogical Tables Illustrative of Modern History*. Oxford: Clarendon Press, 1875.
George, Hereford B. *Battles of English History*. London: Methuen, 1895.
George, Hereford B. *Napoleon's Invasion of Russia*. London: T. Fisher Unwin, 1899.
Gerard, John S.J. *What Was the Gunpowder Plot? The Traditional Story Tested by Original Evidence*. London: Osgood, McIlvaine & Co., 1897.

Green, John Richard. *A Short History of the English People*. London: Macmillan, 1874.
Green, John Richard. *A Short History of the English People*. 17th thousand. London: Macmillan, 1875.
Green, John Richard. *A Short History of the English People*. 49th thousand. London: Macmillan, 1877.
Green, John Richard. *History of the English People*. Vol. 1. London: Macmillan, 1877.
Green, John Richard. *History of the English People*. Vol. 2. London: Macmillan, 1878.
Green, John Richard. *History of the English People*. Vol. 3. London: Macmillan, 1879.
Green, John Richard. *History of the English People*. Vol. 4. London: Macmillan, 1880.
Green, John Richard. *The Making of England*. London: Macmillan, 1881.
Green, John Richard. *The Conquest of England*. London: Macmillan, 1883.
Green, John Richard. *A Short History of the English People*. New ed. London: Macmillan, 1888.
Green, John Richard. *A Short History of the English People*. Illustrated edition by Mrs. J. R. Green and Kate Norgate. Vol. 1. London: Macmillan, 1892.
Green, John Richard. *A Short History of the English People*. Illustrated edition by Mrs. J. R. Green and Kate Norgate. Vol. 2. London: Macmillan, 1893.
Green, John Richard. *A Short History of the English People*. Illustrated edition by Mrs. J. R. Green and Kate Norgate. Vol. 3. London: Macmillan, 1893.
Green, John Richard. *A Short History of the English People*. Illustrated edition by Mrs. J. R. Green and Kate Norgate. Vol. London: Macmillan, 1894.
Green, John Richard. *The Making of England*. Vol. 1. 4th ed. London: Macmillan, 1897.
Green, John Richard. *The Making of England*. Vol. 2. 4th ed. London: Macmillan, 1897.
Green, John Richard. *The Conquest of England*. Vol. 1. 3rd ed. London: Macmillan, 1899.
Green, John Richard. *The Conquest of England*. Vol. 2. 3rd ed. London: Macmillan, 1899.
Green, John Richard. *A Short History of the English People*. New York: American Book Company, [1911].
Grote, Harriet. *Memoir of the life of Ary Scheffer*. London: John Murray, 1860.
Guest, Edwin. *Origines Celticae (a Fragment) and Other Contributions to the History of Britain*. Vol. 1. London: Macmillan, 1883.
Guest, Edwin. *Origines Celticae (a Fragment) and Other Contributions to the History of Britain*. Vol. 2. London: Macmillan, 1883.

Hall, Mrs. Mathew. *The Queens Before the Conquest*. Vol. 1. London: Henry Colburn, 1854.
Hallam, Henry. *The Constitutional History of England from the Accession of Henry II. to the Death of George II*. London: Ward, Lock, & Co. [1859].
Hassall, Arthur. *Life of Viscount Bolingbroke*. London: W. H. Allen, 1889.
Hassall, Arthur. *Louis XIV and the Zenith of the French Monarchy*. London: G. P. Putnam's Sons, 1895.
Hassall, Arthur. *The Making of British Empire (A.D. 1714–1832)*. London: Blackie & Son, 1896.
Hassall, Arthur. *A Handbook of European History 476–1871*. London: Macmillan, 1898.
Hassall, Arthur. *The Balance of Power 1715–1789*. 2nd ed. London: Macmillan, 1898.
Hassall, Arthur. *John Richard Green, Short History of the English People, with Tables and an Analysis*. London: Macmillan, 1917.
Hickson, Mary. Ireland in the Seventeenth Century or the Irish Massacres of 1641-2, Their Causes and Results. Vol. 1. London: Longman, 1884.
Hickson, Mary. Ireland in the Seventeenth Century or the Irish Massacres of 1641-2, Their Causes and Results. Vol. 2. London: Longman, 1884.
Hill, George Birkbeck. *Letters of Samuel Johnson, LL.D.* Vol. 1. Oxford: Clarendon Press, 1892.
Hill, George Birkbeck. *Letters of Samuel Johnson, LL.D.* Vol. 2. Oxford: Clarendon Press, 1892.
Howorth, Henry H. *History of the Mongols from the 9th to the 19th Century*. Vol. 1. London: Longman, 1876.
Howorth, Henry H. *History of the Mongols from the 9th to the 19th Century*. Vol. 2. London: Longman, 1880.
Howorth. Henry H. *History of the Mongols from the 9th to the 19th Century*. Vol. 3. London: Longman, 1888.
Hunt, William. *History of Italy*. London: Macmillan, 1878.
Hunt, William. *Norman Britain*. London: Society for Promoting Christian Knowledge, 1884.
Hunt, William. *The Somerset Diocese, Bath and Wells*. London: Society for Promoting Christian Knowledge, 1885.
Hunt, William. *Bristol*. London: Longman, 1887.
Hunt, William. *The English Church in the Middle Ages*. London: Longman, 1888.
Hunt, William. *The English Church from Its Foundation to the Norman Conquest (597–1066)*. London: Macmillan, 1899.
Hutton, James. *Selections from the Letters and Correspondence of Sir James Bland Burges, Bart., Sometime Under-Secretary of State for Foreign Affairs*. London: John Murray, 1885.

Hutton, William Holden. *The Marquess Wellesley, K.G.* Oxford: Clarendon Press, 1893.
Hutton, William Holden. *King and Baronage (A.D. 1135–1327).* London: Blackie & Son, 1895.
Hutton, William Holden. *Sir Thomas More.* London: Methuen, 1895.
Hutton, William Holden. *William Laud.* London: Methuen, 1895.
Hutton, William Holden. *Philip Augustus.* London: Macmillan, 1896.
Hutton, William Holden. *William Laud.* 2nd ed. London: Methuen, 1896.
Hutton, William Holden. *The Church of the Sixth Century.* London: Longman, 1897.
Hutton, William Holden. *Hampton Court.* London: John. C. Nimmo, 1897.
Hutton, William Holden. *S. John Baptist College.* London: F. E. Robinson, 1898.
Hutton, William Holden. *An Elementary History of the Church in Great Britain.* (London: Rivingtons, 1899.
Hutton, William Holden. *Sir Thomas More.* 2nd ed. London: Methuen, 1900.
Jesse, J. Heneage. *Memoirs of the Life and Reign of King George the Third.* Vol. 1. 2nd ed. London: Tinsley Brothers, 1867.
Jesse, J. Heneage. *Memoirs of the Life and Reign of King George the Third.* Vol. 2. 2nd ed. London: Tinsley Brothers, 1867.
Jesse, J. Heneage. *Memoirs of the Life and Reign of King George the Third.* Vol. 3. 2nd ed. London: Tinsley Brothers, 1867.
Jesse, J. Heneage. *Memoirs of the Life and Reign of King George the Third.* Vol. 1. 2nd ed. London: Tinsley Brothers, 1887.
Kemble, John Mitchell. *The Saxons in England. A History of the English Commonwealth till the Period of the Norman Conquest.* Vol. 1. London: Longman, 1849.
Kitchin G. W. *A History of France.* Vol. 2. 2nd ed. Oxford: Clarendon Press, 1877.
Kitchin G. W. *A History of France.* Vol. 2. 3rd ed. Oxford: Clarendon Press, 1892.
Kitchin G. W. *A History of France.* Vol. 3. 3rd ed. Oxford: Clarendon Press, 1894.
Kitchin G. W. *A History of France.* Vol. 2, 3rd ed. Oxford: Clarendon Press, 1896.
Laughton, John Knox. *State Papers Relating to the Defeat of the Spanish Armada Anno 1588.* Vol. 1. 2nd ed. London: The Navy Record Society, 1895.
Laughton, John Knox. *State Papers Relating to the Defeat of the Spanish Armada Anno 1588.* Vol. 2. 2nd ed. London: The Navy Record Society, 1895.
Lawrance, Hannah. *Historical Memoirs of the Queens of England, from the Commencement of the Twelfth Century.* London: Edward Moxon, 1838.
Lawrance, Hannah. *Historical Memoirs of the Queens of England, from the Commencement of the Twelfth Century.* Vol 2. London: Edward Moxon, 1840.

Lawrance, Hannah. *The History of Woman in England, and Her Influence on Society and Literature, from the Earliest Period*. Vol. 1. London: Henry Colburn, 1843.
Lewis, Lady Theresa. *Extracts from the Journals and Correspondence of Miss Berry from the Year 1783 to 1852*. Vol. 1. 2nd ed. London: Longman, 1866.
Lewis, Lady Theresa. *Extracts from the Journals and Correspondence of Miss Berry from the Year 1783 to 1852*. Vol. 2. 2nd ed. London: Longman, 1866.
Lewis, Lady Theresa. *Extracts from the Journals and Correspondence of Miss Berry from the Year 1783 to 1852*. Vol. 3. 2nd ed. London: Longman, 1866.
Lingard, John. *The History of England from the First Invasion by the Romans to the Accession of William and Mary in 1688*. Vol. 1. 5th ed. London: Charles Dolman, 1849.
Lodge, Richard. *Cardinal Beaufort*. Oxford: Thos. Shrimpton & Son, 1875.
Lodge, Richard. *A History of Modern Europe. From the Capture of Constantinople, 1453, to the Treaty of Berlin, 1878*. London: John Murray, 1885.
Lodge, Richard. *A History of Modern Europe. From the Capture of Constantinople, 1453, to the Treaty of Berlin, 1878*. 2nd edition. London: John Murray, 1887.
Lodge, Richard. *Richelieu*. London: Macmillan, 1896.
Macaulay, Thomas Babington. *The History of England from the Accession of James II*. Vol. 1. London: Longman, 1849.
Macaulay, Thomas Babington. *The History of England from the Accession of James II*. Vol. 1. 4th ed. London: Longman, 1849.
Macaulay, Thomas Babington. *The History of England from the Accession of James II*. Vol. 2. 5th ed. London: Longman, 1849.
Macaulay, Thomas Babington. *The History of England from the Accession of James the Second*. Vol. 3. London: Longman, 1855.
Macaulay Thomas Babington. *The History of England from the Accession of James the Second*. Vol. IV. London: Longman, 1855.
Maitland, Frederic William. *Bracton's Note Book*. Vol. 1. London: C. J. Clay, 1887.
Maitland, Frederic William. *Bracton's Note Book*. Vol. 2. London: C. J. Clay, 1887.
Maitland, Frederic William. *Bracton's Note Book*. Vol. 3. London: C. J. Clay, 1887.
Maitland, Frederic William. *Select Pleas in Manorial and Other Seignorial Courts*. Vol. 1. London: Selden Society, 1889.
Maitland, Frederic William. *Domesday Book and beyond*. Cambridge: Cambridge University Press, 1897.
Maitland, Frederic William. *Roman Canon Law in the Church of England*. London: Methuen, 1898.
Maitland, Frederic William. *Township and Borough*. Cambridge: Cambridge University Press, 1898.

Major, Richard Henry. *The Life of Prince Henry of Portugal, Surnamed the Navigator*. London: Asher, 1868.

Major, Richard Henry. *The Discoveries of Prince Henry the Navigator, and their Results*. London: Sampson, Low, Marsten, Searle, & Rivington, 1877.

May, Thomas Erskine. *The Constitutional History of England since the Accession of George the Third 1760–1860*. Vol. 1. London: Longman, 1861.

May, Thomas Erskine. *The Constitutional History of England since the Accession of George the Third 1760–1860*. Vol. 2. London: Longman, 1863.

May, Thomas Erskine. *The Constitutional History of England since the Accession of George the Third 1760–1860*. Vol. 2. 2nd ed. London: Longman, 1865.

May, Thomas Erskine. *The Constitutional History of England since the Accession of George the Third 1760–1860*. Vol. 1. 3rd ed. London: Longman, 1871.

May, Thomas Erskine. *The Constitutional History of England since the Accession of George the Third 1760–1860*. Vol. 2. 5th ed. London: Longman, 1875.

May, Thomas Erskine. *The Constitutional History of England since the Accession of George the Third 1760–1860*. Vol. 1. 4th ed. London: Longman, 1873.

May, Thomas Erskine. *The Constitutional History of England since the Accession of George the Third 1760–1860*. Vol. 1. 6th ed. London: Longman, 1878.

May, Thomas Erskine. *The Constitutional History of England since the Accession of George the Third 1760–1860*. Vol. 1. 7th ed. London: Longman, 1882.

May, Thomas Erskine. *The Constitutional History of England since the Accession of George the Third 1760–1860*. Vo. 1. 8th ed. London: Longman, 1887.

May, Thomas Erskine. *The Constitutional History of England since the Accession of George the Third 1760–1860*. Vol. 3. 9th ed. London: Longman, 1889.

Medley, Dudley Julius. *A Student's Manual of English Constitutional History*. Oxford: Blackwell, 1894.

Medley, Dudley Julius. *A Student's Manual of English Constitutional History*. 2nd ed. Oxford: Blackwell, 1898.

Merivale, Herman. *Memoirs of Sir Philip Francis, K.C.B.* Vol. 1. London: Longman, 1867.

Merivale, Herman. *Memoirs of Sir Philip Francis, K.C.B.* Vol. 2. London: Longman, 1867.

Milman, Henry Hart. *The History of the Jews. From the Earliest Period down to Modern Times*. Vol. 1. 3rd ed. London: John Murray, 1863.

Milman, Henry Hart. *The History of the Jews. From the Earliest Period down to Modern Times*. Vol. 2. 3rd ed. London: John Murray, 1863.

Milman, Henry Hart. *The History of the Jews. From the Earliest Period Down to Modern Times*. Vol. 3. 3rd ed. London: John Murray, 1863.

Milman, Henry Hart. *History of Latin Christianity*. Vol. 1. 3rd ed. London: John Murray, 1864.

Milman, Henry Hart. *History of Latin Christianity*. Vol. 1. 4th ed. London: John Murray, 1867.

Milman, Henry Hart. *Annals of S. Paul's Cathderal*. London: John Murray, 1868.

Mullinger James Bass. *Cambridge Characteristics in the Seventeenth Century*. London: Macmillan, 1867.

Mullinger James Bass. *The Ancient African Church: Its Rise, Influence, and Decline*. London: E. Johnson & Hamilton, Adams, & Co., 1869.

Mullinger James Bass. *Schools of Charles the Great and the Restoration of Education in the Ninth Century*. London: Longman, 1877.

Mullinger, James Bass. *The University of Cambridge from the Royal Injunctions of 1535 to the Accession of Charles the First*. Vol. 2. Cambridge: Cambridge University Press, 1884.

Mullinger, James Bass. *St. John's College*. London: F. E. Robinson, 1901.

Mullinger, James Bass. *The University of Cambridge*. Volume 3. Cambridge: Cambridge University Press, 1911.

Napier, Mark. *Memorials and Letters Illustrative of the Life and Times of John Graham Claverhouse, Viscount Dundee*. Vol. 1. Edinburgh, Thomas G. Stevenson, 1859.

Napier, Mark. *Memorials and Letters Illustrative of the Life and Times of John Graham Claverhouse, Viscount Dundee*. Vol. 2. Edinburgh, Thomas G. Stevenson, 1862.

Napier, Mark. *Memorials and Letters Illustrative of the Life and Times of John Graham Claverhouse, Viscount Dundee*. Vol. 3. Edinburgh, Thomas G. Stevenson, 1862.

Norgate, Kate. *England under the Angevin Kings*. Vol. 1. London: Macmillan, 1887.

Norgate, Kate. *England under the Angevin Kings*. Vol. 2. London: Macmillan, 1887.

Norgate, Kate. *John Lackland*. London: Macmillan, 1902.

Norgate, Kate. *The Minority of Henry the Third*. London: Macmillan, 1912.

Oliphant. Mrs. *The Makers of Venice. Doges, Conquerors, Painters, and Men of Letters*. London: Macmillan, 1887.

Oliphant, Mrs. *The Makers of Venice. Doges, Conquerors, Painters, and Men of Letters*. New ed. London: Macmillan, 1888.

Oliphant, Mrs. *The Makers of Venice. Doges, Conquerors, Painters, and Men of Letters*. New ed. London: Macmillan, 1889.

Oliphant, Mrs. *The Makers of Venice. Doges, Conquerors, Painters, and Men of Letters*. London: Macmillan, 1891.

Oliphant, Mrs. *The Makers of Venice. Doges, Conquerors, Painters, and Men of Letters*. Extra illustrated ed. London: Macmillan, 1892.

Oman, Charles W. *Warwick the Kingmaker*. London: Macmillan, 1891.

Otté, E. C. *Scandinavian History*. London : Macmillan, 1874.

Owen, Sidney. *India on the Eve of the British Conquest: A Historical Sketch.* London: India Office, 1872.

Owen, Sidney J. *A Selection from the Despatches, Treatises, and Other Papers of the Marquess Wellesley, K.G. during his Government of India.* Oxford: Clarendon Press, 1877.

Owen, Sidney J. *A Selection from the Despatches, Memoranda and Other Papers Relating India of the Field-Marshall the Duke of Wellington, K.G.* Oxford: Clarendon Press, 1880.

Palliser, Mrs. Bury. *History of Lace.* London: Sampson Low, 1865.

Palliser, Mrs. Bury. *History of Lace.* 2nd ed. London: Sampson Low, 1869.

Palliser, Mrs. Bury. *History of Lace.* 3rd ed. London: Sampson Low, 1875.

Palliser, Mrs. Bury. *History of Lace.* Posthumously edited by M. Jourdain and Alice Dryden. London: Sampson Low, 1902.

Pattison, Mark. *Isaac Casaubon 1559-1614.* London: Longman, 1875.

Pattison, Mark. *Milton.* London: Macmillan, 1879.

Pearson, Charles H. *The Early and Middle Ages of England.* London: Bell and Daldy, 1861.

Pearson, Charles H. *History of England during the Early and Middle Ages.* London: Bell and Daldy, 1867.

Plunket, David. *The Life, Letters, and Speeches of Lord Plunket.* Vol. 1. London, Smith, Elder & Co., 1867.

Plunket, David. The Life, Letters, and Speeches of Lord Plunket. Vol. 2. London: Smith, Elder & Co., 1867.

Pollock, Frederick and Frederic William Maitland. *The History of English Law before the Time of Edward I.* Vol. 1. Cambridge: Cambridge University Pres, 1895.

Pollock, Frederick and Frederic William Maitland. *The History of English Law before the Time of Edward I.* Vol. 2. Cambridge: Cambridge University Press, 1895.

Pollock, Frederick and Fredric William Maitland. *The History of English Law before the Time of Edward I.* Vol. 1. 2nd ed. Cambridge: Cambridge University Press, 1898.

Pollock, Frederick and Frederic William Maitland. *The History of English Law before the Time of Edward I.* Vol. 2. 2nd ed. Cambridge: Cambridge University Press, 1898.

Powell, Frederick York. *Early England up to the Norman Conquest.* London: Longman, 1876.

Powell, Frederick York. *Alfred the Great and William the Conqueror.* London: Longman, 1881.

Powell, Frederick York. *Old Stories from British History.* London: Longman, 1882.

Powell, Frederick York. *Old Stories from British History*. 3rd ed. London: Longman, 1885.
Powell, Frederick York. *History of England*. London: Rivingtons, 1885.
Powell, Fredrick York. *Early England up to the Norman Conquest*. 13th ed. London: Longman, 1899.
Powell, Fredrick York. *History of England*. New impression. London: Longman, 1906.
Prendergast, John P. *The Cromwellian Settlement of Ireland*. London: Longman, 1865.
Prendergast, John P. *The Cromwellian Settlement of Ireland*. New York: P. M. Haverty, 1868.
Prendergast, John P. *The Cromwellian Settlement of Ireland*. 2nd ed. London: Longman, 1870.
Prothero, George Walter. *The Life of Simon de Montfort Earl of Leicester with Special Reference to the Parliamentary History of his Time*. London: Longman, 1877.
Ramsay, Sir James H. *Lancaster and York a Century of English History (a.d. 1399–1485)*. Vol. 1. Oxford: Clarendon Press, 1892.
Ramsay, Sir James H. *Lancaster and York a Century of English History (a.d. 1399–1485)*. Vol. 2. Oxford: Clarendon Press, 1892.
Rashdall, Hastings. *John Huss*. Oxford: T. Shrimpton, 1879.
Rashdall, Hastings. *The Universities of Europe in the Middle Ages*. Vol. 1. Oxford: Clarendon Press, 1895.
Rawlinson, George. *The Five Great Monarchies of the Ancient Eastern World*. Vol. 1. London: John Murray, 1862.
Rawlinson, George. *The Five Great Monarchies of the Ancient Eastern World*. Vol. 2. London: John Murray. 1864.
Rawlinson, George. The Five Great Monarchies of the Ancient Eastern World. Vol. 3. London, John Murray, 1865.
Rawlinson, George. *The Five Great Monarchies of the Ancient Eastern World*. Vol. 4. London: John Murray, 1867.
Rogers, James E. Thorold. *A History of Agriculture and Prices in England from the Year after the Oxford Parliament (1259) to the Commencement of the Continental War (1793)*. Vol. 1. Oxford: Clarendon Press, 1866.
Rogers, James E. Thorold. *A History of Agriculture and Prices in England from the Year after the Oxford Parliament (1259) to the Commencement of the Continental War (1793)*. Vol. 2. Oxford: Clarendon Press, 1866.
Rogers, James E. Thorold. *A History of Agriculture and Prices in England from the Year after the Oxford Parliament (1259) to the Commencement of the Continental War (1793)*. Vol. 3. Oxford: Clarendon Press, 1882.
Rogers, James E. Thorold. *Six Centuries of Work and Wages. The History of English Labour*. Vol. 1. London: W. Swan Sonnenschein, 1884.

Rogers, James E. Thorold. *A History of Agriculture and Prices in England from the Year after the Oxford Parliament (1259) to the Commencement of the Continental War (1793).* Vol. 5. Oxford: Clarendon Press, 1887.

Round, J. H. *The Early Life of Anne Boleyn: A Critical Essay*. London: Elliot Stock, 1886.

Round, J. H. *Geoffrey de Mandeville. A Study of the Anarchy*. London: Longman, 1892.

Round, J. H. *Feudal England. Historical Studies on the XIth and XIIth Centuries*. London: Swan Sonnenschein, 1895.

Round J. H. *Studies on the Red Book of the Exchequer*. Printed for private circulation, 1898.

Round J. H. *The Commune of London and Other Studies*. Westminster: Archibald Constable, 1899.

Sampson, Henry. *A History of Advertising from the Earliest Times*. London: Chatto and Windus, 1874.

Scott, Eva. *Rupert Prince Palatine*. Westminster: Archibald Constable, 1899.

Scott, Eva. *Rupert Prince Palatine*. 2nd ed. Westminster: Archibald Constable, 1900.

Seeley J. R. *Lectures and Essays*. London: Macmillan, 1870.

Seeley J. R. *Life and Times of Stein, or Germany and Prussia in the Napoleonic Age*. Vol. 1. Cambridge: Cambridge University Press, 1878.

Seeley J. R. *Life and Times of Stein, or Germany and Prussia in the Napoleonic Age*. Vol. 2. Cambridge: Cambridge University Press, 1878.

Seeley J. R. *Life and Times of Stein, or Germany and Prussia in the Napoleonic Age*. Vol. 3. Cambridge: Cambridge University Press, 1878.

Seeley J. R. *The Expansion of England*. London: Macmillan, 1883.

Seeley, J. R. *Lectures and Essays*. Revised ed. London: Macmillan 1895.

Seeley, J. R. *The Growth of British Policy*. Vol. 2. Cambridge: Cambridge University Press, 1895.

Seeley John Richard. *A Short History of Napoleon the First*. London: Seeley & Co, 1886.

Smith, G. Barnett. *History of the English Parliament. Together with an Account of the Parliament of Scotland and Ireland*. Vol. 1. London: Ward, Lock, Bowden, & Co., 1892.

Smith, G. Barnett. *History of the English Parliament. Together with an Account of the Parliament of Scotland and Ireland*. Vol. 2. London: Ward, Lock, Bowden, 1892.

Smith, G. Barnett. *History of the English Parliament. Together with an Account of the Parliament of Scotland and Ireland*. Vol. 1. 2nd ed. London: Ward, Lock, Bowden, 1894.

Smith, Goldwin. *Irish History and Irish Character*. London: J. H. and Jas. Parker, 1861.

Smith, Goldwin. *Irish History and Irish Character*. 2nd ed. London: J. H. and Jas. Parker, 1862.
Smith, Goldwin. *The United Kingdom. A Political History*. Vol. 1. London: Macmillan, 1899.
Smith, Goldwin. *The United Kingdom. A Political History*. Vol. 2. London: Macmillan, 1899.
Stanhope, Earl. *History of England Comprising the Reign of Queen Anne until the Peace of Utrecht. 1701–1713*. London: John Murray, 1870.
Stanhope, Earl. *History of England Comprising the Reign of Queen Anne until the Peace of Utrecht*. 4th ed. London: John Murray, 1872.
Stirling-Maxwell, Sir William. *Don John of Austria or Passages from the History of the Sixteenth Century 1547–1578*. Vol. I. London: Longman, 1883.
Stirling-Maxwell, Sir William. *Don John of Austria or Passages from the History of the Sixteenth Century 1547–1578*. Vol. 2. London: Longman, 1883.
Stopford Green, Alice. *Henry the Second*. London: Macmillan, 1888.
Stopford Green, Alice. *Town Life in the Fifteenth Century*. Vol. 1. London: Macmillan, 1894.
Stopford Green, Alice. *Town Life in the Fifteenth Century*. Vol. 2. London: Macmillan, 1894.
Stopford Green, Alice. *The Making of Ireland and Its Undoing, 1200–1600*. London: Macmillan, 1908.
Stopford Green, Alice. *The Old Irish World*. London: Macmillan, 1912.
Stopford Green, Alice. *The Making of Ireland and Its Undoing*. 2nd ed. [1909]. London: Macmillan, 1913.
Story, Robert Herbert. *William Carstares: A Character and Career of the Revolutionary Epoch*. London: Macmillan, 1874.
Strickland, Agnes. *Lives of the Queens of England from the Norman Conquest*. Vol. 1. London: Henry Colburn, 1840.
Strickland, Agnes. *Lives of the Queens of England from the Norman Conquest*. Vol. 3. London: Henry Colburn, 1840.
Strickland, Agnes. *Lives of the Queens of England from the Norman Conquest*. Vol. 4. London: Henry Colburn, 1842.
Strickland, Agnes. *Lives of the Queens of Scotland and English Princesses Connected with the Regal Succession of Great Britain*. Edinburgh: William Blackwood, 1850.
Strickland, Agnes. *Lives of the Bachelor Kings of England*. London: Simpkin, Marshall, and Co. 1861.
Strickland, Agnes. *The Lives of the Seven Bishops Committed to the Tower in 1688*. London: Bell and Daldy, 1866.
Strickland, Agnes. *Lives of the Tudor Princesses Including Lady Jane Gray and Her Sisters*. London: Longman, 1868.

Strickland, Agnes. *Lives of the Last Four Princesses of the Royal House of Stuart*. London: Bell and Daldy, 1872.

Stubbs, William. *Registrum Sacrum Anglicanum*. Oxford: University Press, 1858.

Stubbs, William. *Select Charters and Other Illustrations of English Constitutional History from the Earliest Times to reign of Edward the first*. Oxford: Clarendon Press, 1870.

Stubbs, William. *Select Charters and Other Illustrations of English Constitutional History from the Earliest Times to the Reign of Edward the First*. 2nd ed. Oxford: Clarendon Press, 1874.

Stubbs, William. *The Constitutional History of England in Its Origin and Development*. Vol. 1. Oxford: Clarendon Press, 1874.

Stubbs, William. *The Constitutional History of England in Its Origin and Development*. Vol. 2. Oxford: Clarendon Press, 1875.

Stubbs, William. *The Constitutional History of England in Its Origin and Development*. Vol. 1. 2nd ed. Oxford: Clarendon Press, 1875.

Stubbs, William. *Select Charters and Other Illustrations of English Constitutional History from the Earliest Times to the Reign of Edward the First*. 3rd ed. Oxford: Clarendon Press, 1876.

Stubbs, William. *The Early Plantagenets*. London: Longman, 1876.

Stubbs, William. *The Constitutional History of England in Its Origin and Development*. Vol. 2. 2nd ed. Oxford: Clarendon Press, 1877.

Stubbs, William. *The Constitutional History of England in Its Origin and Development*. Vol. 3. Oxford: Clarendon Press, 1878.

Stubbs, William. *The Constitutional History of England in Its Origin and Development*. Vol. 3. 2nd ed. Oxford: Clarendon Press, 1878.

Stubbs, William. *The Constitutional History of England in Its Origin and Development*. Vol. 1. Library Ed. Oxford: Clarendon Press, 1880.

Stubbs, William. *The Constitutional History of England in Its Origin and Development*. Vol. 2. Library Ed. Oxford: Clarendon Press, 1880.

Stubbs, William. *The Constitutional History of England in Its Origin and Development*. Vol. 3. Library Ed. Oxford: Clarendon Press, 1880.

Stubbs, William. *The Constitutional History of England in Its Origin and Development*. Vol. 1. 3rd ed. Oxford: Clarendon Press, 1880.

Stubbs, William. *Select Charters and Other Illustrations of English Constitutional History from the Earliest Times to the Reign of Edward the First*. 4th ed. Oxford: Clarendon Press, 1881.

Stubbs, William. *The Constitutional History of England in Its Origin and Development*. Vol. 1. 4th ed. Oxford: Clarendon Press, 1883.

Stubbs, William. *Seventeen Lectures on the Study of Medieval and Modern History and Kindred Subjects*. Oxford: Clarendon Press, 1886.

Stubbs, William. *Seventeen Lectures on the Study of Medieval and Modern History and Kindred Subjects*. 2nd ed. Oxford: Clarendon Press, 1887.
Stubbs, William. *Select Charters and Other Illustrations of English Constitutional History from the Earliest Times to the Reign of Edward the First*. 6th ed. Oxford: Clarendon Press, 1888.
Stubbs, William. *The Early Plantagenets*. 6th ed. London: Longman, 1889.
Stubbs, William. *Select Charters and Other Illustrations of English Constitutional History from the Earliest Times to the Reign of Edward the First*. 7th ed. Oxford: Clarendon Press, 1890.
Stubbs, William. *The Constitutional History of England in Its Origin and Development*. Vol. 1. 5th ed. Oxford: Clarendon Press, 1891.
Stubbs, William. *Select Charters and Other Illustrations of English Constitutional History from the Earliest Times to the Reign of Edward the First*. 8th ed. Oxford: Clarendon Press, 1895.
Stubbs, William. *The Constitutional History of England in Its Origin and Development*. Vol. 2. 4th ed. Oxford: Clarendon Press, 1896.
Stubbs, William. *The Constitutional History of England in Its Origin and Development*. Vol. 3. 5th ed. Oxford: Clarendon Press, 1896.
Stubbs, William. Registrum Sacrum Anglicanum. 2nd ed. Oxford: Clarendon Press, 1897.
Stubbs, William. *Seventeen Lectures on the Study of Medieval and Modern History and Kindred Subjects*. 3rd ed. Oxford: Clarendon Press, 1900.
Stubbs, William. *The Constitutional History of England*. Vol. 1. 6th ed. Oxford: Clarendon Press, 1903.
Tait, C. W. A. *Analysis of English History, Based on Green's Short History of the English People*. London: Macmillan, 1879.
Tait, C. W. A. *Analysis of English History, Based on Green's Short History of the English People*. 2nd ed. London: Macmillan, 1890.
Taylor, Tom. *Life and Times of Sir Joshua Reynolds*. Vol. 1. London: John Murray, 1865.
Taylor, Tom. *Life and Times of Sir Joshua Reynolds*. Vol. 2. London: John Murray, 1865.
Terry, Charles Sanford. *The Life and Campaigns of Alexander Leslie First Earl of Leven*. London: Longman, 1899.
Thompson, Edith. *History of England*. 4th ed. London: Macmillan, 1874.
Thompson, Edith. *History of England*. Revised ed. London: Macmillan, 1878.
Thompson, Edith. *History of England*. Reprint. London: Macmillan, 1889.
Thompson, Edith. *The Wars of York and Lancaster 1450–1485*. London: David Nutt, 1892.
Thompson, Edith. *History of England*. Revised ed. London: Macmillan, 1901.
Thompson, Edith. *History of England*. Revised ed. London: Macmillan, 1923.
Ward, Adolphus William. *Chaucer*. London: Macmillan, 1879.

Ward, Adolphus William. *The Counter-Reformation*. London: Longman, 1889.
Ward, Adolphus. *Sir Henry Wotton. A Biographical Sketch*. Westminster: Archibald Constable, 1898.
Ward, Adolphus. *Great Britain & Hanover*. Oxford: Clarendon Press, 1899.
Watson, John Selby. *The Life of Richard Porson, M.A.* London: Longman, 1861.
Watson, John Selby. *The Life of William Warburton, D.D. Lord Bishop of Gloucester from 1760 to 1779*. London: Longman, 1863.
Watts, Henry Edward. Spain. *Being a Summary of Spanish History from the Moorish Conquest to the Fall of Granada (711 – 1492 A.D.)*. 3rd ed. London: T. Fisher Unwin, 1893.
Yonge, Charles Duke. *The Life and Administration of Robert Banks, Second Earl of Liverpool, K. G. Late First Lord of the Treasury*. Vol. 1. London: Macmillan, 1868.
Yonge, Charles Duke. *The Life and Administration of Robert Banks, Second Earl of Liverpool, K. G. Late First Lord of the Treasury*. Vol. 2. London: Macmillan, 1868.
Yonge, Charles Duke. *The Life and Administration of Robert Banks, Second Earl of Liverpool, K. G. Late First Lord of the Treasury*. Vol. 3. London: Macmillan, 1868.
Yule, Henry. *The Book of Ser Marco Polo, the Venetian*. Vol. 1. London: John Murray, 1871.
Yule, Henry. *The Book of Ser Marco Polo, the Venetian*. Vol. 1. 2nd ed. London: John Murray, 1875.
Yule, Henry. *The Book of Ser Marco Polo, the Venetian*. Vol. 2. 2nd ed. London: John Murray, 1875.

Index

E. Garritzen, *Reimagining the Historian in Victorian England*,
https://doi.org/10.1007/978-3-031-28461-8

I

Printed by Printforce, the Netherlands